About Island Press

Island Press is the only nonprofit organization in the United States whose principal purpose is the publication of books on environmental issues and natural resource management. We provide solutions-oriented information to professionals, public officials, business and community leaders, and concerned citizens who are shaping responses to environmental problems.

In 1994, Island Press celebrates its tenth anniversary as the leading provider of timely and practical books that take a multidisciplinary approach to critical environmental concerns. Our growing list of titles reflects our commitment to bringing the best of an expanding body of literature to the environmental community throughout North America and the world.

Support for Island Press is provided by The Geraldine R. Dodge Foundation, The Energy Foundation, The Ford Foundation, The George Gund Foundation, William and Flora Hewlett Foundation, The James Irvine Foundation, The John D. and Catherine T. MacArthur Foundation, The Andrew W. Mellon Foundation, The Joyce Mertz-Gilmore Foundation, The New-Land Foundation, The Pew Charitable Trusts, The Rockefeller Brothers Fund, The Tides Foundation, Turner Foundation, Inc., The Rockefeller Philanthropic Collaborative, Inc., and individual donors.

Beyond the Numbers

*Dedicated to the memory
of Irving Mazur, with love.*

Acknowledgments
Many people contributed time, advice, and support to this project.
Special thanks to: José Barzelatto, Heather Boyer, David Devlin-Foltz,
Megan Fraser, Russ Haven, Margaret Hempel, Li Howard,
Jeannine Kerr, Bill LaDue, Kristy Manning, Nancy Olsen,
Tracy Peel, Marty Riche, Charles Savitt, Sharon Stanton Russell,
Susan Sechler, and Steve Viederman.

Cover design: Joyce C. Weston
Interior design: Li Howard and Laurie Ann Mazur
Layout: Li Howard
Copy editing: Marianne Maggini and Tracy Peel

Beyond the Numbers

A Reader on Population, Consumption, and the Environment

....

Edited by
Laurie Ann Mazur

Foreword by Timothy E. Wirth
Counselor, Department of State

ISLAND PRESS
Washington, D.C. • Covelo, California

This publication was made possible through the generous support of the Pew Global Stewardship Initiative, with additional funding provided by Turner Foundation, Inc., and The Rockefeller Philanthropic Collaborative, Inc.

ISBN 1-55963-298-4 (cloth)
ISBN 1-55963-299-2 (paper)
Library of Congress Catalog Card Number: 94-75842

Printed on recycled, acid-free paper ♻
Manufactured in the United States of America
10 9 8 7 6 5 4 3 2

Editor's Addendum

Since this book first went to press a year ago, the international dialogue on population issues has indeed moved "beyond the numbers" — to a more nuanced understanding of the causes and effects of population growth. The change was particularly evident as world leaders gathered in Cairo for the United Nations International Conference on Population and Development in September 1994.

The Cairo conference was a watershed event. Unlike earlier U.N. population meetings, it was characterized by an extraordinary degree of cooperation between the industrialized and developing countries. The conference also marked the U.S. government's return, after a decade of non-engagement, to a leadership role in addressing population growth. An unprecedented level of citizen participation shaped the process, politics, and substance of the conference. And finally, the Cairo conference produced a Program of Action that reflects the new thinking about population and development — and echoes many of the ideas you will find in this book. The Program of Action:

Provides an ethical framework for population policy. The Program of Action reaffirms the basic human right to choose the number and spacing of one's children, and rejects coercion of all forms in population and family planning programs.

Recognizes that unwanted fertility is only one cause of rapid population growth, and that many social and economic factors contribute to the desire for large families. While advocating universal access to family planning services, the Program of Action also seeks to create conditions in which small families are viable and desirable, by promoting maternal and child health, primary education, and sustainable development.

Endorses a comprehensive reproductive health approach to family planning, rather than a narrow emphasis on contraceptive delivery. The Program of Action emphasizes the importance of addressing the full range of reproductive health problems, including the modern pandemic of sexually transmitted diseases.

Recognizes the central role of gender relations. The Program of Action explicitly draws the link between high fertility and the low status of women, condemns gender discrimination, and offers strategies to improve women's access to education, resources, and opportunity. It also urges greater male participation and responsibility in fertility control and the support and care of children.

Addresses the issue of adolescent fertility. Recognizing both the human costs and demographic implications of too-early childbearing, the document underscores the need for programs that help adolescents postpone sexual activity and protect themselves from unwanted pregnancy and disease.

Addresses abortion as a public health concern. While the Program of Action states emphatically that abortion should not be used as a method of family planning, it acknowledges that unsafe abortion is a serious public health problem. The document stresses that, where abortion is legal, it should be safe. It also proposes to reduce unsafe abortion through the provision of quality family planning services.

Recognizes the role of resource consumption in environmental problems. The document examines the interrelationships among population growth, resource consumption, and environmental degradation, and encourages governments to incorporate environmental considerations into population and development policy.

The Cairo conference produced more than agreements in principle; it also prompted substantial new commitments of resources from Germany, the E.C.U., and Japan. In the U.S., annual population assistance has doubled since 1992 to reach almost $600 million in FY 1995, with increases to $1.2 billion promised by the year 2000.

The march of events, however, leaves little time for celebration. The challenge before us now is to translate the Cairo Program of Action from rhetoric to reality. This means taking a hard look at the bilateral and multilateral institutions that exert profound influence on population policy around the globe. It means moving from the theoretical to the practical, by devising population and development programs that really work to improve people's lives. Successful small-scale programs need to be "scaled up" and incorporated into national level programs and the policies of aid agencies. And, perhaps most importantly, it means building support for the Cairo agenda in the U.S. and overseas.

In the U.S., these efforts will take place in a political climate that is indifferent — if not outright hostile — to elements of the Cairo agenda. Many members of Congress want to abolish or severely cut foreign aid, especially population assistance. Opponents of population assistance may seek to steer the debate back to the divisive politics of abortion.

Still, there are reasons for hope. With its ethical appeal and inherent good sense, the Cairo Program of Action has the potential to mobilize a broad coalition of supporters. Indeed, polls show that Americans already support the types of programs recommended in the Cairo document. A public opinion study conducted by the University of Maryland found that strong majorities wanted to increase or maintain foreign aid for child survival programs (91%), development assistance (75%), and family planning (74%). Parlaying that support into a strong, bipartisan constituency for population and aid programs will be a central challenge in the years to come.

The essays collected in this book put forth a bold vision of humane, effective population policies for the 21st century. In Cairo, that vision was endorsed by women and men from all over the world. The Cairo conference was a crucial milestone on the path to a future that is sustainable and just — but it is up to us to take the next step.

Laurie Ann Mazur
March, 1995

Contents

SECTION VI:
POPULATION AND RELIGION

Foreword

BY TIMOTHY E. WIRTH

··················

The publication of *Beyond the Numbers* coincides with a pivotal moment in world history. As nations struggle to define their roles in the post-Cold War era, it is increasingly clear that our definition of national — and global — security must encompass a broad range of non-military issues, particularly population growth, sustainable development, human rights (including gender equity), and the environment.

Those issues will be at the top of the agenda in September 1994, when world leaders gather in Cairo for the third decennial United Nations conference on population. The International Conference on Population and Development (ICPD), as the conference is known, will present a crucial opportunity to reverse trends that threaten our collective future.

The need for action is clear. World population has doubled since 1950 and now stands at 5.6 billion. That number will almost certainly double again, and may even triple, by the end of the next century. Approximately 95 percent of future growth will take place in the developing world, where one billion people now live in abject poverty. We do not know for certain what this means for the world's people. The impact of population growth is not straightforward; its effects can be magnified or mitigated by a host of social, economic, and environmental factors. But we do know this: in many parts of the world, slowing rapid population growth could help spur economic development, ameliorate poverty, achieve a livable environment, and foster greater opportunity for women.

We also know that, as citizens of the industrialized world, we are consuming natural resources at an unprecedented and unsustainable rate. Comprising only a fifth of the world's population, the industrialized nations use two-thirds of all resources consumed and generate four-fifths

State Department counselor Timothy E. Wirth is leading U.S. preparations for the International Conference on Population and Development.

of all pollutants and wastes. It is an open question whether the Earth cannot support its present — much less future — inhabitants at this level of consumption.

These problems are within our capacity to solve, and there is a growing consensus about the most promising solutions. Nonetheless, the international discourse on population and environment issues has long been characterized by bitter conflict. Donor-funded population programs have achieved remarkable success, but they have also been criticized — often justly — for a narrow focus on reducing birthrates and inattention to human rights, women's health, and cultural differences. Developing countries resent the emphasis given to population growth as a cause of environmental problems in the absence of a parallel focus on consumption in the North.

Simmering tensions ignited at the the two previous U.N. population conferences. At the World Population Conference in Bucharest in 1974, the U.S. and other industrial countries advocated programs to slow population growth, while the developing countries countered that "development is the best contraceptive." By 1984, when the U.N. held its second Conference on Population in Mexico City, the tables had turned: The developing countries acknowledged the need for population programs, but President Reagan's U.S. delegation pronounced population "a neutral factor" and scaled back funding for international family planning efforts. And at the 1992 Earth Summit in Rio, a variety of representatives questioned the importance of population growth as an environmental problem.

In the popular media, as well, the debate on population and the environment is often dominated by polar extremes. Last year, for example, a cover story in *The Atlantic Monthly* pitted "Cassandras" — those who believe that population will lead to environmental collapse — against "Pollyannas" — those who are confident that human ingenuity will provide for the burgeoning population. And the debate still rages over whether "development is the best contraceptive" or if family planning programs on their own can achieve significant fertility declines.

These ideological poles exist and are important elements of the debate. However, a focus on the extremes masks considerable and growing consensus in the middle of the spectrum. There is now broad agreement that population growth is part of a constellation of factors which cause environmental degradation. There is increasing recognition that development and family planning can work independently to slow population growth, but that they work best when pursued *together*. It is widely agreed that family planning programs should be provided as part of comprehensive reproductive and primary health initiatives, and that population policy should include development strategies, including education, especially for girls; increased economic opportunity for women; and elimination of legal and social barriers to gender equality. Finally, there is a growing understanding that our efforts to slow population growth abroad must be matched by determined, good faith initiatives to reduce resource consumption at home. Today, the cutting edge of thought on population and consumption issues is not at the extremes, but in the consensus-building middle.

The new consensus forms the cornerstone of the Clinton administration's population policy. Still, much remains to be done. The new consensus has not yet been translated into policy on a grand scale. And our enhanced understanding of the complex dynamics of population growth and environmental degradation magnifies the challenges before us. Our goals — universal access to family planning, health care, education, and economic opportunity — may seem impossible to achieve.

Yet, the transformations of the past few years impel us to reconsider the dimensions of what is possible. If Gorbachev and Yeltsin can discard communism and end the Cold War; if de Klerk and Mandela can work together to end apartheid in South Africa; if Israel and the PLO can break a fifty-year stalemate, should we merely peg our aspirations to what seems possible today?

It is true that budgets are tight — here in the U.S. and elsewhere. But in both the public and private sectors we have assets that haven't been tapped. The challenges we face are so urgent that we must think creatively about how to increase and mobilize resources. For example, economic reform that rewards private initiative, spreads opportunity, and builds freer markets in more parts of the world could bring greater wealth. Sharing ideas and transferring technology could spark growth. Environmentally friendly technologies could lower health costs, improve productivity, reduce waste, and raise profits. Coordinating international assistance could bring more results for the same amount of aid dollars. And we are all hopeful that a more peaceful world will free up funds for economic and social development.

There are plenty of sound reasons for hope. For all the suffering still in the world, we can still say that in the last 50 years we have made more progress in alleviating human misery than in the previous two millennia. Life expectancy in the developing countries grew by one-third, death rates for infants and children were cut by half, and real incomes more than doubled. If we could do all that while burdened with the political and economic costs of the Cold War, how much greater should be the goals we set for ourselves now?

The Cairo conference offers an unparalleled opportunity to develop an action plan to meet those goals. In the U.S., the signs are encourag-

ing: The broad consensus described above provides a solid basis for policymaking, and the Clinton administration is poised to take a leadership
role on global population and environment issues.

Still, we will not succeed without the informed support of the
American people. That is where *Beyond the Numbers* comes in. This volume includes essays by many of the architects of the new thinking on
population and environment issues. In its pages you will encounter plenty of lively debate and controversy. But you will also find the elements of
an emerging consensus — a blueprint, perhaps, for a new world order
that is sustainable and just.

LAURIE ANN MAZUR

Beyond the Numbers: An Introduction and Overview

....

The numbers are simply astounding. World population now totals 5.6 billion and is expected to double within the next century. Every year, the world gains another 91 million inhabitants — the equivalent of another New York City every month, another China every decade.[1] Nearly all of that growth will take place in the developing countries.

But, as the title of this volume suggests, numbers tell only part of the story. As John Muir once said, "When we try to pick out anything by itself, we find it hitched to everything else in the universe." So it is with population issues: Both the causes and the effects of population growth are "hitched" to a complex tangle of social, political, and economic factors. Moreover, population touches on many deeply divisive subjects, including human rights, the unequal distribution of resources, abortion, foreign aid, and the status of women. Not surprisingly, these issues have ignited conflict between the industrialized North and the developing South and among environmentalists, religious institutions, feminists, and family planning professionals. Complexity and conflict have helped cause a political and conceptual gridlock that confounds creative solutions.

The purpose of this reader is to illuminate the contours of these complex issues and to explore areas of debate beyond the polarized extremes — to look, in essence, beyond the numbers. To that end, we have solicited essays and reprinted articles by a broad array of activists, academics, and policymakers. The authors of *Beyond the Numbers* do not all speak the

Laurie Ann Mazur is an independent writer and consultant to nonprofit organizations.

same language; they vary widely in their perspectives, interpretive approaches, and voice. Still, we have not represented every facet of the debate, nor have we included voices from every part of the world. In general, the perspectives collected here are clustered near the center of the spectrum of opinion. As State Department counselor Timothy E. Wirth observes in his foreword, the cutting edge of thought on population issues is not at the extremes, but in the "consensus-building middle."

Our framework for examining these issues is population, consumption, and the environment. This orientation reflects the fact that we — the editor and publisher of this book — are American environmentalists. But we have come to realize that no single discipline is equal to the task of interpreting the complexities of population growth. Our environmentalism serves as a (green-tinted) lens through which we view these issues and not, we hope, as a set of blinders that obscures other points of view.

THE CONVENTIONAL WISDOM IS WRONG

If there is one generalization to be made from the wealth of fact and opinion presented here, it is this: The conventional wisdom about population, consumption, and the environment is often wrong.

For example, the conventional wisdom holds that population growth is a result of high birth rates. But, as Carl Haub and Martha Farnsworth Riche point out in "Population by the Numbers: Trends in Population Growth and Structure," (page 95) it is declining mortality, not rising fertility, that is causing the current population surge. Over the last 40 years, fertility rates have fallen in most parts of the world. But, because death rates have dropped even more steeply, the absolute number of births has gone up. The children born in the 1960s and 1970s survived in greater numbers than any previous generation, so there are now more people of reproductive age than ever. Although couples are choosing to have fewer children, the sheer number of men and women of childbearing age has caused the total number of births to soar. This "population momentum" means that even if births and deaths were brought into balance tomorrow, population would continue to grow for quite some time.

The conventional wisdom also holds population growth responsible for increasing migration from the impoverished countries of the South to the industrialized countries of the North. But as Hania Zlotnik notes in "International Migration: Causes and Effects,"(page 359) the poor countries with the highest rates of population growth are not the ones that send the most migrants to the developed world. Indeed, until the late 1980s, most of those migrants came from *other developed countries*. And the regions that currently send the most migrants to developed countries — Latin America and East and Southeast Asia — have relatively low rates of population growth, while the region with the highest rate of growth —

sub-Saharan Africa — sends the fewest migrants to developed countries. Rapid population growth *per se* is not a cause of migration, although, as we'll see, it is one of several interconnected factors that can degrade the quality of life and thus create an incentive to migrate.

POPULATION AND THE ENVIRONMENT

Perhaps the most pervasive piece of misguided conventional wisdom holds that rapid population growth leads inevitably to environmental decline. Intuitively, this proposition makes sense: More humans consume more resources and generate more waste. A quick look at the data appears to support this equation. Carbon dioxide emissions, for example, follow a steep upward curve that closely matches population growth rates for the last quarter century. However, as Mark Sagoff points out in "Population, Nature, and the Environment," (page 33) vast differences in consumption mean that some populations have a far greater environmental impact than others. With only 25 percent of the world's people, the industrialized nations of the North generate nearly three-quarters of all carbon dioxide emissions, accounting for about half of the manmade "greenhouse" gases in the atmosphere.[2] So in terms of global climate change, consumption in the North poses a greater threat than population growth in the South.

When a community or nation is described as "overpopulated," the implication is that its numbers have grown too large in relation to the stock of available resources. But resources are distributed so inequitably that it is often impossible to determine their capacity to sustain a given population. For example, one researcher writes of a Philippine village where poor farmers cultivate land on a sloping hillside. As the community's population has grown, the farmers have placed greater demands on the land, degrading the quality of the soil. But the entire village could be amply sustained on a small portion of the huge sugar cane estate that surrounds the hill. While population growth has made the situation worse, poverty and inequitable resource distribution are at the heart of the problem.[3] Thus, says Robert Repetto of the World Resources Institute, "It is misleading to describe the resource degradation that results when marginal farmers misuse marginal lands as a consequence of population pressures, when, in reality, it is a consequence of the gross inequality in access to resources between the rich and poor."[4]

On a global level, inequity is even more striking. The 25 percent of the world's population that lives in the developed countries lays claim to 85 percent of all forest products consumed, 72 percent of steel production, and 75 percent of energy use.[5] Developed countries also generate about 75 percent of the global burden of pollutants and wastes.[6] Of course, the current inequitable distribution of resources need not be a given. And as living standards rise in the developing world, per capita environmental impact will increase. This scenario could cause even greater environmental devastation, unless appropriate technologies and policies are in place to prevent it.

Technology can indeed help moderate the effects of population growth. It has been estimated that simply by employing energy efficiency measures, the North American economy could do everything it now does, with currently available technologies and at equivalent or lower costs, using half as much energy.[7] Halving our energy consumption would ameliorate many environmental problems, from acid rain to climate change.

Technology should perhaps be understood to include not only the means of production, but also what has been called the "technologies of distribution." These include economic, social, and political arrangements that assure people access to education, opportunity, and the shared powers and responsibilities of citizenship.[8] Improved technologies of distribution could greatly enhance the planet's capacity to support human life. As M.I.T. economist Lester Thurow has written:

> If the world's population had the productivity of the Swiss, the consumption habits of the Chinese, the egalitarian instincts of the Swedes, and the social discipline of the Japanese, then the planet could support many times its current population without privation for anyone. On the other hand, if the world's population had the productivity of Chad, the consumption habits of the United States, the inegalitarian instincts of India, and the social discipline of Argentina, then the planet could not support anywhere near its current numbers. [9]

NO LIMITS?

Does this mean, then, that with the right "technologies" of production and distribution, there are no limits to the number of people the planet can sustain? That is the position taken by neoconservative economists, most notably Julian Simon. In *The Ultimate Resource*, Simon declared that "There is no meaningful limit to our capacity to keep growing forever."[10] When markets function well, economists say, human populations are unlikely to go crashing headlong into resource limits. Any sort of scarcity will trigger "warning signals" in the form of higher prices, which will set in motion a range of adaptive behaviors — substitution, recycling, conservation — that will prevent resource depletion. In "Population, Living Standards and Sustainability: An Economic View," (page 76) David Horlacher and Landis MacKellar offer a more qualified, yet still optimistic, version of this hypothesis. Horlacher and MacKellar also note that as population grows and living standards rise, so does demand for environmental quality and resources to protect the environment.

Economists find plenty of evidence to support these views. They are particularly fond of reminding doomsayers that, in the two centuries since Thomas Malthus predicted that human numbers would soon outstrip the food supply, global food production has generally kept well ahead of population growth. Between 1950 and 1984, the Green Revolution helped expand world grain production 2.6-fold, raising the per capita grain harvest by 40 percent.[11] Economists can also summon innumerable cases of markets responding to scarcity. For example, after the oil crisis of the 1970s, automakers designed a new generation of fuel-efficient cars and gasoline consumption per mile fell by 29 percent between 1973 and 1988.[12]

Needless to say, environmentalists are less sanguine about the Earth's infinite bounty. Because the scale of current resource consumption and population growth is unprecedented, they argue, we are in effect playing a high-stakes game of chance with the limits of sustainability. Since the outcome of the game cannot be known, and since the stakes are so high, environmentalists invoke the precautionary principle as a reason to slow population growth and curb consumption.

Environmentalists have seen population growth and consumption overwhelm their hard-won gains in pollution control. As environmental writer Paul Harrison reports, emissions reductions from automotive fuel-efficiency in the U.S. have been completely offset by a doubling of the number of cars on the road since 1970. Thus, says Harrison, "Population and consumption will go on raising the hurdles that technology must leap."[13]

While environmentalists acknowledge that properly functioning markets can prevent resource depletion, they observe that markets are often dysfunctional or nonexistent. For example, government subsidies to logging, grazing, and mining in the American West have kept prices artificially low, effectively muting any warning signals that resource limits are dangerously near. Moreover, there are no markets for many commonly held resources, such as the global atmosphere and biological diversity. There is now an international effort to protect those assets, but the process moves slowly and is hindered by a lack of understanding of the complex systems it seeks to protect.

At the heart of the environmentalists' perspective on population growth is the concept of "carrying capacity." The planet's carrying capacity is, in essence, its ability to sustain life. Carrying capacity cannot simply be measured in terms of grain output, fish harvests, or timber production. It refers to the health of the ecological systems and processes that filter wastes, regenerate soils, and replenish aquifers. Those systems, which we have belatedly begun to understand and respect, form the very basis of life on Earth.

If we evaluate environmental quality from the perspective of ecosystem health rather than by production of crops and renewable resources, there are many causes for alarm. Climate change, the ozone hole, and species loss are all red-alerts that ecological systems are under severe strain. Even this century's spectacular increases in food production are less impressive when viewed in this light. The Green Revolution boosted food output but left a legacy of poisoned groundwater, degraded and salinated soil, and lost biodiversity. That legacy can now be quantified even by conventional indicators. As Sandra Postel reports in "Carrying Capacity: The Earth's Bottom Line" (page 48), many measures of the planet's ability to support human life — availability of fresh water, acreage of irrigated farmland, grain production, and world fish catch — have stagnated or declined in recent years.

Human ingenuity may yet devise the technologies of production and distribution needed to sustain the expanding human population. But, as Mark Sagoff suggests, we may salvage the *environment* — the systems and resources that sustain life — without saving *nature* — those parts of the biosphere that have not been modified by human intervention or shaped to meet human needs. It is this concern that gives mainstream environmentalists their (sometimes deserved) reputation for caring more about trees and owls than about people.[14] But reverence for nature is deeply embedded in the human tradition. Nature is our Garden of Eden, a well of spiritual sustenance, an irreplaceable source of perspective and humility. It is also being destroyed at a breathtaking pace. Environmentalists feel an obligation to act as stewards of nature to insure that it survives for future generations.

THE CHICKEN AND EGG OF DEMOGRAPHY

Just as population growth can offset improvements in environmental quality, it can stall social and economic development — or so the conventional wisdom goes. But here, too, the relationship is more complex than it appears.

Clearly, there is a link of some kind between development and slower population growth. As Haub and Riche explain, most preindustrial societies are characterized by stable populations and high rates of fertility and mortality. Then, in the early stages of development, improvements in public health reduce mortality and a population explosion follows as births outnumber deaths. Some time later, fertility falls to "replacement level" of roughly two children per couple, and population size stabilizes. This is known as the "demographic transition," which demographer Nathan Keyfitz calls "the most universally observed and least readily explained phenomena of modern times."[15]

Does development precede or follow population stabilization? This question remains the "chicken and egg" of demography, and evidence has been rallied to support both points of view. Many contend that birth rates remain high where children are needed for social and economic security; where parents rely on children for support in old age, for example, or where childbearing is a woman's only means of attaining status and security. Moreover, where infant and child mortality rates are high, parents tend to have many children in order to insure that some survive. In policy terms, this is known as the "demand" argument, because it suggests that social and economic development are necessary preconditions to create demand for family planning. In contrast, "supply-siders" believe that if family planning services are made available, fertility will decline and development will follow. The supply versus demand debate has at times raged with great intensity, because different answers drive different policy approaches and funding priorities.

Demand-siders usually cite the successful demographic transition in Europe, where populations stabilized and prospered without interventions to reduce fertility — in fact, without the aid of modern contraceptives. But others believe the European experience has limited relevance for the impoverished nations of the developing world. They note that both the scale and pace of current population growth dwarfs the earlier increase, and that Europeans managed their transition by emigrating and colonizing the "new world."

Certainly, the developing countries today are in a very different situation than their European predecessors. As Haub and Riche explain, mortality rates have fallen steeply due to advances in public health, implemented mostly by multilateral agencies. But fertility remains high because neither supply of nor demand for family planning is firmly in place. In "Seeking Common Ground: Unmet Need and Demographic Goals," (page 158) Steven Sinding, John Ross, and Allan Rosenfield calculate that at least 100 million women in the developing world wish to have no more children but lack access to modern contraception. And, as many authors in this volume report, social and economic factors — especially the low status of women — still serve as powerful inducements to have large families.

Some economists have argued that population growth can serve as an engine of development, by improving economies of scale for infrastructure construction, for example. But this is not currently the case in the developing countries of the South, where governments are struggling to provide services and employment for ever-expanding populations. The United Nations Population Fund estimates that there are now a half-billion people in developing countries who are unemployed or underemployed — a number that is equivalent to the entire workforce of the industrialized countries. To accommodate their growing populations, developing countries must create some 30 million new jobs every year just to maintain current employment levels.[16] Few consider it likely that the debt-ridden (and sometimes corrupt) regimes of the South can generate increases of that magnitude. Thus, population growth and underdevelopment create a downward cycle in which "population growth prevents the development that would slow population growth," says Nathan Keyfitz.[17] A report by the National Research Council concluded that "slower population growth would be beneficial to economic development for most developing countries."[18]

The downward cycle of population growth and underdevelopment joins with environmental stress to create what James Grant, executive director of the United Nations Children's Fund, terms the Poverty-Population-Environment (PPE) spiral. As we have seen, poverty contributes to population growth by maintaining the demand for high fertility. Population growth, in turn, can perpetuate poverty by impeding development. Environmental stress is both a cause and effect of poverty and population growth. As in the case of the Philippine villagers cited above, the poor are forced to mine their resources unsustainably through overgrazing, farming steep slopes, and denuding forests for fuel. The result, writes Grant, "has been the drawing of the poorest into a cycle by which poverty forces growing numbers of people into environmentally vulnerable areas and the resulting environmental stress becomes yet another cause of their continued poverty...."[19]

Ominously, PPE problems can also contribute to war and social instability. As Thomas Homer-Dixon, Jeffrey Boutwell, and George Rathjens report in "Environmental Change and Violent Conflict,"(page 391) PPE problems are already a significant cause of violent conflict, insurgencies, and refugee movements in many parts of the world. These conflagrations, and their attendant toll in human misery, are expected to intensify as more people are drawn into the PPE spiral's downward vortex.

DEVELOPMENT CAN MAKE MATTERS WORSE

Ironically, some development strategies may actually make PPE problems worse by failing to account for fundamental inequities between rich and poor. Again, the example of world food production is instructive. The Green Revolution, as noted earlier, produced enormous increases in per capita food production. But that bounty was not evenly distributed to each *capita*. Global food markets are plagued by surpluses, yet more than one billion of the world's people are malnourished. Why?

Some analysts place part of the blame on the Green Revolution's emphasis on cash crops and capital-intensive inputs, such as fertilizer and pesticides. Capital-intensive agriculture means that wealthier farmers are more able to benefit from the new technologies, exacerbating the gap between rich and poor. As Mark Sagoff observes, "The vast increases in yields-per-acre of recent decades depend on technologies that do nothing for subsistence farmers who cannot afford to purchase them." And, because Green Revolution techniques make agriculture more profitable, rich farmers and corporations buy up the best, flattest land, while poor peasants are forced onto environmentally fragile marginal lands. In India, the Philippines, and Indonesia — all major clients of the Green Revolution — half of the rural population lives below the poverty line.[20] Thus, argues Fatima Vianna Mello in "Sustainable Development — For and By Whom?" (page 71), any discussion of population and sustainable development must confront "the destructive character of the reigning development model, which aims at gaining profits and satisfying market forces, and not at meeting the basic human needs of the population."

Structural adjustment policies are perhaps the most widely criticized elements of the reigning development model. These policies, which are imposed by the World Bank and International Monetary Fund as conditions for receiving financial assistance, typically include a shift from domestic food production to production for export, wage controls, and cuts in social spending. Although structural adjustment policies are intended to foster economic growth, they have often had the opposite effect. In most countries where they have been implemented, structural adjustment policies have intensified poverty, hastened the plundering of natural resources, and exacerbated environmental problems. The 33 African countries that received structural adjustment loans in the 1980s experienced a steady decline in gross domestic product, food production, and social spending, and a 17 percent increase in people living in poverty. Government spending on education declined from $11 billion to $7 billion during that period, and primary school enrollment fell from 80 percent in 1980 to 69 percent in 1990.[21] Structural adjustment in Africa, in the words of the British charity Oxfam, has "had the effect of forcing the region to struggle up a downward escalator."[22]

Some current development strategies harm women and children by reinforcing patterns of gender bias. According to Jodi Jacobson, director of the Health and Development Policy Project, gender bias "boils down to [the] grossly unequal allocation of resources — whether of food, credit, education, jobs, information, or training."[23] In "Investing in Women: The Focus of the '90s," (page 209) Nafis Sadik recounts the tragic consequences of gender bias. In most parts of the world, she observes, women are more likely than men to be malnourished, poor, and illiterate; they have fewer opportunities to earn income, own little real property, and have less access to education.

Development programs can perpetuate gender bias. As Jacobson points out, women perform the lion's share of work in the subsistence economies of the developing world. In those cultures, women are the primary breadwinners, growing crops to feed their children and contributing a greater proportion of their earnings to family welfare. However, in the eyes of government statisticians and development experts, subsistence agriculture is not considered "productive" labor. Therefore, cash crops, which are generally controlled by men, receive the bulk of credit and other resources that improve productivity. Because men contribute a proportionally smaller share of their earnings to the family budget, development programs that favor male-controlled cash crops can drive women and children deeper into poverty. In Africa, a World Bank report found that "it is not uncommon for children's nutrition to deteriorate while wrist watches, radios, and bicycles are acquired by the adult male household members."[24]

Gender bias impacts on the population and environment components of the PPE spiral as well. Jacobson writes, "Gender bias is also the single most important cause of rapid population growth. Where women have little access to productive resources and little control over family income, they depend on children for social status and economic security."[25] And, because they are often responsible for securing food, fuel, and fresh water, women tend to have a greater interest in preserving croplands, forests, and other resources for perpetual use. Men, on the other hand, are more likely to regard natural resources as commodities that can be converted into cash.[26] Sustainable development initiatives thus invite failure when they ignore the role of women.

A HISTORY OF CONFLICT

What can we conclude about population growth, the environment, and development? Certainly, population growth is one of several interrelated factors that can cause environmental degradation and hinder development. It is reasonable, then, to ask how we might slow the rate of population growth, while addressing the other elements of the PPE spiral. This question has generated many different answers, and much bitter debate, over the last several decades. Newcomers to these issues are often mystified by the vehemence of that debate, which is undoubtedly incomprehensible without an understanding of its history. Here, then, is a very abbreviated version of that history and the conflicts it engendered.

As Peter J. Donaldson and Amy Ong Tsui explain in "The International Family Planning Movement" (page 111), some of the earliest advocates of family planning were turn-of-the-century social reformers, feminists, and — for very different reasons — eugenicists. Reformers and feminists saw family planning as a means of improving the lot of the working class and increasing women's self-determination. Eugenicists promoted family planning as a means of increasing fertility among those with allegedly superior genetic stock, and diminishing the population of "undesirables" — the physically and mentally handicapped and criminals.

In the 1950s, when the surge in population catalyzed widespread concern, an organized multilateral effort was launched to reduce fertility. In the 1950s, 1960s, and early 1970s, donor-funded population programs were implemented in most developing countries. The U.S. took the lead in funding these programs, for a variety of strategic and humanitarian reasons. These early programs operated on the assumption that there was a great latent demand for family planning in the developing countries, and that modern contraceptives — notably the birth control pill and intrauterine device (IUD) — would serve as "magic bullets" to bring down growth rates. That assumption was at least partly correct; since the early 1960s, the percentage of women in the developing countries using contraceptives has risen from 18 to 50 percent, according to Donaldson and Ong Tsui.

But there are some grim chapters in the history of population programs. As Mahmoud Fathalla explains in "From Family Planning to Reproductive Health," (page 143) some programs have employed abusive or coercive tactics. Perhaps the most notorious case was in India, where several million forced sterilizations were performed following a "national emergency period" in 1976. This is an extreme example, but, as Ruth Macklin shows in "Ethical Issues in Reproductive Health," (page 191) many programs use less heavy-handed inducements which can also violate the fundamental right to reproductive autonomy.

Since the 1970s, population programs have drawn fire from feminists and women's health advocates. Women's groups charged that many programs focused too narrowly on the demographic bottom line, without proper regard for human welfare and especially women's needs. As Adrienne Germain and Jane Ordway write in "Population Control and Women's Health: Balancing the Scales," (page 135) population programs "have viewed women as producers of too many babies, and measured their accomplishments in numbers of contraceptive or sterilization 'acceptors' and statistical estimates of 'births averted.'" And family planning programs in developing countries generally offer poor quality of care. Clients are given little or no choice among contraceptive methods and are rarely warned about side effects. (See "Quantifying Quality," by Barbara Mensch, page 174.) Not surprisingly, low-quality care is a major impediment to contraceptive usage and continuation, report Anrudh Jain and Judith Bruce in "Quality: The Key to Success" (page 171).

But not even high-quality reproductive health care would enable all women to control their fertility in cultures where women lack fundamental rights. "In many societies," writes Nafis Sadik, "a young woman is still trapped within a web of traditional values which assign a very high value to childbearing and almost none to anything else she can do." Nor do women always have the final say about whether to use contraception. In

"Family and Gender Issues for Population Policy," (page 242) Cynthia Lloyd notes that in patriarchal cultures, when husbands and wives disagree about family planning, it is usually the husband who prevails.

Women of color in the U.S. have rallied against abuses closer to home. In "Reflections on African-American Resistance to Population Policies and Birth Control," (page 281) Denese Shervington analyzes the legacy of distrust left by "racially motivated population policies" in the U.S. Shervington notes that eugenics laws and other programs have often been used to justify attacks on the reproductive autonomy of women of color. For example, in the 1960s, 65 percent of the women sterilized in North Carolina were African American, though they constituted a much smaller percentage of the population.

The 1970s saw a backlash against population programs from other quarters as well. Just as feminists analyzed fertility in the broader context of women's lives, thinkers from the South analyzed population programs in the larger economic and political context. Some came to view donor-funded population programs as a substitute for meaningful economic assistance and as a means of preserving the inequitable global regime of haves and have nots. Population programs were also accused of "cultural imperialism" — exporting western values along with birth control devices.

These concerns erupted at the U.N.-sponsored World Conference on Population, held in Bucharest in 1974. Developing country delegates strongly objected to the conference's plan of action, which set population control targets without addressing economic development. The then newly formed Group of 77 non-aligned nations called instead for a New International Economic Order to counteract the inequities of the world economy. The supply versus demand debate flared in Bucharest: Developed countries favored supply-side family planning programs, while the G-77 expressed the demand-side view that "development is the best contraceptive."[27]

By the 1980s, the political winds had shifted, and the harshest criticism of population programs blew from the political right. As Sharon Camp relates in "The Politics of U.S. Population Assistance," (page 122) the election of Ronald Reagan galvanized the anti-abortion movement in the U.S. Abortion opponents lobbied successfully to curtail both public and private support for abortion and contraception, both in the U.S. and overseas. At the same time, neoconservative economic doctrine took hold in policymaking circles. The Reagan administration echoed Julian Simon's conviction that population growth can spur development, and that technological innovations can extend natural resources indefinitely.

The 1984 World Population Conference in Mexico City was a mirror image of the Bucharest debate. By that time, most developing countries acknowledged the need to slow population growth. But in a stunning policy reversal, Reagan's U.S. delegation declared population a "neutral factor" in development. The U.S. announced its so-called "Mexico City policy," which denied U.S. funding to any organization that provided abortion services, counseling, or referral — even with money from other sources. The Mexico City policy dealt a serious blow to the International Planned Parenthood Federation, which was forced to scale back its programs worldwide.

According to Sharon Camp, the U.S. government's retreat from leadership of world population efforts bears at least part of the blame for the slower progress made toward population stabilization in the last decade. During the 1970s, she notes, world contraceptive use grew by 53 percent and average family size declined 22 percent. But during the 1980s, contraceptive use grew by less than 20 percent and family size declined less than eight percent.

There may be other explanations for the slackening pace of fertility change. In some countries, it may reflect a backlash against coercive family planning programs. In India, for example, the forced sterilization program brought down Indira Gandhi's government and is blamed for setting back the progress of fertility decline by about a decade.[28]

And there may be limits to what family planning programs can achieve until social changes boost demand. It is true that such programs can have a significant impact even without socioeconomic development. In Bangladesh, for example, aggressive family planning initiatives helped reduce fertility rates by 21 percent between 1970 and 1991, although Bangladesh remains an impoverished agrarian nation with a traditional culture. Contraceptive use among married Bangladeshi women rose from three to 40 percent during that time.[29] However, the total fertility rate is still a long way from replacement level, at 4.9 children per woman.[30] In a recent study of a family planning program in the Matlab region of Bangladesh, the Population Reference Bureau questioned "whether fertility can decline to replacement level in the absence of significant improvements in living conditions."[31]

THE 1990S: FROM CONFLICT TO CONSENSUS?

The 1990s have witnessed a resurgence of interest in population issues. In the U.S., much of the renewed interest comes from environmentalists worried about the planet's carrying capacity. But environmentalists have not been warmly received by many participants in the debate. Developing country governments resent the emphasis given to population growth as a cause of environmental degradation, when consumption in the North clearly deserves much of the blame. Feminists and women's health advocates fear that environmental concerns could reinvigorate numbers-driven population programs that ignore human rights and women's health.

And the Catholic Church, a leading opponent of abortion and "artificial" contraception, has also expressed concern about coercive population programs. (The role of the Catholic Church is explored in Section VI in essays by Frances Kissling and Maura Anne Ryan.) At the 1992 Earth Summit in Rio, developing country delegates, feminists, and the Vatican all launched separate efforts to downplay the role of population growth in environmental problems.

At the same time, there are many hopeful signs that gridlock is yielding to consensus. The controversies of the 1970s offered many valuable lessons, both about the causes of PPE problems and about how best to remedy them. Criticism from the South, for example, has kindled a deeper understanding of the economic and personal contexts in which decisions about childbearing are made. The supply versus demand debate has lost some of its fervor, as studies have confirmed the benefits of fostering development and providing family planning *simultaneously*. As demographers John Bongaarts, W. Parker Mauldin, and James Phillips have written, "Investments in socioeconomic development and family planning programs have much more than simply additive effects on fertility. Instead, they operate synergistically, with one reinforcing the other."[32]

In recent years, there has been increased dialogue between feminists and women's health advocates and the population establishment which they have criticized. Many in the population community have come to see the ways in which gender bias harms women and impedes population stabilization. Among feminists, there are some who believe that all population policies are antithetical to women's well being. But there are many more who support policies that incorporate voluntary family planning, high-quality reproductive health care, and improved opportunities for girls and women. As Ruth Dixon-Mueller writes in "Women's Rights and Reproductive Choice: Rethinking the Connections," (page 227) family planning "gives women the means to shape their lives in ways undreamed of by those who have never questioned the inevitability of frequent childbearing...." But Dixon-Mueller stresses that the full benefits of reproductive autonomy are made manifest only when "opportunity structures," such as education and employment, are in place for women. "Thus," she writes, "it makes little sense to promote policies encouraging contraceptive use and smaller families without simultaneously addressing other legal, social, and economic constraints on women's rights."

The key issue, for many feminists, is that of *motivation*. Policies that are motivated primarily by a desire to bring down birth rates, many believe, will trample women's rights and health. But policies that seek first and foremost to improve the conditions of women's lives will have the beneficial side effect of reducing fertility. Women must be the subjects, rather than the objects, of population policy — or population policy simply won't work. This view is now promoted by many in the population estab-lishment, but some advocates for women are concerned that the new commitment to women's issues has not yet been translated into changed programs and funding streams. "They are talking the talk," says Jane Ordway of the International Women's Health Coalition, "but will they walk the walk?"[33]

The population and environment debate has become more temperate of late. As Donella Meadows observes in "Seeing the Population Issue Whole," (page 23) some economists "are recognizing that one does not have to pitch out the market completely to impose upon it standards of social justice and environmental stewardship." Also, Meadows notes that environmentalists "are admitting that nature can be preserved without dashing the hopes of human beings, and they are learning the hard way that equating the human population with an out-of-control cancer is not an auspicious beginning for a political discussion."

Religious groups have also become much more involved in the popula-tion and environment debate. In the past, much attention has focused on the role of religious groups as an obstacle to solving population problems — most notably, on the opposition to abortion and contraception by evangelical Protestants and Catholics. But many participants in the debate have come to realize that the world's religious traditions have much to offer: Indeed, religious teachings could serve as the moral com-pass in our journey toward a sustainable future. In Section VI, we offer a sampling of religious perspectives on PPE issues. Although these essays represent only a segment of the world's faiths, their wisdom may be uni-versal. In "Population Policies and Christian Ethics," (page 310) Reverend James Martin-Schramm argues that justice must serve as the touchstone for ethical population policy. Maura Anne Ryan concurs and adds that the pursuit of justice "is a challenge which truly undertaken involves self-critique, sacrifice, and the renunciation of privilege in all its forms." In "American Religious Groups and Population Policy," (page 303) L. Anathea Brooks and Teresa Chandler report that most American faith groups have affirmed their support for environmental stewardship and humane population policy.

ELEMENTS OF A SOLUTION

The emerging consensus is more than just an agreement in principle. There is also broad agreement on solutions. The bad news, for anyone who harbors fantasies of a "magic bullet" that will solve PPE problems, is that these are enormously complex problems that elude simplistic reme-dies. The good news is that there is a broad range of interventions that can have a significant impact. These interventions are each beneficial in and of themselves and are thus ethically unassailable. Better still, they have been shown to have a mutually reinforcing, positive synergism when implemented together. Just one example: Family planning can help lower child mortality rates, and lower child mortality in turn increases the demand for family planning. Therefore, the effect of implementing all of the solutions outlined below would be greater than the sum of its parts. These are the central elements of a solution:

Satisfy unmet need for contraception. As noted above, at least 100 million women — one in six married women in the developing countries outside China — have an unmet need for contraception. Approximately one in four births in the developing world (excluding China) is unwanted, and even more are unplanned.[34] Another measure of unmet need is the estimated 50 million abortions that are performed each year, nearly half of which are illegal and often unsafe. (See "Abortion and the Global Crisis in Women's Health," by Jodi Jacobson, page 177.) Universal access to modern contraception (including safe, legal abortion) would go a long way toward eliminating unwanted fertility. Steven Sinding, John Ross, and Allan Rosenfield contend that satisfying unmet need would decrease fertility by as much, or more, than is called for in most countries' demographic targets.

Improve the quality of reproductive health care and family planning services. Better reproductive health care is essential to halt the pandemic of disease and death that stalks so many of the world's women. Every year, 500,000 women die of maternity-related causes.[35] According to Mahmoud Fathalla, the maternal mortality rate in developing countries is fifteen times higher than in developed countries. Complications from unsafe abortion are leading causes of maternal mortality worldwide. And, as Christopher Elias notes in "AIDS: An Agenda for Population Policy," (page 185) the soaring incidence of sexually transmitted disease and reproductive tract infections demands that family planning programs take a broader approach to reproductive health. Incorporating reproductive health care into family planning programs can also have an important demographic effect. High-quality family planning programs which offer a range of reproductive health services have been shown to attract and sustain clients more effectively than programs that focus more narrowly on contraceptive delivery.

Improve child survival. Here again, the gap between North and South is appalling. Mahmoud Fathalla reports that in developing countries, the infant mortality rate is almost six times higher, and the child mortality rate is seven times higher, than in developed countries. In addition to its toll in human misery, high infant and child mortality prevents the transition to lower fertility, notes Judith Bruce in "Population Policy Must Encompass More Than Family Planning Services," (page 150). Child mortality encourages high fertility in several ways: It encourages excess births to insure that some children will survive; it instills a sense of fatalism and inability to plan for the future; and it discourages parents' investments in children's health — which in turn perpetuates high mortality rates. According to UNICEF, we already have the medical and technical capacity to drastically reduce infant and child mortality by the end of the century.

Increase access to education, especially for girls. Education of girls is one of the most reliable determinants of fertility decline. Judith Bruce cites a recent United Nations study which shows that women who have completed seven years of school have on average three children less than their unschooled counterparts. Educated women marry later, begin childbearing later, and space their births more widely.[36] As a result, the children of women with even a few years of schooling are more likely to survive than the children of uneducated women.[37] Education also enhances women's authority at home and in the community and broadens their personal and occupational horizons.

Expand life choices for young women. In many parts of the world, young women are forced by social pressures, coercion, and even violence into early marriage and childbearing. Adolescent childbearing can have a devastating effect on the health and lives of young mothers and their children. And because of the phenomenon of population momentum, it also has a significant demographic effect. The Population Council estimates that if first-time mothers in the developing countries were on average five years older, world population would ultimately stabilize at 6.1 billion, rather than at 7.3 billion.[38] But young women must have alternatives to early and frequent childbearing. Judith Bruce recommends "making young women the special focus of formal and non-formal education, community development programs, and participation in income generation and small-scale credit. [39]

Improve the status of women. Gender bias against women is a colossal human tragedy and waste, and it confounds every effort to slow population growth. The low status of women contributes to high fertility in several ways. For example, where sons are preferred over daughters, parents will have more children in order to produce the desired number of sons. Where resources such as food, education, and health care are distributed unequally to sons and daughters, parents can afford to have more children than where resources are distributed equitably. Where men bear a smaller share of the burdens of child rearing and support, they have less interest than women in limiting their fertility — yet unequal marital power relations often mean that men's fertility preferences prevail.[40] Finally, where women are denied education, secure livelihoods, property ownership, credit, and the full legal and social rights of citizenship, they will continue to rely on childbearing as a source of status and security. Thus, population policy must promote measures to insure an equal sharing of the responsibilities of childbearing and work to eliminate gender bias in education, development programs, employment, and the law.

Reduce consumption in the developed countries. No program to protect the global environment can succeed without altering the way resources are used in the developed world. As Alan Durning writes in "The Conundrum of Consumption," (page 40), "The global environment cannot support 1.1 billion of us living like American consumers, much less 5.5 billion people, or a future population of at least eight billion." And unless we citizens of the developed countries are working to clean our own house, so to speak, we lack any moral credibility in our efforts to encourage lower fertility and wise resource use in the developing world.

Efforts to reduce consumption must work on at least four levels. First, they must examine the infrastructure of life in industrialized societies and promote sustainable technologies such as solar energy, public transportation, and waste recycling. Second, they must insure that markets reflect environmental reality. "The marketplace," Timothy E. Wirth has said, "…is distorted by subsidies and out-of-date accounting systems that fail to incorporate either present environmental costs or tomorrow's economic needs." Wirth recommends adopting a new formula for calculating Gross National Product that incorporates natural assets as well as eliminating federal subsidies that encourage unsustainable use of water, timber, grazing lands, and minerals.[41]

Third, we must grapple with the culture of consumerism. We cannot build a sustainable world with an economy predicated on planned obsolescence, in an advertising-saturated society that encourages us to define our identity through our choice of consumer goods. This change will require no less than a cultural paradigm shift: as Alan Durning puts it, toward "sufficiency rather than excess." We have much to learn about sufficiency from indigenous cultures, notes Russel Barsh in "Indigenous Peoples, Population, and Sustainability," (page 290). In those cultures, as in preindustrial Europe, communitarian values and ecological awareness have traditionally held consumption in check.

Finally, we will not substantially reduce consumption unless we rethink our emphasis on growth — the relentless expansion of the quantity of materials we rip from the Earth. Truly sustainable development would emphasize instead qualitative improvements in the way resources are used and distributed.

Support economic and social development for the world's poorest citizens. The downward spiral of PPE problems will not be broken as long as one-fifth of humanity lives in abject poverty. Poverty and inequity are not part of the natural order of the universe; they are tractable and soluble. But alleviating poverty will require fundamental changes in the world economy: a fair and equitable system of global trade; responsible and responsive government in the North and South; an easing of the crushing burden of debt now shouldered by developing countries; and a renewed investment in social development, including education and health. It will also require a careful evaluation of current development strategies' effect on women and the very poor. Finally, the failed policies of structural adjustment must be replaced by development programs that are participatory, equitable, and responsive to local populations.

WHAT'S STOPPING US?

Why haven't these measures been implemented on a grand scale? Money, as always, is a major stumbling block. For too long, equally worthy programs have been forced to compete with each other for scarce resources. Family planning and reproductive health, for example, have been regarded as separate policy options, rather than as two halves of a single, mutually reinforcing whole. Scarce resources also lead to antagonism in the debate over funding priorities. Family planning providers are understandably fearful that their already inadequate resources will be further diluted if other social and health initiatives are funded from family planning budgets. Clearly, the answer is not to slice the existing pie into smaller pieces, but to make a larger pie.

The resources needed are meager in relation to their anticipated benefits. James Grant estimates that we could reduce child mortality by one-third, halve maternal mortality and child malnutrition, provide primary school education for 80 percent of the world's children, make family planning services available to all who want them, and provide clean water and sanitation for all communities — if the world's nations collectively contributed $25 billion a year over current spending levels.[42] To put that figure in perspective, consider that world *advertising* expenditures totaled $256 billion in 1990.[43]

But the scarcest resource is not money, writes Donella Meadows, "It is our willingness to listen to each other and learn from each other and to seek the truth rather than seek to be right." Indeed, much could be gained if the participants in the population debate stopped fighting each other and combined forces. This is not to suggest that differences can, or should, be set aside; each constituency in the debate has legitimate concerns that cannot be ignored. Women must be assured that they will be regarded as subjects, not objects in the design of population policy. Family planning advocates must be assured that their crucial efforts will not be sacrificed to other goals. Developing countries must be assured that population programs are not offered as a substitute for economic and social development. When these participants feel they are not being heard — or that their concerns are being given lip service rather than sincere attention — their positions become more intractable and consensus seems even more remote.

Another scarce resource is political will. As James Grant observes, "Political vision often appears to be circumscribed by opinion polls, and to extend only as far as the next election, whereas the widely acknowledged problems which threaten our own and our children's futures require vision and action on a different scale in both place and time."[44] Still, there are encouraging signs. In the U.S., the Clinton administration has voiced its commitment to assume a leadership role on population and environment issues. Vice President Al Gore has suggested that these problems merit nothing less than a "Global Marshall Plan."[45] The 1994 International Conference on Population and Development offers a once-in-a-decade opportunity to launch such a plan.

But, as Timothy E. Wirth notes in his foreword, the U.S. cannot take the lead in solving the problems of poverty, population, and the environment without the informed support of the American people. Opinion polls show that the seeds of such support exist, although in dormant form.[46] For that support to blossom, we must first comprehend our stake in building a sustainable and equitable future. We must transcend the limits of conventional wisdom and look beyond the numbers — to the human lives and dreams the numbers represent.

Population, Consumption, Development, and the Environment

....

DONELLA H. MEADOWS

Seeing the Population Issue Whole

....

The debate has been going on for almost two hundred years, since the Reverend Thomas Robert Malthus, in reaction to a group of optimistic French writers, penned his famous dictum in his 1798 *Essay on the Principle of Population*: "Taking the population of the world at any number ... the human species would increase in the ratio of — 1,2,4,8,16,32,64,128,256,512,&c, and subsistence as — 1,2,3,4,5,6,7,8,9,10,&c."

Since then, the label *Malthusian* has been attached to those who believe that the human population could push or is pushing against the earth's resources. Their opponents, the anti-Malthusians, hold that this fear is not only exaggerated but dangerous. At best, they believe, it expresses too little faith in the adaptive, creative potential of humankind. At worst, they say, it allows some people to declare other people too numerous, a threat to the planet — with horrendous social consequences.

The optimists sometimes are called *cornucopians* or Marxists, since Marx was one of the harshest critics of Malthus. Other labels for the two sides have been *antinatalists* or *ecofreaks* for the Malthusians and *pronatalists* or *technotwits* for the anti-Malthusians.

Whatever the labels, since Malthus wrote, the human population has grown by a factor of six, and total human energy use by a factor of one hundred or so. Human life expectancy has increased nearly everywhere.

Donella H. Meadows is an adjunct professor in the Environmental Studies Program at Dartmouth College. This chapter is reprinted from the June, 1993 issue of The Economist.

The forest cover of the earth has been cut by a third and the area of undisturbed wetlands by half. The composition of the atmosphere has been altered by human-generated pollution. Hundreds of millions of people have starved to death; thousands of species have gone extinct. Mines and oil wells have been depleted — and new ones have been discovered. The economy has gone on growing.

One reason the argument continues is that history offers such mixed evidence. If you are part of the richest 20 percent of the world's population, you can easily read the past as an uninterrupted human triumph over the limits of the earth. If you are among the desperately poor, you might well agree with Malthus. As he put it, in *A Summary View of the Principle of Population* (1880), "The pressure arising from the difficulty of procuring subsistence is not to be considered as a remote one which will be felt only when the earth refuses to produce any more, but as one which actually exists at present over the greatest part of the globe."

Or, as ecologist Garrett Hardin put it, writing in the November 1972 issue of the *Bulletin of the Atomic Scientists*, "Malthus has been buried again. (This is the 174th year in which that redoubtable economist has been interred. We may take it as certain that anyone who has to be buried 174 times cannot be wholly dead.)"

If a debate persists with passion for nearly two centuries, it must be true not only that the evidence is complex enough to support both sides, but also that each side is actively sifting the evidence, accumulating only that which supports preconceived notions. If people are doing that, there must be more to the argument than a scientific disagreement. There must be emotional investment as well. The protagonists in the Malthusian debate are not so much searching for truth as they are acting on commitments to see the world their way, and refusing to see it otherwise.

That shouldn't be surprising, science historian Thomas Kuhn would say, because even supposedly scientific debates have their ideological content. All human beings develop ego involvement with their own beliefs. They do so especially when the beliefs are fundamental, when they touch on the nature of humanity, the purpose of existence, the question of how we relate to nature and to each other. For centuries, scientists could not look objectively at the idea that the earth was not at the center of the universe. Many still have trouble with the idea that the earth is finite.

At least that's how I see it, as a person who has been active in the Malthusian debate but who has become less interested in winning than in understanding the intransigent nature of the discussion. I assume the argument resists resolution partly because the issues it raises are so complex and partly because they are so emotional. What I wonder is, what would we see if we were willing to approach the question of human population growth and planetary limits purely scientifically? What if we could divest ourselves of hopes, fears, and ideologies long enough to entertain all arguments and judge them fairly?

What we would see, I think, is that all sides are partly right and mostly incomplete. Each is focusing on one piece of a very complex system. Each is seeing its piece correctly. But because no side is seeing the whole, no side is coming to wholly supportable conclusions.

In short, to resolve the Malthusian conundrum and to find a way of thinking and acting that can guide a growing population to a sufficient and supportable standard of living within the earth's limits, we need all points of view. We need to treat them all with respect. We need to integrate them.

There are more than two points of view. The argument is not simply pro- and anti-Malthus. To begin what I hope will be a more comprehensive discussion, I will describe four sides of the debate here, with the understanding that many people put elements of these four together in their own unique combinations, and that there are other points of view as well. To avoid traps of labeling, I will use colors to characterize each side — though even colors carry emotional loads, as you will see.

Because of space limitations, I will have to simplify what are in each case self-consistent, sophisticated human worldviews. I will try to do so fairly (though probably no one can do that) so that the wisdom as well as the weakness in all these views will be apparent.

THE BLUES

The Blue view of the Malthusian question focuses on the possibility of keeping capital growing faster than population, so everyone can be better off. Progress, as defined by this view, comes from the accumulation of productive capital, from the building of infrastructure (roads, dams, ports) to make that capital more effective, and from the education of humans to make them more skilled and inventive in producing output from capital.

An important part of the Blue model is the assumption that capital grows most efficiently in a market system, where it is privately owned, where those who make it grow are directly rewarded, and where government interferes minimally.

Blues see living demonstrations of the workability of their view all around them. The world's most vibrant, diverse, productive, and innovative economies are those where industry is strong and where people reap material incentives for hard work, cleverness, or willingness to sacrifice in the present to invest for the future. Singapore, South Korea, and Taiwan are examples of successful development under the Blue model — and the United States, Japan, or Europe represents a vision for all the world's people of where that model can lead.

The Blues focus on raising the total level of output, not on the distribution of that output. They assume that concentrations of wealth are necessary to spur investment and that wealth will "trickle down" to enrich everyone.

Some Blues worry about population growth as a drain on investment — if too much is needed for consumption, schools, health care, and the like, then not enough will be available to plow back into the economy. Others, at the anti-Malthusian extreme, don't see even very rapid population growth as a problem at all. They see every new mouth as equipped with two new hands. The problem, in their eyes, is not how to slow the multiplication of people but how to multiply capital fast enough to put tools and machines in those hands so the new arrivals can earn their own way and even produce a surplus.

Blues see the human economy with great clarity. They see the natural world, from which raw materials come and to which wastes and pollution flow, only dimly — as a set of opportunities, a cost of production, or a source of government regulation — not as a complex system in its own right and certainly not as limited or vulnerable one.

Insofar as Blues admit that raw materials or the earth's ability to absorb pollution may be constrained, they assume that human technology and the market system can adjust. If a resource becomes scarce, if a pollution stream becomes unbearable, a cost will rise, prices will incorporate that new cost, and a technical change will bring about more efficiency or a substitute or an abatement process. There have been many examples of this kind of adjustment, from beneficiation technologies that yield metals from poorer ores to catalytic converters that have reduced the average pollution emission per car by as much as 85 percent.

Blues assume that human beings are basically competitive, individualistic, and motivated by material gain. They believe that there are real differences in merit and competence among people. Justice in this model means appropriate rewards for productivity. Injustice means rewarding those who are not productive with goods or services taken from those who are. One of the strongest assumptions in the Blue model is that promoting the good of individuals and companies will add up to the good of the entire system.

THE REDS

Reds are quiet these days, subdued by the collapse of the former Soviet Union. But their way of looking at the world has by no means disappeared; nor, would they say, has it been invalidated. What the Reds see more clearly than anyone else is the way societies systematically enrich those who already are rich, leaving the poor behind.

Reds do not assume that the enriched ones reap just rewards of superior productivity, while the left-behind ones fail because they are unwilling to work or invest. They point out many social processes, from interest payments to differential educations to the distribution of political power, that reward those who already have won and condemn many to lives of continuous losing.

Reds want to fix these inequities and oppressions. They envision a community of people working together to control resources and produce goods. It is a community that respects every person as a full member, sharing both work and output. Red assumptions about human nature are, of course, quite different from those of Blues. Reds believe that people care about the welfare of others and that no one can be truly happy while others are miserable. They believe that people respond to opportunities to serve the larger society, not just to material rewards. Justice to a Red means meeting the needs of all and never discriminating against the least fortunate.

In the Red view, labor — not capital and not natural resources — is the most critical factor of production. Therefore, people should be rewarded for their labor. Some Reds harbor a deep streak of resentment toward people who earn through rents or dividends or other payments related to ownership rather than work.

Most Reds do not trust the free market alone to add up to the common good. They see the need for social control of the economy to keep it functional and equitable. The Soviet Union, modern Reds would say, was not a real example of their philosophy at work — it was too large to manage centrally, and there was too much greed and corruption at the top. For examples of their model at its best, most Reds would point to cooperatives and worker-controlled industries all over the world, or to the mixed socialist-capitalist economies of Scandinavia.

Most Reds, like Blues, see development in terms of large-scale industry but with factories controlled by representatives of their workers. Historically, Reds have not been much concerned about population growth or the environment. They have assumed that people with tools, land, education, and political empowerment will regulate their own numbers. Like Blues, Reds see economic growth as good in itself and as a key to solving social problems. They have not until recently focused on natural resources. The possibility of a limited earth is not easy to accommodate within the Red philosophy.

THE GREENS

If Blues turn their attention to the growth of capital and technology and Reds are especially conscious of labor and patterns of distribution, Greens keep their eyes on resource depletion and pollution. They see not capital, not labor, but materials and energy as the most critical factors of production. They are worried about the size of the economic system relative to the size that nature can support. Whereas both Blues and Reds strive to make economies grow bigger, those who see the world through Green lenses fear that economies and population can grow too big to be sustainable.

Progress, according to Greens, should bring people to a state of sufficiency, not one of constant material growth. The key word is *enough* — enough food, clothing, shelter, education, and health care, and also enough clean water, green trees, and unspoiled natural beauty. The major threats to achieving this vision are production methods that waste resources and populations and economies that stress ecosystems.

The path to development, in this view, is to reduce excessive human demands for both production and reproduction. That means stabilizing or even diminishing populations, moderating material wealth, and choosing technologies that enhance, rather than destroy, the natural world. Greens are as technologically optimistic as Blues and Reds but only when it comes to technologies they like. They believe solar energy can work but not nuclear power. They think materials can be recycled almost indefinitely but not taken from the earth indefinitely. Greens are less likely than Blues or Reds to call upon a generalized technology to solve all problems. No technology, say the Greens, will allow continuous expansion of population and production on a limited planet.

From the Green point of view, both the market and social-equity measures may be necessary in an ideal world, but they will do no good if they are not contained within a mind-set of harmony with the environment. This mind-set would admit that human beings are both communal and individualistic, both greedy and altruistic, but it also would assume that humans have evolved in relationship with nature, that they require continued contact with nature to be truly happy, and that functioning ecosystems are needed to support a functioning economy. Both justice and pragmatism, from the Green point of view, must ensure the welfare of all species, not just *Homo sapiens.*

Green thinkers favor incentives for small families and disincentives for large ones. They favor either adjustments to the market, so that real environmental costs are contained within prices, or strong regulatory measures to prevent public and private actors from destroying resources. Most Greens are not as fond of coercion as Blues think they are, but, as with all these points of view, there is a range of opinion even within a single camp. Some Greens would be quite willing to use the police power of society to protect nature from greedy or senseless depredation. They would argue that crimes against nature are, directly or indirectly, crimes against humanity as well.

Blues tend to see Greens as Reds in disguise. Reds tend to see Greens as elitists who live at the ends of long, winding roads and who do not care about the struggle for economic justice. In fact, Greens do not fit on the Red-Blue spectrum at all.

THE WHITES

The White view combines some aspects of all the previously mentioned colors (which is why I have dubbed it white), but it rejects their centralist, we-will-tell-you-how-to-behave tone. Whites see any policy as worthwhile only if it comes out of the wisdom and efforts of the people. Their emphasis is not on revolutionary redistribution or population control or building factories or planting trees but on empowering people to take control of their own lives. They care less about what should be done and more about who decides.

This model sees progress as local self-reliance. An important concept to the Whites is appropriate technology — technology that uses tools that can be manufactured and maintained at the local level, that uses nearby resources and skills, and that yields products needed close at hand. The best agents for development in this view might be facilitators (something on the order of Peace Corps workers) who are familiar with modern technologies (vaccines, for instance, or how to hybridize plant varieties). A facilitator should know how to tap outside resources and knowledge, when necessary, but should come from, live with, and feel himself to be one of the people.

From the White point of view, all other ideologies originate from citified, intellectual people who confuse their compassion for the poor with their own personal agendas. Environment, class conflict, and the free market are abstractions that show their worth only when applied to specific questions like how to make a particular field grow more grain, or how to allocate the water from a new irrigation system, or how to design a biogas energy system or get a fair price for fish.

According to Whites, big loans for big industry are likely to trample real development; very small loans are what is needed. Market incentives are fine as long as they help little businesses compete instead of reinforcing the power of big businesses. Land reform, family planning, health care, education, reforestation, and prevention of soil erosion are all okay as long as they are planned and controlled by local people. They are not okay if they are promoted by central authorities.

Many White organizations are strongly environmental — such as the rubber tappers in Brazil — because people often have considerable knowledge of local resources and a direct stake in preserving them. But Whites see the environment as a *working* environment, one within which and from which people live, not one that is kept pristine for the admiration of tourists.

The White view of human nature is very positive. It assumes that even the most common people (and especially the most common people) can be entrepreneurial, communal, industrious, moderate in their material demands, and gentle to their environment — if they are free to be that way. Justice in this view means removing the obstacles that keep people from taking control of their own lives.

THE CLASH OF MODELS; THE CONSOLIDATION OF MODELS

I have simplified each of these views greatly, but not as much as they simplify each other. Each side has a tendency to define itself by its own more moderate beliefs and to see the other sides at their extremes.

I could go on to describe how the various parties call upon their own stables of biased experts, how they use the same words in different ways, how they seize upon different indices of good or bad performance, how they commit egregious logical errors. But I want to dwell here not on the ways these protagonists differentiate themselves from one another, but on how their arguments overlap — and are coming together.

Reds the world over are experimenting with the undeniable efficiencies of the market. Some Blues are recognizing that one does not have to pitch out the market completely to impose upon it standards of social justice and environmental stewardship. White development leaders I know are asking themselves how they can scale up, how they can use the skills of large-scale management without losing the ability to listen to the people. Greens are admitting that nature can be preserved without dashing the hopes of human beings, and they are learning the hard way that equating the human population with an out-of-control cancer is not an auspicious beginning for a political discussion. The earth itself is making clear to people from the Himalayas to the World Bank that their visions of development have to rest upon a foundation of a healthy, functioning planet.

The Greens are correct: Population growth that causes people to level forests and overgraze lands exacerbates poverty. The Reds are correct: The helplessness of poverty creates the motivation for parents to have many children, as their only hope of providing for themselves. The Blues are right: Economic development can bring down birthrates. The Whites are right: Development schemes work only when they are not imposed from on high. The Greens are right: Family planning alone is not enough to stabilize populations. The Reds are right: Populations stabilize when all people have a real economic stake in their society.

Capital can be the scarcest factor of production at some times and places, labor at other times and places, materials and energy and pollution-absorption capacity at still others. The limits the Greens point out really are there. So are the injustices the Reds want us to see. So are the market and technical responses the Blues have faith in. And so is the wisdom of the people that the Whites respect.

The earth almost certainly is more resilient than the Greens think it is, and less so than the Blues think it is. The collapse of industrial society because of populations and economies growing past the earth's carrying capacity is not inevitable. It also is not impossible.

What would we do about population, development, and environment if we allowed ourselves to see from all these viewpoints at once? My guess is that at first we would not so much change policy as alter the way we advocate, choose, and implement policy. We would be less doctrinaire and more open to learning. Instead of seeing the results we want to see from the social and technical experiments going on in the world, we would try to see what actually is happening. We would proceed more experimentally. We probably would learn that different continents, cultures, and communities require different blends of the policies advocated by the various points of view, but that in general all the policies being advocated are needed to one extent or another.

Most parts of the world need more productive capital. The places that already are overcapitalized need more efficient, resource-conserving, nonpolluting, elegant capital. Every nation on earth needs technologies that produce the same final result from less energy, material, and labor. True ingenuity and productivity should be rewarded quickly and unambiguously; at the same time, the inequities that stifle ingenuity and productivity should be removed.

Population growth should be slowed and finally stopped everywhere. The way to do that is through both fertility-control technologies and people-based, empowering economic development.

Mindless, wasteful consumption also should be slowed and stopped. How to do that is hard for us in the rich parts of the world to imagine, because it is the change that will affect us most. I believe it is also the change that will free us and delight us most. First, it will involve rethinking our economics enough to get our prices and indices of success (such as the GNP) to reflect real costs and human values. Second, it will require greater sophistication about the goal of growth, which will be seen as a means, not an end. Some kinds of growth are needed and welcome; others are excessive and destructive. We have to be able to differentiate. Finally, I think, we have to distinguish our material needs, which are undeniable, from our nonmaterial needs. The latter, which also are undeniable, involve everything from self-esteem to salvation and cannot be met by material accumulation. Getting that straight — seeing, for example, that a car is a means of transportation, not a means of self-importance — is key not only to "saving the earth" but to real happiness.

In the final analysis, if we were to admit the relevance of all points of view, we would see that we need to pay as close attention to the earth's energy and material flows as we do to our economy's money flows. We need to keep resource accounts, like bank accounts, and never commit the foolishness of spending down our capital while calling it income. If we did that, we would discover that our planet is enormously bountiful but not infinite. We would see how to achieve our human dreams without destroying either the resources or the natural magnificence that will allow future generations to achieve their dreams.

The scarcest resource is not oil, metals, clean air, capital, labor, or technology. It is our willingness to listen to each other and learn from each other and to seek the truth rather than seek to be right. Because we have not done that, another resource has become critically scarce: time. With the world population growing now by 95 million a year, most of whom are born in poor nations; with forests, soil, water, and ecosystems being degraded around the world; with people to educate and factories to build and new technologies to develop, there is no time to continue the Malthusian argument fruitlessly for another two hundred years.

MARK SAGOFF

Population, Nature,
and the Environment

....

In 1989, when Jessica Tuchman Mathews of the World Resources
Institute proposed an agenda for international environmental policy
in the 1990s, she observed that "for the first time in its history,
mankind is rapidly—if inadvertently—altering the basic physiology of
the planet." Pointing to the effects of deforestation and soil degradation,
changes in the composition of the atmosphere, and other examples of
environmental deterioration, Ms. Mathews observed, "Individuals and
governments alike are beginning to feel the cost of substituting for (or
doing without) the goods and services once freely provided by healthy
ecosystems." "Nature's bill," she continued,

> is presented in many different forms: the cost of commercial fertil-
> izer needed to replenish once naturally fertile soils; the expense of
> dredging rivers that flood their banks because of soil erosion hun-
> dreds of miles upstream; the loss in crop failures due to the indis-
> criminate use of pesticides that inadvertently kill insect pollinators;
> or the price of worsening pollution, once filtered from the air by
> vegetation. Whatever the immediate cause for concern, the value
> and absolute necessity for human life of functioning ecosystems is
> finally becoming apparent.

Ms. Mathews saw population growth as central to most of the environmental trends she described. She did not mean by this that the human

*Mark Sagoff is the director of the Institute for Philosophy and Public Policy at the
University of Maryland. This chapter is reprinted from the* Report from the
Institute for Philosophy and Public Policy, *Fall, 1993.*

population was about to exceed some absolute limit imposed by the carrying capacity of the earth: "The planet," she wrote, "may ultimately be able to accommodate the additional five or six billion people projected to be living here by the year 2100." However, she warned, "it seems unlikely that the world will be able to do so unless the means of production change dramatically."

ASSESSING THE ROLE OF POPULATION GROWTH

By now, we are accustomed to arguments that rapid population growth inherently poses a threat to the natural environment. It seems obvious, for example, that the many billions of people who will be added to the world population over the next several decades will present greater demands for energy; they will need to burn more coal, gas, oil, or wood to satisfy their ordinary needs. Moreover, the demand for energy, and therefore the adverse environmental impact, may be all the greater if economic growth provides higher standards of living for the world's poor. This plausible view of events has led Paul Ehrlich among others to argue that population growth is the salient cause of environmental deterioration. The "causal chain of the deterioration," Ehrlich wrote in 1968, "is easily followed to its source . . . too many people."

In presenting this view, however, most analysts are also careful to say how complex the relationship is between population levels and the resource base. As physicist John Holdren observes, "I know of no analyses that have even begun to quantify in any comprehensive way the role played by population growth . . . in the growth of associated environmental damages." There are simply too many variables in addition to population that must be taken into account. "Policies, technologies, and institutions determine the impact of population growth," Ms. Mathews writes. "These factors can spell the difference between a highly stressed, degraded environment and one that can provide for many more people."

Consider, for example, the often-cited stories about Nepalese fuelwood gatherers who deplete forests at non-renewable rates, or African farmers who overcultivate their land, causing the soil to erode. These stories, usually presented as evidence of the harm created by overpopulation, are actually more compelling as illustrations of the baneful environmental consequences of poverty. Nepalese peasants do not have the same access as their better-off compatriots to natural gas and petroleum, and so they must destroy the forests for fuel. Subsistence farmers cannot afford soil conservation measures, and so they exhaust the land. Moreover, as World Bank official R. Paul Shaw explains, the "connections among population growth, poverty, and environment are usually exacerbated by powerful economic and political factors":

Distortionary pricing policies, urban bias in development expenditures . . . half-hearted agrarian reforms and failure to establish rights of land tenure, mismanagement of common lands, protectionism, massive international indebtedness, intergroup conflict, ethnocide, and genocide have all contributed to the vicious downward spiral of poverty and environmental degradation. The poor have been exploited, shifted, and marginalized to the extent that they often have no choice but to participate in the denigration of resources, with full knowledge that they are mortgaging their own future.

Thus, it would clearly be a mistake to blame deforestation or soil erosion solely on the existence of "too many people."

Analysts also point out that the developed world's consumption and capital are often more responsible for resource depletion in the poorer countries than are the growing populations of those countries. Disparities between the consumption levels of developed and developing nations are profound. According to a 1991 report by the U.N. Population Fund, developed countries, which contain only 25 percent of world population, account for 75 percent of world energy use and consume 85 percent of all forest products and 72 percent of steel production. Informed estimates blame developed countries for about 75 percent of the 2,500 billion tons of waste generated in 1985 worldwide. At present, the per capita energy consumption of the United States is 250 times as great as that of many poor countries.

These disparities suggest that the developed world, despite its stable population levels, threatens the resource base and biospheric system much more than the developing world with its rising population. Here again, the example of deforestation is instructive. While local populations can nibble around the edges of rain forests, it takes enormous capital investments to deforest on a major scale. The massive highway project that opened up the Brazilian rain forest to grand-scale exploitation could not have been built by the local peasants; it was financed by the World Bank. As a U.S. official, quoted by R. Paul Shaw, describes the situation: "Yes, rapidly growing numbers of peasants contribute to tropical deforestation, but on a global scale their activities are probably more akin to picking up branches and twigs after commercial chain saws have done their work."

It is undeniable that rapid population growth can exacerbate, or work in tandem with other factors in bringing about, an array of environmental problems, including those associated with poverty, urbanization, energy use, and waste disposal. Yet the connection between population growth and environmental deterioration is difficult if not impossible to quantify, and much of the "common wisdom" concerning it is probably overstated.

FOOD AND FARMLAND

This cautious approach to the question of how population levels affect the environment is consistent with the views of ecologist Barry Commoner. Mr. Commoner has long argued that population growth is only one element influencing environmental quality, and that population levels interact with other factors in a variety of ways in different circumstances. This, in turn, leads him to conclude that environmental and resource concerns cannot serve as the principal reasons for instituting population programs. It is possible, he believes, to "produce bountiful harvests, productive machinery, rapid transportation, and decent human dwellings sufficient to support the world population without despoiling the environment." But in order to do this, we must, as Jessica Mathews suggests, change the "means of production," developing technologies that will enable us to meet human needs without destroying the earth.

The example of food and farmland offers some support for Mr. Commoner's position, but also poses challenges to it. From a global point of view, persistent and growing surpluses, rather than episodes of scarcity, have been the recent bane of agricultural markets; indeed, food production worldwide has increased substantially during the past thirty years. Yet more than a billion people are malnourished because they have no money to pay for food, even at today's historically low prices.

Thus, it is not the earth's capacity to produce food that is inadequate. There is food enough to sustain our growing human population—but it does not reach those in need. The reason is that the vast increases in yields-per-acre of recent decades depend on technologies that do nothing for subsistence farmers who cannot afford to purchase them. Early users of these technologies, by raising yields and cutting costs, manage to drive less efficient farmers out of business, and as a result of this "technological treadmill," more and more arable land is taken out of production. In the United States, farm subsidies have also played a role in reducing cropland. Although acreage devoted to subsistence farming around the world is on the rise, this increase is offset by decreases in acreage used by large-scale growers.

Because economies of scale place the major growers at a great advantage, farmers in developing countries are hard-pressed to compete with large producers from abroad, even for the urban markets in their own countries. These farmers also confront political, social, and economic barriers in planting and harvesting. In Somalia, where farmers are now producing a crop sufficient to feed the country, it is clear that the political and social climate is as much a factor in food production as weather and soil conditions.

The dilemma in developing countries—whether to improve local production or to permit global markets to supply grain and other commodities—has no easy solution. The decision depends on a thousand variables, including growing conditions, economic systems, and the integration of a locality into global trading networks. Ultimately, however, the question of food is a question of justice and equity; it relates not to the earth's capacity to feed a growing population, but to the ability of markets and social systems to make food available to the poor. Thus, while technology has made it possible for us to produce more food, it has not created the conditions necessary to feed the ever-rising numbers of the world's people.

ENERGY PRODUCTION

The example of energy production might lend more unequivocal support to technological optimists such as Mr. Commoner. Lester Brown of the Worldwatch Institute recently noted that the sun sends more than enough energy our way to provide clean, low-priced fuel for economic growth, provided we can capture that energy and use it efficiently. Some analysts favor intensive development of wind energy and photovoltaic arrays, while others hope for a breakthrough in fusion technology. Such innovations may well allow us to convert more and more of the earth's surface and its resources to sustainable human use.

However, this optimistic scenario can occur only under favorable political, economic, and cultural circumstances. Those nations without the wealth to purchase technology from abroad or the resources to develop it themselves may confront overwhelming difficulties in coping with population growth. This is true even of countries that are rich in natural resources—Nigeria is an example—but which suffer from serious political and economic problems.

What is more, even if we are able to develop technologies that replace fossil fuels with solar or fusion energy, this alone will do nothing to protect Nature or to spare our ecological and evolutionary heritage. Indeed, it may simply enable us to continue doing what technology always has done—to subdue Nature and transform it to our use. In order to grasp this point, however, we must first make a distinction between Nature and the environment.

NATURE VERSUS THE ENVIRONMENT

Nature and the environment are best understood as distinct concepts. Nature is the object of religious, aesthetic, and cultural contemplation and appreciation; in the nineteenth century, it was also the province of natural history, which attended to natural facts without inquiring into their practical usefulness to human beings. The environment, in contrast, is a concept of more recent origin. It is the object of the economic and biological sciences that attempt to predict, control, and "price" flows of materials and resources (from genetic materials to biospheric systems) in order to maximize long-run benefits from their use.

John McPhee, Edward Abbey, and many other Nature writers—who eulogize the earth's vanishing natural heritage—tend to define Nature and technology as opposites. As essayist Noel Perrin remarks, Nature comprises "everything on this planet that is at least partially under the control of some other will than ours." Moreover, many of us believe that human beings, by "conquering" Nature and imposing our will upon it, contaminate it. This sense of "contamination" is of religious origin; as we use technology to control and manipulate Nature, we reenact the crime that expelled us from paradise. Ecology book titles such as *The Exploited Eden* and *Leaving Eden* illustrate this view.

Yet the environment is, in fact, a kind of technology. It is the "plumbing" and "infrastructure" we find, as distinct from that which we build ourselves. Its value is instrumental—not religious, moral, cultural, or aesthetic. The environment is what Nature becomes when we view it as a life-support system and as a collection of materials. It is "natural capital" as distinct from "human capital"; it is a collection of "services" that often come "free" in the sense that because nobody owns them, nobody can charge a fee for their use.

Human beings cannot live very long or very well in Nature in the sense of uncultivated wilderness. Indeed, during earlier centuries, the conquest and subjugation of Nature—not its preservation and protection—seemed to be the path to human survival. When the Pilgrims landed in what they described as the American wilderness, half of them died during the first year for want of food, houses, and "improvements."

The landscape that greets a visitor to Plymouth Rock today—vacation homes, shopping malls, highways—shows what a growing population does to Nature under even the most favorable circumstances. Cape Cod—like Ocean City, Atlantic City, and other tourist meccas along the Atlantic coast—attracts thousands and thousands of visitors. It is an environment people will pay $1,000 a week to visit—but one in which very little of Nature is left.

To a large extent, human beings have prospered by pushing Nature back and putting a largely man-made environment in its place—cities where there were forests, farms where there were fields. Today, a widely shared moral and aesthetic commitment to retain the last vestiges of creation has led us to try to stop this process—to protect rain forests, for example, from the economic forces that would replace them with ranches and farms. But this commitment to preserving Nature stems primarily from a belief in its intrinsic value, rather than from a desire to preserve natural resources for our future use. Indeed, biologists such as Daniel Janzen have argued that the single greatest force favoring conservation is not the utility of Nature but its opposite—the *inviability* of rain forests and other environments for agricultural purposes. "When genetic engineering gives us crop plants and animals that thrive in the various tropical rain forest habitats," Mr. Janzen writes, "it is 'good-bye, rain forest.'"

FUTURE IMPACT

We have seen that the developed world, despite its stable population levels, poses a greater threat to the resource base and biospheric system than does the developing world. But this state of affairs is unlikely to last. If incomes in less developed countries continue to grow at about three percent annually, forty years from now these countries might possibly produce more than half of global waste loadings (though still less per capita than the richer nations), and the world economy will be five times as large as it is today.

The impact of this growing population may be far greater on Nature than on the environment. Indeed, the very technological improvements and break-throughs that optimists such as Mr. Commoner count on may only increase the speed with which natural landscapes are transformed into economically successful man-made and man-managed ones. As these technologies become more efficient, they might succeed in bringing the earth increasingly under human control. The more sustainable technologies we develop to make ourselves independent of natural processes—biotechnology, fusion energy, and so on—the less prudence will require us to protect the natural world. But then, as we have seen, there are other, more powerful reasons—religious, aesthetic, cultural—that we might choose to do so.

In view of the doubling of world population that demographers expect over the next fifty years, there surely will be more than enough people on earth to destroy Nature and imperil the environment. Indeed, even the five or six billion of us alive today could do that. Accordingly, we must look to changes in our political, economic, and social relationships—as well as to more benign technologies—if we are to keep any remnants of Nature intact for future generations. We shall also have to enter into and enforce more agreements such as the Montreal Protocol on ozone to protect common atmospheric and other resources. Only in conjunction with such initiatives can population policies help us to sustain functioning ecosystems and forestall the death of Nature.

ALAN THEIN DURNING

The Conundrum of Consumption

. . . .

For Sidney Quarrier of Essex, Connecticut, Earth Day 1990 was Judgment Day — the day of ecological reckoning. While tens of millions of people around the world were marching and celebrating in the streets, Sidney was sitting at his kitchen table with a yellow legal pad and a pocket calculator. The task he set himself was to tally up the burden he and his family had placed on the planet since Earth Day 1970.[1]

Early that spring morning he began tabulating everything that had gone into their house — oil for heating, nuclear-generated electricity, water for showers and watering the lawn, cans of paint, appliances, square footage of carpet, furniture, clothes, food, and thousands of other things — and everything that had come out — garbage pails of junk mail and packaging, newspapers and magazines by the cubic meter, polluted water, and smoke from the furnace. He listed the resources they had tapped to move them around by car and airplane, from fuel and lubricants to tires and replacement parts. "I worked on that list most of the day," Sid remembers. "I dug out wads of old receipts, weighed trash cans and the daily mail, excavated the basement and shed, and used triangulation techniques I hadn't practiced since graduate school to estimate the materials we used in the roofing job."[2]

Manufacturing and delivering each of the objects on his list, Sid knew, had required additional resources he was unable to count. National sta-

---

Alan Thein Durning is the director of Northwest Environment Watch. This chapter is adapted from How Much Is Enough? The Consumer Society and the Fate of the Earth *(New York: W.W. Norton & Company, 1992).*

tistics suggested, for example, that he should double the energy he used in his house and car to allow for what businesses and government used to provide him with goods and services. He visualized a global industrial network of factories making things for him, freighters and trucks transporting them, stores selling them, and office buildings supervising the process. He wondered how much steel and concrete his state needed for the roads, bridges, and parking garages he used. He wondered about resources used by the hospital that cared for him, the air force jets and police cars that protected him, the television stations that entertained him, and the veterinary office that cured his dog.

As his list grew, Sid was haunted by an imaginary mountain of discarded televisions, car parts, and barrels of oil — all piling up toward the sky on his lot. "It was a sober revisiting of that period...It's only when you put together all the years of incremental consumption that you realize the totality." That totality hit him like the ton of paper packaging he had hauled out with the trash over the years: "The question is," Sid said, "Can the earth survive the impact of Sid, and can the Sids of the future change?"[3]

That *is* the question. Sidney Quarrier and his family are no gluttons. "During those years, we lived in a three bedroom house on two-and-a-half acres in the country, about 35 miles from my job in Hartford," Sidney recounts. "But we have never been rich," he insists. "What frightened me was that our consumption was typical of the people here in Connecticut."[4]

Sid's class — the American middle class — is the group that, more than any other, defines and embodies the contemporary international vision of the good life. Yet the way the Quarriers lived for those 20 years is among the world's premier environmental problems, and may be the most difficult to solve.

Only population growth rivals high consumption as a cause of ecological decline, and at least population growth is now viewed as a problem by many governments and citizens of the world. Consumption, in contrast, is almost universally seen as good — indeed, increasing it is the primary goal of national economic policy. The consumption levels exemplified in the two decades Sid Quarrier reviewed are the highest achieved by any civilization in human history. They manifest the full flowering of a new form of human society: the consumer society.

This new manner of living was born in the United States, and the words of an American best capture its spirit. In the age of U.S. affluence that began after World War II, retailing analyst Victor Lebow declared: "Our enormously productive economy ... demands that we make consumption our way of life, that we convert the buying and use of goods into rituals, that we seek our spiritual satisfaction, our ego satisfaction, in consumption ... We need things consumed, burned up, worn out,

replaced, and discarded at an ever increasing rate." Most citizens of western nations have responded to Lebow's call, and the rest of the world
appears intent on following.[5]

The life-style made in the United States is emulated by those who can
afford it around the world, but many cannot. The economic fault lines that
fracture the globe defy comprehension. The world has 202 billionaires and
more than three million millionaires. It also has 100 million homeless people who live on roadsides, in garbage dumps, and under bridges. The value
of luxury goods sales worldwide — high-fashion clothing, top-of-the-line
autos, and the other trappings of wealth — exceeds the gross national
products of two thirds of the world's countries. Indeed, the world's average
income, about $5,000 a year, is below the U.S. poverty line.[6]

THE ECOLOGICAL CLASS SYSTEM

The world has three broad ecological classes: the consumers, the middle income, and the poor. These groups, ideally defined by their per
capita consumption of natural resources, emissions of pollution, and disruption of habitats, can be distinguished in practice through two proxy
measures: their average annual incomes and their life-styles.

The world's poor — some 1.1 billion people — includes all households that earn less than $700 a year per family member. They are mostly
rural Africans, Indians, and other South Asians. They eat almost exclusively grains, root crops, beans, and other legumes, and they drink mostly
unclean water. They live in huts and shanties, they travel by foot, and
most of their possessions are constructed of stone, wood, and other substances available from the local environment. This poorest fifth of the
world's people earns just two percent of world income.[7]

The 3.3 billion people in the world's middle-income class earn
between $700 and $7,500 per family member and live mostly in Latin
America, the Middle East, China, and East Asia. This class also includes
the low-income families of the former Soviet bloc and of western industrial nations. With notable exceptions, they eat a diet based on grains and
water, and lodge in moderate buildings with electricity for lights, radios,
and, increasingly, refrigerators and clothes washers. (In Chinese cities,
for example, two thirds of households now have washing machines and
one fifth have refrigerators.) They travel by bus, railway, and bicycle, and
maintain a modest stock of durable goods. Collectively, they claim 33 percent of world income.[8]

The consumer class — the 1.1 billion members of the global consumer
society — includes all households whose income per family member is
above $7,500. Though that threshold puts the lowest ranks of the con-
sumer class scarcely above the U.S. poverty line, they — rather, we — still
enjoy a life-style unknown in earlier ages. We dine on meat and

processed, packaged foods, and imbibe soft drinks and other beverages from disposable containers. We spend most of our time in climate-controlled buildings equipped with refrigerators, clothes washers and dryers, abundant hot water, dishwashers, microwave ovens, and a plethora of other electric-powered gadgets. We travel in private automobiles and airplanes, and surround ourselves with a profusion of short-lived, throwaway goods. The consumer class takes home 64 percent of world income — 32 times as much as the poor.[9]

The consumer class counts among its members most North Americans, West Europeans, Japanese, Australians, and the citizens of Hong Kong, Singapore, and the oil sheikdoms of the Middle East. Perhaps half the people of Eastern Europe and the Commonwealth of Independent States are in the consumer class, as are about one fifth of the people in Latin America, South Africa, and the newly industrializing countries of Asia, such as South Korea.[10]

The gaping divide in material consumption between the fortunate and unfortunate stands out starkly in their impacts on the natural world. The soaring consumption lines that track the rise of the consumer society are, from another perspective, surging indicators of environmental harm. The consumer society's exploitation of resources threatens to exhaust, poison, or unalterably disfigure forests, soils, water, and air. We, its members, are responsible for a disproportionate share of all the global environmental challenges facing humanity.

The consumer class's use of fossil fuels, for example, causes an estimated two thirds of the emissions of carbon dioxide from this source. (Carbon dioxide is the principal greenhouse gas.) The poor typically are responsible for the release of a tenth of a ton of carbon apiece each year through burning fossil fuels; the middle-income class, half a ton; and the consumers, 3.5 tons. In the extreme case, the richest tenth of Americans pump 11 tons into the atmosphere annually.[11]

Parallel class-by-class evidence for other ecological hazards is hard to come by, but comparing industrial countries, home to most of the consumers, with developing countries, home to most of the middle-income and poor, gives a sense of the orders of magnitude. Industrial countries, with one fourth of the globe's people, consume 40-86 percent of the earth's various natural resources.

From the crust of the earth, we take minerals; from the forests, timber; from the farms, grain and meat; from the oceans, fish; and from the rivers, lakes, and aquifers, fresh water. The average resident of an industrial country consumes three times as much fresh water, 10 times as much energy, and 19 times as much aluminum as someone in a developing country. The ecological impacts of our consumption even reach into the local environments of the poor. Our appetite for wood and minerals, for example, motivates the road builders who open tropical rain forests to poor settlers, resulting in the slash-and-burn forest clearing that is condemning countless species to extinction.

High consumption translates into huge impacts. In industrial countries, the fuels burned release perhaps three fourths of the sulfur and nitrogen oxides that cause acid rain. Industrial countries' factories generate most of the world's hazardous chemical wastes. Their military facilities have built more than 99 percent of the world's nuclear warheads. Their atomic power plants have generated more than 96 percent of the world's radioactive waste. And their air conditioners, aerosol sprays, and factories release almost 90 percent of the chlorofluorocarbons that destroy the earth's protective ozone layer.[12]

The furnishings of our consumer life-style — things like automobiles, throwaway goods and packaging, a high-fat diet, and air conditioning — can only be provided at great environmental cost. Our way of life depends on enormous and continuous inputs of the very commodities that are most damaging to the earth to produce: energy, chemicals, metals, and paper. In the United States, those four industries are all in the top five of separate industry-by-industry rankings for energy intensity and toxic emissions, and similarly dominate the most-wanted lists for polluting the air with sulfur and nitrogen oxides, particulates, and volatile organic compounds.[13]

Thus from global warming to species extinction, we consumers bear a huge burden of responsibility for the ills of the earth. Yet our consumption too seldom receives the attention of those concerned about the fate of the planet, who focus on other contributors to environmental decline. Consumption is the neglected variable in the global environmental equation. In simplified terms, an economy's total burden on the ecological systems that undergird it is a function of three variables: the size of the population, average consumption, and the broad set of technologies — everything from dinner plates to communications satellites — the economy uses to provide goods and services. Generally, environmentalists work on regulating and changing technologies, and family planning advocates concentrate on slowing population growth.

There are good reasons for emphasizing technology and population. Technologies are easier to replace than cultural attitudes. Family planning has enormous human and social benefits aside from its environmental pluses. Yet the magnitude of global ecological challenges requires progress on all three fronts. Environmental economist Herman Daly of the World Bank points out, for example, that simply stopping the growth in rates of global pollution, ecological degradation, and habitat destruction — not reducing those rates, as is clearly necessary — would require within four decades a twentyfold improvement in the environmental performance of current technology. And that assumes both that industrial countries immediately halt the growth of their per-capita resource consumption, allowing the developing countries to begin catching up, and that world population no more than doubles in that period.[14]

Changing technologies and methods in agriculture, transportation, urban planning, energy, and the like could radically reduce the environmental damage caused by current systems, but a twentyfold advance is farfetched. Autos that go three or four times as far on a tank of fuel are feasible; ones that go 20 times as far would defy the laws of thermodynamics. Bicycles, buses, and trains are the only vehicles that can reduce the environmental costs of traveling by that much, and to most in the consumer class they represent a lower standard of living. Clothes dryers, too, might run on half as much energy as the most efficient current models, but the only way to dry clothes with one twentieth the energy is to use a clothesline — another retrogressive step, in the eyes of the consumer society. So technological change and population stabilization cannot suffice to save the planet without their complement in the reduction of material wants.

THE HUMAN COSTS OF CONSUMPTION

Ironically, high consumption is a mixed blessing in human terms too. People living in the nineties are on average four-and-a-half times richer than their great-grandparents were at the turn of the century, but they are not four-and-a-half times happier. Since 1940, Americans alone have used up as large a share of the earth's mineral resources as did everyone before them combined. Yet this historical epoch of titanic consumption appears to have failed to make the consumer class any happier. Regular surveys by the National Opinion Research Center of the University of Chicago reveal, for example, that no more Americans report they are "very happy" now than in 1957. The "very happy" share of the population has fluctuated around one third since the mid-fifties, despite near-doublings in both gross national product and personal consumption expenditures per capita.[15]

Worse, two primary sources of human fulfillment — social relations and leisure — appear to have withered or stagnated in the rush to riches. Thus many of us in the consumer society have a sense that our world of plenty is somehow hollow — that, hoodwinked by a consumerist culture, we have been fruitlessly attempting to satisfy with material things what are essentially social, psychological, and spiritual needs.[16]

Of course, the opposite of overconsumption — destitution — is no solution to either environmental or human problems. It is infinitely worse for people and bad for the natural world too. Dispossessed peasants slash-and-burn their way into the rain forests of Latin America, hungry nomads turn their herds out onto fragile African rangeland, reducing it to desert, and small farmers in India and the Philippines cultivate steep slopes, exposing them to the erosive powers of rain. Perhaps half the world's billion-plus absolute poor are caught in a downward spiral of ecological and economic impoverishment. In desperation, they knowingly abuse the land, salvaging the present by savaging the future.[17]

HOW MUCH IS ENOUGH?

If environmental destruction results when people have either too little or too much, we are left to wonder, How much is enough? What level of consumption can the earth support? When does having more cease to add appreciably to human satisfaction? Is it possible for all the world's people to live comfortably without bringing on the decline of the planet's natural health? Is there a level of living above poverty and subsistence but below the consumer life-style — a level of sufficiency? Could all the world's people have central heating? Refrigerators? Clothes dryers? Automobiles? Air conditioning? Heated swimming pools? Airplanes? Second homes?

Many of these questions cannot be answered definitively, but for each of us in the consumer society, asking is essential nonetheless. Unless we see that more is not always better, our efforts to forestall ecological decline will be overwhelmed by our appetites. Unless we ask, we will likely fail to see the forces around us that stimulate those appetites, such as relentless advertising, proliferating shopping centers, and social pressures to "keep up with the Joneses." We may overlook forces that make consumption more destructive than it need be, such as subsidies to mines, paper mills, and other industries with high environmental impacts. And we may not act on opportunities to improve our lives while consuming less, such as working fewer hours to spend more time with family and friends.

Still, the difficulty of transforming the consumer society into a sustainable one can scarcely be overestimated. We consumers enjoy a life-style that almost everybody else aspires to, and why shouldn't they? Who would just as soon not have an automobile, a big house on a big lot, and complete control over indoor temperature throughout the year? The momentum of centuries of economic history and the material cravings of 5.5 billion people lie on the side of increasing consumption.

We may be, therefore, in a conundrum — a problem admitting of no satisfactory solution. Limiting the consumer life-style to those who have already attained it is not politically possible, morally defensible, or ecologically sufficient. And extending that life-style to all would simply hasten the ruin of the biosphere. The global environment cannot support 1.1 billion of us living like American consumers, much less 5.5 billion people, or a future population of at least eight billion. On the other hand, reducing the consumption levels of the consumer society, and tempering material aspiration elsewhere, though morally acceptable, is a quixotic proposal. It bucks the trend of centuries. Yet it may be the only option.

If the life-supporting ecosystems of the planet are to survive for future generations, the consumer society will have to dramatically curtail its use of resources — partly by shifting to high-quality, low-input durable goods and partly by seeking fulfillment through leisure, human relationships, and other nonmaterial avenues. We in the consumer society will have to live a technologically sophisticated version of the life-style currently practiced lower on the economic ladder.

If our grandchildren are to inherit a planet as bounteous and beautiful as we have enjoyed, we in the consumer class must — without surrendering the quest for advanced, clean technology — eat, travel, and use energy and materials more like those on the middle rung of the world's economic ladder. If we can learn to do so, we might find ourselves happier as well, for in the consumer society, affluence has brought us to a strange pass. Who would have predicted a century ago that the richest civilizations in history would be made up of polluted tracts of suburban development dominated by the private automobile, shopping malls, and a throwaway economy? Surely, this is not the ultimate fulfillment of our destiny.

In the final analysis, accepting and living by sufficiency rather than excess offers a return to what is, culturally speaking, the human home: to the ancient order of family, community, good work, and good life; to a reverence for skill, creativity, and creation; to a daily cadence slow enough to let us watch the sunset and stroll by the water's edge; to communities worth spending a lifetime in; and to local places pregnant with the memories of generations. Perhaps Henry David Thoreau had it right when he scribbled in his notebook beside Walden Pond, "A man is rich in proportion to the things he can afford to let alone."[18]

SANDRA POSTEL

Carrying Capacity: The Earth's Bottom Line

....

I t takes no stretch of the imagination to see that the human species is now an agent of change of geologic proportions. We literally move mountains to mine the earth's minerals, redirect rivers to build cities in the desert, torch forests to make way for crops and cattle, and alter the chemistry of the atmosphere in disposing of our wastes. At humanity's hand, the earth is undergoing a profound transformation — one with consequences we cannot fully grasp.

It may be the ultimate irony that in our efforts to make the earth yield more for ourselves, we are diminishing its ability to sustain life of all kinds, humans included. Signs of environmental constraints are now pervasive. Cropland is scarcely expanding any more, and a good portion of existing agricultural land is losing fertility. Grasslands have been overgrazed and fisheries overharvested, limiting the amount of additional food from these sources. Water bodies have suffered extensive depletion and pollution, severely restricting future food production and urban expansion. And natural forests — which help stabilize the climate, moderate water supplies, and harbor a majority of the planet's terrestrial biodiversity — continue to recede.

These trends are not new. Human societies have been altering the earth since they began. But the pace and scale of degradation that started about mid-century — and continues today — is historically new. The central conundrum of sustainable development is now all too apparent:

Sandra Postel is vice president for research at the Worldwatch Institute. This chapter is reprinted from The State of the World 1994, *by Lester Brown and the staff of the Worldwatch Institute (New York: W.W. Norton & Company, 1994).*

population and economies grow exponentially, but the natural resources that support them do not.

Biologists often apply the concept of "carrying capacity" to questions of population pressures on an environment. Carrying capacity is the largest number of any given species that a habitat can support indefinitely. When that maximum sustainable population level is surpassed, the resource base begins to decline — and sometime thereafter, so does the population.

A simple but telling example of a breach of carrying capacity involved the introduction of 29 reindeer to St. Matthew Island in the Bering Sea in 1944. Under favorable conditions, the herd expanded to 6,000 by the summer of 1963. The following winter, however, the population crashed, leaving fewer than 50 reindeer. According to a 1968 study by biologist David R. Klein of the University of Alaska, the large herd had overgrazed the island's lichens, its main source of winter forage, and the animals faced extreme competition for limited supplies during a particularly severe winter. Klein concluded that "food supply, through its interaction with climatic factors, was the dominant population regulating mechanism for reindeer on St. Matthew Island."[1]

Of course, human interactions with the environment are far more complicated than those of reindeer on an island. The earth's capacity to support humans is determined not just by our most basic food requirements but also by our levels of consumption of a whole range of resources, by the amount of waste we generate, by the technologies we choose for our varied activities, and by our success at mobilizing to deal with major threats. In recent years, the global problems of ozone depletion and greenhouse warming have underscored the danger of overstepping the earth's ability to absorb our waste products. Less well recognized, however, are the consequences of exceeding the sustainable supply of essential resources — and how far along that course we may already be.

As a result of our population size, consumption patterns, and technology choices, we have surpassed the planet's carrying capacity. This is plainly evident by the extent to which we are damaging and depleting natural capital. The earth's environmental assets are now insufficient to sustain both our present patterns of economic activity and the life-support systems we depend on. If current trends in resource use continue and if world population grows as projected, by 2010 per capita availability of rangeland will drop by 22 percent and the fish catch by 10 percent. Together, these provide much of the world's animal protein. The per capita area of irrigated land, which now yields about a third of the global food harvest, will drop by 12 percent. And cropland area and forestland per person will shrink by 21 and 30 percent, respectively.[2]

The days of the frontier economy — in which abundant resources were available to propel economic growth and living standards — are over. We have entered an era in which global prosperity increasingly depends on using resources more efficiently, distributing them more equitably, and reducing consumption levels overall. Unless we accelerate this transition, powerful social tensions are likely to arise from increased competition for the scarce resources that remain. The human population will not crash wholesale as the St. Matthew Island reindeer did, but there will likely be a surge in hunger, crossborder migration, and conflict — trends already painfully evident in parts of the world.[3]

Wiser and more discriminating use of technology offers the possibility of tremendous gains in resource efficiency and productivity, helping us get more out of each hectare of land, ton of wood, or cubic meter of water. In this way, technology can help stretch the earth's capacity to support humans sustainably. Trade also has an important, though more limited role. Besides helping spread beneficial technologies, it enables one country to import ecological capital from another. Trade can thus help surmount local or regional scarcities of land, water, wood, or other resources.

In these ways, technology and trade can buy time to tackle the larger challenges of stabilizing population, reducing excessive consumption, and redistributing wealth. Unfortunately, past gains in these two areas have deluded us into thinking that any constraint can be overcome, and that we can therefore avoid the more fundamental tasks. And rather than directing technology and trade toward sustainable development, we have more often used them in ways that hasten resource depletion and degradation.

The roots of environmental damage run deep. Unless they are unearthed soon, we risk exceeding the planet's carrying capacity to such a degree that a future of economic and social decline will be impossible to avoid.

DRIVING FORCES

Since mid-century, three trends have contributed most directly to the excessive pressures now being placed on the earth's natural systems — the doubling of world population, the quintupling of global economic output, and the widening gap in the distribution of income. The environmental impact of our population, now numbering 5.5 billion, has been vastly multiplied by economic and social systems that strongly favor growth and ever-rising consumption over equity and poverty alleviation; that fail to give women equal rights, education, and economic opportunity — and thereby perpetuate the conditions under which poverty and rapid population growth persists; and that do not discriminate between means of production that are environmentally sound and those that are not.

Of the three principal driving forces, the growing inequality in income between rich and poor stands out in sharpest relief. In 1960, the richest

20 percent of the world's people absorbed 70 percent of global income; by 1989 (the latest year for which comparable figures are available), the wealthy's share had climbed to nearly 83 percent. The poorest 20 percent, meanwhile, saw their share of global income drop from an already meager 2.3 percent to just 1.4 percent. The ratio of the richest fifth's share to the poorest's thus grew from 30 to one in 1960 to 59 to one in 1989. (See Table 1.)[4]

This chasm of inequity is a major cause of environmental decline: it fosters overconsumption at the top of the income ladder and persistent poverty at the bottom. By now, ample evidence shows that people at either end of the income spectrum are far more likely than those in the middle to damage the earth's ecological health — the rich because of their high consumption of energy, raw materials, and manufactured goods, and the poor because they must often cut trees, grow crops, or graze cattle in ways harmful to the earth merely to survive from one day to the next.[5]

Families in the western United States, for instance, often use as much as 3,000 liters of water a day — enough to fill a bathtub 20 times. Overdevelopment of water there has contributed to the depletion of rivers and aquifers, destroyed wetlands and fisheries, and, by creating an illusion of abundance, led to excessive consumption. Meanwhile, nearly one out of every three people in the developing world — some 1.2 billion people in all — lack access to a safe supply of drinking water. This contributes to the spread of debilitating disease and death, and forces women and children to trek many hours a day to collect a few jugs of water to meet their family's most basic needs.[6]

Disparities in food consumption are revealing as well. (See Table 2.) As many as 700 million people do not eat enough to live and work at their full potential. The average African, for instance, consumes only 87 per-

TABLE 1: Global Income Distribution, 1960-1989

Year	Share of Global Income Going to Richest 20 Percent	Share of Global Income Going to Poorest 20 Percent	Ratio of Richest to Poorest
1960	70.2	2.3	30 to 1
1970	73.9	2.3	32 to 1
1980	76.3	1.7	45 to 1
1989	82.7	1.4	59 to 1

Source: United Nations Development Programme, *Human Development Report* 1992 (New York: Oxford University Press, 1992).

cent of the calories needed for a healthy and productive life. Meanwhile, diets in many rich countries are so laden with animal fat as to cause increased rates of heart disease and cancers. Moreover, the meat-intensive diets of the wealthy usurp a disproportionately large share of the earth's agricultural carrying capacity since producing one kilogram of meat takes several kilograms of grain. If everyone in the world required as much grain for their diet as the average American does, the global harvest would need to be 2.6 times greater than it is today — a highly improbable scenario.[7]

Economic growth — the second driving force — has been fueled in part by the introduction of oil onto the energy scene. Since mid-century, the global economy has expanded fivefold. As much was produced in two-and-a-half months of 1990 as in the entire year of 1950. World trade, moreover, grew even faster: exports of primary commodities and manufactured products rose elevenfold.[8]

The extent to which the overall scale of economic activity damages the earth depends largely on the technologies used and the amount of resources consumed in the process. Electricity generated by burning coal

---

TABLE 2: Grain Consumption Per Person in Selected Countries 1990

Country	Grain Consumption Per Person (kilograms)
Canada	974
United States	860
Soviet Union	843
Australia	503
France	465
Turkey	419
Mexico	309
Japan	297
China	292
Brazil	277
India	186
Bangladesh	176
Kenya	145
Tanzania	145
Haiti	100
World Average	323

Sources: Worldwatch Institute estimate, based on U.S. Department of Agriculture, *World Grain Database* (unpublished printout) (Washington, D.C.: 1992); Population Reference Bureau, *1990 World Population Data Sheet* (Washington, D.C.: 1990).

may contribute as much to economic output as an equal amount generated by wind turbines, for example, but burning coal causes far more environmental harm. A similar comparison holds for a ton of paper made from newly cut trees and a ton produced from recycled paper.

Unfortunately, economic growth has most often been of the damaging variety — powered by the extraction and consumption of fossil fuels, water, timber, minerals, and other resources. Between 1950 and 1990, the industrial roundwood harvest doubled, water use tripled, and oil production rose nearly sixfold. Environmental damage increased proportionately.[9]

Compounding the rises in both poverty and resource consumption related to the worsening of inequality and rapid economic expansion, population growth has added greatly to pressures on the earth's carrying capacity. The doubling of world population since 1950 has meant more or less steady increases in the number of people added to the planet each year. Whereas births exceeded deaths by 37 million in 1950, the net population gain in 1993 was 87 million — roughly equal to the population of Mexico. [10]

Aside from the late fifties, when the massive famine caused by China's Great Leap Forward led to a sharp drop in annual population growth, the only interval of sustained reductions in the yearly addition to population during the last 40 years occurred in the seventies. Fairly widespread improvements in living standards then and the introduction of family planning programs in a number of countries caused birth rates to drop. This translated into a decline in the growth rate. If the trend of the seventies had continued along roughly the same path, world population would have stabilized in 2030 at 6.7 billion.[11]

Instead, the population growth rate stopped declining in the late seventies and remained stalled through much of the eighties. This initiated another period of record-setting additions to world population. The U.N. medium population projection now shows world population reaching 8.9 billion by 2030 — 2.2 billion more people than if the slowdown of the seventies had continued — and levelling off at 11.5 billion around 2150.[12]

Rarely do the driving forces of environmental decline operate in isolation; more often they entangle, like a spider's web. Where people's livelihoods depend directly on the renewable resource base around them, for example, poverty, social inequity, and population growth fuel a vicious cycle in which environmental decline and worsening poverty reduce options for escaping these traps. This is plainly evident in the African Sahel, where traditional agricultural systems that depended on leaving land fallow for a time to restore its productivity have broken down under population pressures.[13]

On Burkina Faso's Mossi Plateau, for instance, some 60 percent of the arable land is under cultivation in a given year, which means it is not lying idle long enough to rejuvenate. The reduced organic content and moisture-storage capacity of the soil lowers crop productivity and makes farmers more vulnerable to drought. In addition, with firewood in scarce supply in many Sahelian countries, families often use livestock dung for fuel, which also robs the land of nutrients. The result is a lowering of the land's carrying capacity, reduced food security, greater poverty, and continued high population growth.[14]

To take another example, the U.S. government protects domestic sugar producers by keeping sugar prices at three to five times world market levels. Because of the lost market opportunity, low-cost sugarcane growers in the Philippines produce less, putting cane-cutters out of work. The inequitable distribution of cropland in the Philippines combines with rapid population growth to leave the cutters little choice but to migrate into the hills to find land to grow subsistence crops. They clear plots by deforesting the upper watershed, causing increased flooding and soil erosion, which in turn silts up reservoirs and irrigation canals downstream. Poverty deepens, the gap between the rich and poor widens, and the environment deteriorates further.[15]

THE RESOURCE BASE

The outer limit of the planet's carrying capacity is determined by the total amount of solar energy converted into biochemical energy through plant photosynthesis minus the energy those plants use for their own life processes. This is called the earth's net primary productivity (NPP), and it is the basic food source for all life.

Prior to human impacts, the earth's forests, grasslands, and other terrestrial ecosystems had the potential to produce a net total of some 150 billion tons of organic matter per year. Stanford University biologist Peter Vitousek and his colleagues estimate, however, that humans have destroyed outright about 12 percent of the terrestrial NPP and now directly use or co-opt an additional 27 percent. Thus, one species — *Homo sapiens* — has appropriated nearly 40 percent of the terrestrial food supply, leaving only 60 percent for the millions of other land-based plants and animals.[16]

It may be tempting to infer that, at 40 percent of NPP, we are still comfortably below the ultimate limit. But this is not the case. We have appropriated the 40 percent that was easiest to acquire. It may be impossible to double our share, yet theoretically that would happen in just 60 years if our share rose in tandem with population growth. And if average resource consumption per person continues to increase, that doubling would occur much sooner.

Perhaps more important, human survival hinges on a host of environmental services provided by natural systems — from forests' regulation of the hydrological cycle to wetlands' filtering of pollutants. As we destroy, alter, or appropriate more of these natural systems for ourselves, these environmental services are compromised. At some point, the likely result is a chain reaction of environmental decline — widespread flooding and erosion brought on by deforestation, for example, or worsened drought and crop losses from desertification, or pervasive aquatic pollution and fisheries losses from wetlands destruction. The simultaneous unfolding of several such scenarios could cause unprecedented human hardship, famine, and disease. Precisely when vital thresholds will be crossed, no one can say. But as Vitousek and his colleagues note, those "who believe that limits to growth are so distant as to be of no consequence for today's decision makers appear unaware of these biological realities."[17]

How have we come to usurp so much of the earth's productive capacity? In our efforts to feed, clothe, house, and otherwise satisfy our evergrowing material desires, we have steadily converted diverse and complex biological systems to more uniform and simple ones that are managed for human benefit. Timber companies cleared primary forests and replaced them with monoculture pine plantations to make pulp and paper. Migrant peasants torched tropical forests in order to plant crops merely to survive. And farmers plowed the prairie grasslands of the U.S. Midwest to plant corn, creating one of the most productive agricultural regions in the world. Although these transformations have allowed more humans to be supported at a higher standard of living, they have come at the expense of natural systems, other plant and animal species, and ecological stability.

Continuing along this course is risky. But the flip side of the problem is equally sobering. What do we do when we have claimed virtually all that we can, yet our population and demands are still growing?

This is precisely the predicament we now face. Opportunities to expand our use of certain essential resources — including cropland, rangeland, fisheries, water, and forests — are severely limited, and a good share of the resources we have already appropriated, and depend on, are losing productivity. And unlike energy systems, where we can envisage a technically feasible shift from fossil fuels to solar-based sources, there are no identifiable substitutes for these essential biological and water resources.

Between 1980 and 1990, cropland area worldwide expanded by just two percent, which means that gains in the global food harvest came almost entirely from raising yields on existing cropland. Most of the remaining area that could be used to grow crops is in Africa and Latin America; very little is in Asia. The most sizable near-term additions to the cropland base are likely to be a portion of the 76 million hectares of savanna grasslands in South America that are already accessible and potentially cultivable, as well as some portion of African rangeland and forest. These conversions, of course, may come at a high environmental price, and will push our 40-percent share of NPP even higher.[18]

Moreover, a portion of any cropland gains that do occur will be offset by losses. As economies of developing countries diversify and as cities expand to accommodate population growth and migration, land is rapidly being lost to industrial development, housing, road construction, and the like. Canadian geographer Vaclav Smil estimates, for instance, that between 1957 and 1990, China's arable land diminished by at least 35 million hectares — an area equal to all the cropland in France, Germany, Denmark, and the Netherlands combined. At China's 1990 average grain yield and consumption levels, that amount of cropland could have supported some 450 million people, about 40 percent of its population.[19]

In addition, much of the land we continue to farm is losing its inherent productivity because of unsound agricultural practices and overuse. The Global Assessment of Soil Degradation, a three-year study involving some 250 scientists, found that more than 550 million hectares are losing topsoil or undergoing other forms of degradation as a direct result of poor agricultural methods. (See Table 3.)[20]

TABLE 3: Human-Induced Land Degradation Woldwide, 1945-Present

Region	Over-grazing	Defores-tation	Agricultural Mismanage-ment	Other[1]	Total	Degraded Area as Share of Total Vegetated Land
			(million hectares)			(percent)
Asia	197	298	204	47	746	20
Africa	243	67	121	63	494	22
South America	68	100	64	12	244	14
Europe	50	84	64	22	220	23
North & Cent. Amer.	38	18	91	11	158	8
Oceania	83	12	8	0	103	13
World	679	579	552	155	1,965	17

[1]Includes exploitation of vegetation for domestic use (133 million hectares) and bioindustrial activities, such as pollution (22 million hectares).

Source: Worldwatch Institute, based on "The Extent of Human-Induced Soil Degradation," Annex 5 in L.R. Oldeman et al., World Map of the Status of Human-Induced Soil Degradation (Wageningen, Netherlands: United Nations Environment Programme and International Soil Reference and Information Centre, 1991).

TABLE 4: Population Size and Availability of Renewable Resources, Circa 1990 With Projections for 2010

	Circa 1990	2010	Total Change	Per Capita Change
	(million)		(percent)	
Population	5,290	7,030	+33	-
Fish Catch (tons)[1]	85	102	+20	-10
Irrigated Land (hectares)	237	277	+17	-12
Cropland (hectares)	1,444	1,516	+ 5	-21
Rangeland and Pasture (hectares)	3,402	3,540	+ 4	-22
Forests (hectares)[2]	3,413	3,165	- 7	-30

[1]*Wild catch from fresh and marine waters, excludes aquaculture*
[2]*Includes plantations; excludes woodlands and shrublands.*

Sources: Population figures from U.S. Bureau of the Census, Department of Commerce, *International Data Base,* unpublished printout, November 2, 1993; 1990 irrigated land, cropland, and rangeland from U.N. Food and Agriculture Organization (FAO), *Production Yearbook 1991* (Rome: 1992); fish catch from M. Perotti, chief, Statistics Branch, Fisheries Department, FAO, Rome, private communication, November 3, 1993; forests from FAO, *Forest Resources Assessment 1990* (Rome: 1992 and 1993) and other sources documented in endnote 30. For explanation of projections, see text.

On balance, unless crop prices rise (which in turn depends on economic conditions in developing countries and on whether purchasing power rises sufficiently to push up the demand for food), it appears unlikely that the net cropland area will expand much more quickly over the next two decades than it did between 1980 and 1990. Assuming a net expansion of 5 percent, which may be optimistic, total cropland area would climb to just over 1.5 billion hectares. Given the projected 33-percent increase in world population by 2010, the amount of cropland per person would decline by 21 percent. (See Table 4).

Pasture and rangeland cover some 3.4 billion hectares of land, more than twice the area in crops. The cattle, sheep, goats, buffalo, and camels that graze them convert grass, which humans cannot digest, into meat and milk, which they can. The global ruminant livestock herd, which numbers about 3.3 billion, thus adds a source of food for people that does not subtract from the grain supply, in contrast to the production of pigs, chickens, and cattle raised in feedlots.[21]

Much of the world's rangeland is already heavily overgrazed and cannot continue to support the livestock herds and management practices that exist today. According to the Global Assessment of Soil Degradation, overgrazing has degraded some 680 million hectares since mid-century.

This suggests that 20 percent of the world's pasture and range is losing productivity and will continue to do so unless herd sizes are reduced or more sustainable livestock practices are put in place.[22]

During the eighties, the total range area increased slightly, in part because land deforested or taken out of crops often reverted to some form of grass. If similar trends persist over the next two decades, by 2010 the total area of rangeland and pasture will have increased four percent, but it will have dropped 22 percent in per capita terms. In Africa and Asia, which together contain nearly half the world's rangelands and where many traditional cultures depend heavily on livestock, even larger per capita declines could significantly weaken food economies.

Fisheries — another natural biological system that humans depend on — add calories, protein, and diversity to human diets. The annual fish catch from all sources, including aquaculture, totalled 97 million tons in 1990, about 5 percent of the protein humans consume. Fish account for a good portion of the calories consumed overall in many coastal regions and island nations.[23]

The world fish catch has climbed rapidly in recent decades, expanding nearly fivefold since 1950. But it peaked at just above 100 million tons in 1989. Although catches from both inland fisheries and aquaculture (fish farming) have been rising steadily, they have not offset the decline in the much larger wild marine catch, which fell from a historic peak of 82 million tons in 1989 to 77 million in 1991, a drop of 6 percent.[24]

With the advent of mechanized hauling gear, bigger nets, electronic fish detection aids, and other technologies, almost all marine fisheries have suffered from extensive overexploitation. Under current practices, considerable additional growth in the global fish catch overall looks highly unlikely.

TABLE 5: Net Imports of Forest Products, Selected East Asian Countries, 1961-91

Country	1961	1971	1981	1991
(thousand cubic meters)[1]				
Japan	8,800	45,000	50,000	70,100
South Korea	500	2,900	4,700	12,700
Taiwan	0	1,200	6,700	8,800
China	200	200	3,100	6,800
Hong Kong	900	1,300	2,000	2,500
Singapore	200	1,600	1,400	1,100

[1]All forest products are expressed in equivalent units of wood fiber content.

Sources: Worldwatch Institute, based on sources documented in endnote 47.

Indeed, the U.N. Food and Agriculture Organization (FAO) now estimates that all 17 of the world's major fishing areas have either reached or exceeded their natural limits, and that 9 are in serious decline.[25]

FAO scientists believe that better fisheries management might allow the wild marine catch to increase by some 20 percent. If this could be achieved, and if the freshwater catch increased proportionately, the total wild catch would rise to 102 million tons; by 2010, this would nonetheless represent a 10-percent drop in per capita terms.[26]

Fresh water may be even more essential than cropland, rangeland, and fisheries; without water, after all, nothing can live. Signs of water scarcity are now pervasive. Today, 26 countries have insufficient renewable water supplies within their own territories to meet the needs of a moderately developed society at their current population size — and populations are growing fastest in some of the most water-short countries, including many in Africa and the Middle East. Rivers, lakes, and underground aquifers show widespread signs of degradation and depletion, even as human demands rise inexorably.[27]

Water constraints already appear to be slowing food production, and those restrictions will only become more severe. Agricultural lands that receive irrigation water play a disproportionate role in meeting the world's food needs: the 237 million hectares of irrigated land account for only 16 percent of total cropland but more than a third of the global harvest. For most of human history, irrigated area expanded faster than population did, which helped food production per person to increase steadily. In 1978, however, per capita irrigated land peaked, and it has fallen nearly 6 percent since then.[28]

There is little to suggest that this disturbing trend will turn around soon. The rising cost of building new irrigation projects, growing competition for scarce water, mounting concern about the social and environmental effects of large dams, and the steady loss of perhaps two million hectares of irrigated land each year due to salinization suggest that the pace of net irrigation expansion is not likely to pick up any time soon. Indeed, it may slow further. Assuming, perhaps optimistically, that the irrigation base spreads at an average rate of two million hectares a year for the next two decades, irrigated area would climb to 277 million hectares — a 17-percent gain over 1990, but a 12-percent loss in per capita terms.[29]

Forests and woodlands, the last key component of the biological resource base, contribute a host of important commodities to the global economy — logs and lumber for constructing homes and furniture, fiber for making paper, fruits and nuts for direct consumption, and, in poor countries, fuelwood for heating and cooking. More important even than these benefits, however, are the ecological services forests perform — from conserving soils and moderating water cycles to storing carbon, protecting air quality, and harboring millions of plant and animal species.

Today forests cover 24 percent less area than in 1700 — 3.4 billion hectares compared with an estimated 4.5 billion about 300 years ago. Most of that area was cleared for crop cultivation, but cattle ranching, timber and fuelwood harvesting, and the growth of cities, suburbs, and highways all claimed a share as well. Recent assessments suggest that the world's forests declined by about 130 million hectares between 1980 and 1990, an area larger than Peru.[30]

Most of this clearing occurred in the tropics, which lost 154 million hectares of natural forests but gained 18 million hectares of plantations. Among temperate countries, substantial losses occurred in China (some 13 million hectares). Other temperate areas, however, including the former Soviet Union, experienced a net gain. For the world as a whole, if the average net loss of 3.7 percent per decade continues, by 2010 the world's forested area will shrink by an additional seven percent; per capita forest area will drop by an astonishing 30 percent.[31]

Tragically, much tropical forest is being cleared in order to cultivate soils that cannot sustain crop production for more than a few years. Yet the species extinguished in the process are gone forever. And the associated environmental destruction will have repercussions for generations.

REDIRECTING TECHNOLOGY

Advances in technology — which is used broadly here to mean the application of knowledge to an activity — offer at least a partial way out of our predicament. The challenge of finding ways to meet the legitimate needs of our growing population without further destroying the natural resource base certainly ranks among the greatest missions humanity has ever faced. In most cases, "appropriate" technologies will no longer be engineering schemes, techniques, or methods that enable us to claim more of nature's resources, but instead systems that allow us to benefit more from the resources we already have. As long as the resulting gains are directed toward bettering the environment and the lives of the less fortunate instead of toward increased consumption by the rich, such efforts will reduce human impacts on the earth.

The power of technology to help meet human needs was a critical missing piece in the world view of Thomas Malthus, the English curate whose famous 1798 essay postulated that the growth of human population would outstrip the earth's food-producing capabilities. His prediction was a dire one — massive famine, disease, and death. But a stream of agricultural advances combined with the productivity leaps of the Industrial Revolution made the Malthusian nightmare fade for much of the world.

Without question, technological advances have steadily enhanced our capacity to raise living standards. They not only helped boost food production, the main concern of Malthus, they also increased our access to sources of water, energy, timber, and minerals. In many ways, however, technology has proved to be a double-edged sword. Take, for example, the chlorofluorocarbons that at first appeared to be ideal chemicals for so many different uses. It turned out that once they reached the upper atmosphere they began destroying the ozone layer, and thus threatened life on the planet.

Likewise, the irrigation, agricultural chemicals, and high-yielding crop varieties that made the Green Revolution possible also depleted and contaminated water supplies, poisoned wildlife and people, and encouraged monoculture cropping that reduced agricultural diversity. Huge driftnets boosted fish harvests but contributed to overfishing and the depletion of stocks. And manufacturing processes that rapidly turn timber into pulp and paper have fueled the loss of forests and created mountains of waste paper.

As a society, we have failed to discriminate between technologies that meet our needs in a sustainable way and those that harm the earth. We have let the market largely dictate which technologies move forward, without adjusting for its failure to take proper account of environmental damages. Now that we have exceeded the planet's carrying capacity and are rapidly running down its natural capital, such a correction is urgently needed.

Meeting future food needs, for instance, now depends almost entirely on raising the productivity of land and water resources. Over the last several decades, remarkable gains have been made in boosting cropland productivity. Between 1950 and 1991, world grain production rose 169 percent despite only a 17-percent increase in the area of grain harvested. An impressive 131-percent increase in average grain yield — brought about largely by Green Revolution technologies — allowed production to expand so greatly. If today's grain harvest were being produced at 1950's average yield, we would need at least twice as much land in crops as today — and pressure to turn forests and grasslands into cropland would have increased proportionately.[32]

Whether technological advances continue to raise crop yields fast enough to meet rising demand is, at the moment, an open question. Given the extent of cropland and rangeland degradation and the slowdown in irrigation expansion, it may be difficult to sustain the past pace of yield increases. Indeed, per capita grain production in 1992 was 7 percent lower than the historic peak in 1984. Whether this is a short-term phenomenon or the onset of a longer-term trend will depend on what new crop varieties and technologies reach farmers' fields and if they can overcome the yield-suppressing effects of environmental degradation. Another factor is whether agricultural policies and prices encourage farmers to invest in raising land productivity further.[33]

Currently, yields of the major grain crops are still significantly below their genetic potential, so it is possible that scientists will develop new crop varieties that can boost land productivity. They are working, for example, on a new strain of rice that may offer yield gains within a decade. And they have developed a wheat variety that is resistant to leaf rust disease, which could both increase yields and allow wheat to be grown in more humid regions.[34]

Gains from biotechnology may be forthcoming soon as well. According to Gabrielle Persley of the World Bank, rice varieties bioengineered for virus resistance are likely to be in farmers' fields by 1995. Wheat varieties with built-in disease and insect resistance, which could reduce crop losses to pests, are under development. And scientists are genetically engineering maize varieties for insect resistance, although no commercial field applications are expected until sometime after 2000. It remains to be seen whether these and other potential gains materialize and whether they collectively increase yields at the rates needed. The recent cutback in funding for international agricultural research centers, where much of the work on grain crops takes place, is troubling.[35]

Parallelling the need to raise yields, however, is the less recognized challenge of making both existing and future food production systems sustainable. A portion of our current food output is being produced by using land and water unsustainably. Unless this is corrected, food production from these areas will decline at some point.

For instance, in parts of India's Punjab, the nation's breadbasket, the high-yielding rice paddy-wheat rotation that is common requires heavy doses of agricultural chemicals and substantial amounts of irrigation water. A recent study by researchers from the University of Delhi and the World Resources Institute in Washington, D.C., found that in one Punjab district, Ludhiana, groundwater pumping exceeds recharge by one third and water tables are dropping nearly one meter per year. Even if water use were reduced to 80 percent of the recommended level, which would cause yields to drop an estimated 8 percent, groundwater levels would still decline by a half-meter per year. Given the importance of the Punjab to India's food production, the authors' conclusion is sobering, to say the least: "Unless production practices are developed that dramatically reduce water use, any paddy production system may be unsustainable in this region."[36]

Indeed, in many agricultural regions — including northern China, southern India (as well as the Punjab), Mexico, the western United States, parts of the Middle East, and elsewhere — water may be much more of a constraint to future food production than land, crop yield potential, or most other factors. Developing and distributing technologies and practices that improve water management is critical to sustaining the food production capability we now have, much less increasing it for the future.

Water-short Israel is a front-runner in making its agricultural economy more water-efficient. Its current agricultural output could probably not have been achieved without steady advances in water management — including highly efficient drip irrigation, automated systems that apply water only when crops need it, and the setting of water allocations based on predetermined optimum water applications for each crop. The nation's success is notable: between 1951 and 1990, Israeli farmers reduced the amount of water applied to each hectare of cropland by 36 percent. This allowed the irrigated area to more than triple with only a doubling of irrigation water use.[37]

Whether high-tech, like the Israeli systems, or more traditional, like the vast canal schemes in much of Asia, improvements in irrigation management are critical. At the same time, technologies and methods to raise the productivity of rainfed lands are urgently needed. Particularly in dry regions, where land degradation and drought make soil and water conservation a matter of survival, improvements on many traditional methods could simultaneously raise local food production, reduce hunger, and slow environmental decline.[38]

In the Burkina Faso province of Yatenga, for example, farmers have revived a traditional technique of building simple stone lines across the slopes of their fields to reduce erosion and help store moisture in the soil. With the aid of Oxfam, a U.K.-based development organization, they improved on the earlier technique by constructing the stone walls along contour lines, using a simple water-tube device to help them determine a series of level points. The technique has raised yields by up to 50 percent, and is now being used on more than 8,000 hectares in the province.[39]

Matching the need for sustainable gains in land and water productivity is the need for improvements in the efficiency of wood use and reductions in wood and paper waste in order to reduce pressures on forests and woodlands. A beneficial timber technology is no longer one that improves logging efficiency — the number of trees cut per hour—but rather one that makes each log harvested go further. Raising the efficiency of forest product manufacturing in the United States, the world's largest wood consumer, roughly to Japanese levels would reduce timber needs by about a fourth, for instance. Together, available methods of reducing waste, increasing manufacturing efficiency, and recycling more paper could cut U.S. wood consumption in half; a serious effort to produce new wood-saving techniques would reduce it even more.[40]

With the world's paper demand projected to double by the year 2010, there may be good reason to shift production toward "treeless paper" — that made from nonwood pulp. Hemp, bamboo, jute, and kenaf are among the alternative sources of pulp. The fast-growing kenaf plant, for example, produces two to four times more pulp per hectare than southern pine, and the pulp has all of the main qualities needed for making most grades of paper. In China, more than 80 percent of all paper pulp is made from nonwood sources. Treeless paper was manufactured in 45 countries in 1992, and accounted for 9 percent of the world's paper supply. With proper economic incentives and support for technology and market development, the use of treeless paper could expand greatly.[41]

These are but a few examples of the refocusing of technology that is needed. A key policy instrument for encouraging more sustainable and efficient means of production is the institution of environmental taxes, which would help correct the market's failure to include environmental harm in the pricing of products and activities. In addition, stronger criteria are needed within development institutions and aid agencies to ensure that the projects they fund are ecologically sound and sustainable.

The many past gains from technological advances might make concerns about resource constraints seem anachronistic. But as Dartmouth College professor Donella Meadows and her coauthors caution in their 1992 study *Beyond the Limits*, "the more successfully society puts off its limits through economic and technical adaptations, the more likely it is in the future to run into several of them at the same time." The wiser use of technology can only buy time — and precious time it is — to bring consumption and population growth down to sustainable levels and to distribute resources more equitably.[42]

THE ROLE OF TRADE

Consider two countries, each with a population of about 125 million. Country A has a population density of 331 people per square kilometer, has just 372 square meters of cropland per inhabitant (one seventh the world average), and imports almost three fourths of its grain and nearly two thirds of its wood. Country B, on the other hand, has a population density less than half that of Country A and nearly five times as much cropland per person. It imports only one tenth of its grain and no wood. Which country has most exceeded its carrying capacity?[43]

Certainly it would be Country A — which, as it turns out, is Japan — a nation boasting a real gross domestic product (GDP) of some $18,000 per capita. Country B, which from these few indicators seems closer to living within its means, is Pakistan — with a real GDP per capita of only $1,900. By any economic measure, Japan is far and away the more successful of the two, so how can questions of carrying capacity be all that relevant?[44]

The answer, of course, lies in large part with trade. Japan sells cars and computers, and uses some of the earnings to buy food, timber, oil, and other raw materials. And that is what trade is supposed to be about: selling what one can make better or more efficiently, and buying what others have a comparative advantage in producing. Through trade, countries with scarce resources can import what they need from countries with a greater abundance. If those reindeer on St. Matthew Island had been able to import lichens, their numbers might not have crashed so drastically.

Imports of biologically based commodities like food and timber are, indirectly, imports of land, water, nutrients, and the other components of ecological capital needed to produce them. Many countries would not be able to support anything like their current population and consumption levels were it not for trade. To meet its food and timber demands alone, the Netherlands, for instance, appropriates the production capabilities of 24 million hectares of land — 10 times its own area of cropland, pasture, and forest.[45]

In principle, there is nothing inherently unsustainable about one nation relying on another's ecological surplus. The problem, however, is the widespread perception that all countries can exceed their carrying capacities and grow economically by expanding manufactured and industrial goods at the expense of natural capital — paving over agricultural land to build factories, for example, or clear-cutting forest to build new homes. But all countries cannot continue to do this indefinitely. As economist Herman Daly observes, "One country's ability to substitute man-made for natural capital depends on some other country's making the opposite (complementary) choice."[46]

In other words, globally the ecological books must balance. Many economists see no cause for worry, believing that the market will take care of any needed adjustments. As cropland, forests, and water grow scarce, all that is necessary, they say, is for prices to rise; the added incentives to conserve, use resources more productively, alter consumption patterns, and develop new technologies will keep output rising with demand. But once paved over for a highway or housing complex, cropland is unlikely to be brought back into production — no matter how severe food shortages may become. Moreover, no mechanism exists for assuring that an adequate resource base is maintained to meet needs that the marketplace ignores or heavily discounts — including those of vital ecosystems, other species, the poor, or the next generation.

Trade in forest products illuminates some of these trends. East Asia, where the much-touted economic miracles of Japan and the newly industrializing countries have taken place, has steadily and rapidly appropriated increasing amounts of other nations' forest resources. In Japan, where economic activity boomed after World War II, net imports of forest products rose eightfold between 1961 and 1991. (See Table 5.) The nation is now the world's largest net importer of forest products by far. Starting from a smaller base, South Korea's net imports have more than quadrupled since 1971, and Taiwan's have risen more than sevenfold.[47]

China is a big wild card in the global forest picture. According to He Bochuan, a lecturer at Sun Yat-sen University in Guangdong, China's consumption of raw wood — some 300 million cubic meters per year — exceeds the sustainable yield of its forests and woodlands by 30 percent. During the last decade alone its net forest product imports more than doubled. With one fifth of the world's population, economic growth rates that have averaged around 12 percent in recent years, only about 13 percent of its land covered by trees, and its own limited stocks undergoing depletion, China could fast become the leading importer of wood. If its per capita use of forest products were to rise to the level of Japan's today, China's total demand would exceed Japan's by nine times — and its import needs would place enormous pressure on forests worldwide.[48]

Thus, like technology, trade cuts both ways. It can help overcome local or regional carrying capacity constraints by allowing countries to bring in resources to meet their needs. But it can also foster unsustainable consumption levels by creating the illusion of infinite supplies. And because trade allows environmental damage to be done in a land far from where the products are used, it encourages what Stanford University biologists Gretchen Daily and Paul Ehrlich call "discounting over distance" — placing a lower value on environmental harm done far from home.[49]

East Asia's appetite for wood, for instance, has done just that — depleting and degrading Southeast Asia's species-rich tropical forests. Although logging is directly responsible for only a small portion of tropical deforestation (clearing for agriculture is the biggest cause), the construction of logging and access roads provides peasants, migrants, and land speculators a gateway into the forest that often initiates a chain reaction culminating in the clearing and burning of large forest tracts.[50]

The last several decades have seen the rise and fall of one Southeast Asian timber exporter after another. The wave pattern began with the Philippines during the sixties, followed by Indonesia and Thailand during a good bit of the seventies, and then by Malaysia during the eighties — with most of the shipments going to Japan. Having experienced extensive forest losses, Thailand and the Philippines are now net importers. Indonesia, with a bigger forest base, remains a significant net exporter. But the largest exporter of tropical wood products is now Malaysia — which shipped the equivalent of nearly 26 million cubic meters in 1991. With timber cutting in Malaysian forests estimated to be four times the sustainable yield, a decline in exports appears inevitable there as well.[51]

Like technology, trade is neither inherently good nor bad. One of its strengths is its ability to spread the benefits of more efficient and sustainable technologies and products, whether they be advanced drip irrigation systems, nontimber products from tropical forests, or the latest paper recycling techniques. Trade can also generate more wealth in developing countries, which conceivably could permit greater investments in environmental protection and help alleviate poverty. So far, however, the potential gains from trade have been overwhelmed by its more negative facets — in particular, by its tendency to foster ecological deficit-financing and unsustainable consumption.

In light of this, it is disturbing, to say the least, that negotiators involved in the eight-year-long Uruguay Round of the General Agreement on Tariffs and Trade (GATT) seem barely interested in the role trade plays in promoting environmental destruction. While the reduction of government subsidies and other barriers to free trade — the main concern of the GATT round — could make international markets more efficient and increase the foreign exchange earnings of developing countries, that offers no guarantee that trade will be more environmentally sound or socially equitable.

Since environmental taxes, regulations, or other means of internalizing environmental costs would help redirect products and activities toward those that are less damaging to the earth's natural systems, they would also help make trade — which is merely the exchange of those goods and services — more sustainable. For this to work, however, countries would need to adopt such measures more or less simultaneously to avoid placing some of them at a competitive disadvantage. And developing countries would likely participate only if they received substantial financial and technical assistance from wealthier ones, which for decades have benefited from having the environment absorb the damages caused by their activities. Unfortunately, the reluctance of rich countries to agree at the 1992 Earth Summit to any sizable transfer of funds to poorer nations does not bode well for such an initiative.[52]

There is talk that the next series of GATT negotiations may be a "green round" that would address the trade-environment nexus more directly, although probably not as broadly as is needed. Moreover, with short-term considerations such as slow economic growth and high unemployment taking precedence over long-term concerns, a coordinated effort to make trade more sustainable through cost-internalizing measures is not high on the agenda. If action is delayed too long, however, the future will arrive in a state of ecological impoverishment that no amount of free trade will be able to overcome.

LIGHTENING THE LOAD

Ship captains pay careful attention to a marking on their vessels called the Plimsoll line. If the water level rises above the Plimsoll line, the boat is too heavy and is in danger of sinking. When that happens, rearranging items on the ship will not help much. The problem is the total weight, which has surpassed the carrying capacity of the ship.[53]

Economist Herman Daly sometimes uses this analogy to underscore that the scale of human activity can reach a level that the earth's natural systems can no longer support. The ecological equivalent of the Plimsoll line may be the maximum share of the earth's biological resource base that humans can appropriate before a rapid and cascading deterioration in the planet's life-support systems is set in motion. Given the degree of resource destruction already evident, we may be close to this critical mark. The challenge, then, is to lighten our burden on the planet before "the ship" sinks.

More than 1,600 scientists, including 102 Nobel laureates, underscored this point in collectively signing a "Warning to Humanity" in late 1992. It states that "No more than one or a few decades remain before the chance to avert the threats we now confront will be lost and the prospects for humanity immeasurably diminished.... A new ethic is required — a new attitude towards discharging our responsibility for caring for ourselves and for the earth.... This ethic must motivate a great movement, convincing reluctant leaders and reluctant governments and reluctant peoples themselves to effect the needed changes."[54]

A successful global effort to lighten humanity's load on the earth would directly address the three major driving forces of environmental decline — the grossly inequitable distribution of income, resource-consumptive economic growth, and rapid population growth — and would redirect technology and trade to buy time for this great movement. Although there is far too much to say about each of these challenges to be comprehensive here, some key points bear noting.

Wealth inequality may be the most intractable problem, since it has existed for millennia. The difference today, however, is that the future of both rich and poor alike hinges on reducing poverty and thereby eliminating this driving force of global environmental decline. In this way, self-interest joins ethics as a motive for redistributing wealth, and raises the chances that it might be done.

Important actions to narrow the income gap include greatly reducing Third World debt, much talked about in the eighties but still not accomplished, and focusing foreign aid, trade, and international lending policies more directly on improving the living standards of the poor. If decision makers consistently asked themselves whether a choice they were about to make would help the poorest of the poor — that 20 percent of the world's people who share only 1.4 percent of the world's income — and acted only if the answer were yes, more people might break out of the poverty trap and have the opportunity to live sustainably.[55]

Especially in poorer countries, much could be gained from greater support for the myriad grassroots organizations working for a better future. These groups constitute a powerful force for achieving sustainable development in its truest form — through bottom-up action by local people. In an October 1993 address at the World Bank, Kenyan environmentalist Wangari Maathai noted that among the great benefits of the Green Belt Movement, the tree planting campaign she founded, was the understanding it gave people that "no progress can be made when the environment is neglected, polluted, degraded and over-exploited. Many people have also come to appreciate that taking care of the environment is not the responsibility of only the Government but of the citizens as well. This awareness is empowering and brings the environment close to the people. Only when this happens do people feel and care for the environment."[56]

A key prescription for reducing the kinds of economic growth that harm the environment is the same as that for making technology and trade more sustainable — internalizing environmental costs. If this is done through the adoption of environmental taxes, governments can avoid imposing heavier taxes overall by lowering income taxes accordingly. In addition, establishing better measures of economic accounting is critical. Since the calculations used to produce the gross national product do not account for the destruction or depletion of natural resources, this popular economic measure is extremely misleading. It tells us we are making progress even as our ecological foundations are crumbling. A better beacon to guide us toward a sustainable path is essential. The United Nations and several individual governments have been working to develop such a measure, but progress has been slow.[57]

Besides calling on political leaders to effect these changes, individuals in wealthier countries can help lighten humanity's load by voluntarily reducing their personal levels of consumption. By purchasing "greener products" for necessities and reducing discretionary consumption, the top one billion can help create ecological space for the bottom one billion to consume enough for a decent and secure life.

In September 1994, government officials will gather in Cairo for the International Conference on Population and Development, the third such gathering on population. This is a timely opportunity to draw attention to the connections between poverty, population growth, and environmental decline — and to devise strategies that simultaneously address the root causes. Much greater efforts are needed, for instance, to raise women's social and economic status and to give women equal rights and access to resources. Only if gender biases are rooted out will women be able to escape the poverty trap and to choose to have fewer children. In the realm of family planning, an essential step is to meet the needs of more than 100 million couples who want to limit or plan their families but who lack access to the means to do so safely and effectively. To succeed, such programs must also meet women's reproductive health needs as they perceive them.[58]

Progress in these areas was set back during the eighties when population related issues were politicized under the Reagan and Bush administrations and when the United States stopped funding the U.N. Population Fund and the International Planned Parenthood Federation. Fortunately, funding has been restored under the Clinton administration. And at a May 1993 preparatory meeting for the Cairo conference, State Department Counselor Tim Wirth indicated a major course correction in U.S. policy when he noted that "advancing women's rights and health and promoting family planning are mutually reinforcing objectives," and that "all barriers which deprive women of equal opportunity must be removed."[59]

The challenge of living sustainably on the earth will never be met, however, if population and environment conferences are the only forums in which it is addressed. Success hinges on the creativity and energy of a wide range of people in many walks of life. The scientists' Warning to Humanity ends with a call to the world's scientists, business and industry leaders, the religious community, and people everywhere to join in the urgent mission of halting the earth's environmental decline.

Everyone is aboard the same ship. The Plimsoll line carries the same meaning for all. And time appears short to accomplish the challenging task of lightening the human load.

FATIMA VIANNA MELLO
translated by Joanna Berkman

Sustainable Development For and By Whom?

....

To hold population growth responsible for environmental degradation conceals the central problem: the failure of the current development model itself.

Instead of blaming population growth for environmental problems, the link between population and environment should be analyzed from a point of view which takes into consideration the benefits which the population, if it had access to and control over natural resources, could bring about for the environment. Extractivist workers in Amazonia, for example, are attempting to protect the forest from large property owners and international capitalists. By fighting to maintain their way of life, which depends on the continued existence of the forest, extractivists are keeping the forest alive. Environmental protection must also acknowledge the enormous wisdom and know-how of women, for it is women who give continuity to life, dealing on a daily basis with land, water, food and garbage. Women are the ones most interested in a healthy environment, since they and their children are the first victims of pollution.

Unfortunately, the perspective of those with the most to offer and the most to lose — the population itself — has not prevailed in the predominant debate about population, development, and the environment. All seem to agree that this debate be framed in terms of the "population

---

Fatima Vianna Mello is a historian, an activist in the Brazilian feminist health movement, and a member of FASE, a Brazilian development non-governmental organization. This chapter is reprinted from Brazil Network Contato, *January 31, 1992.*

problem," where the explicit question is the size and growth of the population in the Third World. In this way, those who identify population as the greatest environmental threat do so without distinguishing exactly which sectors of the population do most of the polluting, and where.

Emphasizing only population size and growth leaves aside the division of the population into social classes. Different social classes utilize natural resources for different ends and in different ways, and they have extremely unequal standards of living and levels of consumption. This means that some human beings appropriate resources for themselves in the form of profits, while the majority is denied access to those resources.

The quantitative approach defined by the present development model begs fundamental questions about the distribution of population and of wealth. In Brazil, the accelerated process of urbanization during the last 40 years resulted from the extremely high concentration of land in very few hands and from expropriation of rural workers. Those trends combined with processes of industrialization, internationalization and expansion of the economy, which were nourished at the expense of an abundant, cheap work force.

The quantitative approach also fails to consider the division of the population between women and men. Women's bodies have simply been chosen as the target of population control advocates, with no attempt to understand the mechanisms of patriarchal culture which, among other things, created the sexual division of labor (and wealth) and placed most contraceptive responsibilities upon women and their bodies.

This debate is being conducted in the most simplistic, mechanistic and limited form possible, driving all analytical energies towards one predefined goal: to control population growth in the so-called developing countries. By making population limitation in the Third World the main environmental solution, the issues of unequal distribution and consumption of wealth become nonquestions. In spite of its oversimplicity, the argument that population growth produces pervasive and detrimental environmental impacts through increased resource consumption and deterioration is becoming the consensus in the U.S. While there are different levels of consistency and emphasis in this analysis, it is found in publications of the most diverse organizations — from environmentalists to multilateral organizations, from the population lobby to U.N. agencies. This approach has become a priority not only on the agenda for the U.N. Conference on Environment and Development (UNCED), but also in the long-term analyses and proposals referred to as "sustainable development."

POPULATION GROWTH AND SUSTAINABLE DEVELOPMENT

The term "sustainable development" has been widely adopted. The central argument is that to attain the objective of greatest sustainability, the rate of population growth must decrease. Quite frequently, the argument asserts that the largest factor in "unsustainability" is the pressure exerted by population growth on natural resources in developing countries and that, therefore, a "sustainable" future depends fundamentally on controlling the growth of this population.

Isabel Carvalho's analysis of the values assumed by the ideology of "sustainable development" dismantles the arguments that have led to this unrealistic consensus. Examining the criteria of sustainability — what we want to sustain, for what purpose, and for whom — Carvalho, a Brazilian researcher, notes: "The fact of advocating for a more productive society, at a lower socio-environmental cost, does not necessarily imply the choice of a more orderly, just or participatory society. We could arrive at a high level of efficiency, with new, clean technologies, and even with a diminution in the level of absolute poverty, without significantly altering the degree of political participation and ethics in social relations."

In the case of tropical forests, it is often argued that one of the main causes of destruction, the "unsustainable practices" supposedly perpetrated by this burgeoning population, is the human demand for survival. An August, 1989, Zero Population Growth fact sheet illustrates this position, stating that "as increasing human numbers accelerate the demand for land and timber from the forests, this balancing act is becoming ever more difficult....Throughout the tropics, developing nations are struggling to meet the food needs of their rapidly growing populations, placing enormous pressures on their forests."

But to which human beings does this argument refer? In the Brazilian Amazon, there is an enormous difference among humans. This difference is the origin of the conflict between social classes which have opposite interests in the Amazon, the most striking evidence of which is the unacceptable violence by large landowners against rubber tappers, Indians, extractivist workers and other riverside inhabitants. The human demand for survival by the resident population is mentioned as the greatest factor in the destruction of the Amazon forest. But at no time is it pointed out that the development model practiced by megaranchers and large national and multinational corporations requires expansion, accumulation, and exploitation of nature. There is no mention whatsoever of Brazil's external debt and of its devastating consequences from a social and environmental point of view.

With some exceptions, including the Consortium for Action to Protect the Earth (CAPE), which states that in developing countries, "national and multilateral policies and unequal distribution of land and resources tend to deepen poverty among populations," the responsibility of multilateral organizations in financing socioeconomic disasters is not duly noted by the great majority of participants in the debate. In Brazil, sectors that have been heavily financed by the World Bank — especially the energy and agricultural sectors — are examples of the ecological disasters that characterize these "development projects."

To name population growth as an obstacle to sustainable development is to emphasize the putative unsustainable activities of the population. This avoids all analysis of the destructive character of the reigning development model, which aims at gaining profits and satisfying market forces, and not at meeting the basic human needs of the population.

TOWARD A DEMOCRATIC DEVELOPMENT MODEL

Current arguments assert the need to diminish population growth in order for citizens to better enjoy the benefits of development. Can it really be true that population size and growth are the greatest obstacles to implementing a development scheme that meets citizens' needs?

Using a kind of mathematical logic, some arguments try to prove with graphs and projections that the pressure on natural resources increases in exact proportion to the numerical growth of the population. However, in Brazil the birth rate is falling but the rate of environmental destruction continues unabated or is worsening, which clearly proves the inconsistency of the arguments that relate environmental destruction with high fertility rates. Clearly, it is not the sheer number of people, but their consumption patterns that determine environmental impact. The average North American, for example, consumes dozens of times more energy than an Indian or Brazilian. Consumption levels in the North are the result not just of individual behavior patterns, but more importantly, of technical standards inherent in capitalist development, which create and feed these consumption levels. This technological standard, together with the rule of accumulation, is viable largely through the social exclusion of the majority of the populations of both the South and North.

The tenuous connection that has been constructed between population growth and pressure on natural resources has become a form of specious conventional wisdom, and merely serves to cover up the central issue, which turns out to be the failure of the development model itself. To advocate a decrease in population growth as a means to better distribute wealth and to protect natural resources blames the poor people of the world for the tremendous socio-environmental imbalances that exist. It is a stance that serves the status quo and fights neither social inequalities nor the domination of the South by the North.

We need a new concept of development, one that puts the majority of the population and its needs as the first priority and as the point that defines all other policies. Those who have always been excluded from decision-making processes and from access to wealth, both of which are concentrated in the hands of a minority, need to be protagonists in this debate. The angle of vision must change from the question of population growth to questions of administration. Who controls natural resources? Who decides how resources will be used? Who has control over information and knowledge? And for every woman, the most basic question of all, who controls her body?

F. LANDIS MACKELLAR AND DAVID E. HORLACHER

Population, Living Standards And Sustainability: An Economic View

....

The dominant feature of the global socioeconomic landscape is the demographic contrast between the well-off populations of Europe, North America, and Japan and the poor populations of Asia, Africa, the Middle East, and Latin America. The population of the developed, industrialized world is small in absolute numbers, relatively old in average age, and is growing at a very slow rate. That of the under-developed, mostly agricultural and commodity-producing world is large in absolute numbers, relatively young in age, and is growing very rapidly. The World Bank estimates that per capita gross domestic product (GDP— the value of goods and services produced) in the more-developed countries (MDCs) is over 20,000 U.S. dollars (USD), whereas in the less-developed countries (LDCs), it is only about USD 1,000. Ethical imperatives, as well as concerns for global security — the lessening of social tensions within poor countries and of migratory pressures between poor and well-off countries — demand that substantial progress be made towards narrowing this gap.

Demographers expect that narrowing the gap will hasten population stabilization. The theory of demographic transition predicts that, as living standards rise, first mortality declines and then, somewhat later, fertility declines. During the transition itself, population grows rapidly because the fall in fertility precedes the fall in mortality. [See "Population By The Numbers," by Carl Haub and Martha Farnsworth

F. Landis MacKellar is the Andrew W. Mellon Visiting Scholar at the Population Reference Bureau and David E. Horlacher is the Hepburn Professor of Economics at Middlebury College.

Riche, page 95.] Demographic trends since World War II have confirmed the theory of transition and the world, taken as a whole, is now entering its later phases. Nonetheless, future growth trends are already, to a large extent, embedded in the current population. Demographers can predict with virtual certainty that world population will reach six billion before the end of this century and project with considerable confidence that by 2025 it will have surpassed eight billion. Virtually all of this increase in world population will occur in LDCs.

Rapid population growth in LDCs might pose a threat to global sustainability in one of two ways.[1] First, it might impede economic growth and mire LDCs in poverty, perpetuating or even widening the gap in living standards between LDCs and MDCs. Research on the effect of population growth on economic development has not reached particularly strong conclusions,[2] so we do not pursue this line of reasoning here. Second, and paradoxically, if it does not impede economic growth, then satisfying the consumption demands of rapidly growing and increasingly affluent Third World populations might endanger global environmental sustainability. Many experts have argued that the earth cannot even sustain all of the planet's current — much less future — inhabitants at the level of consumption which now prevails in the MDCs. It has widely been concluded, based on such reasoning, that pressure on the global ecosystem can be relieved only by implementing a combination of aggressive family planning policies to slow LDC population growth, and policies to stabilize or reduce consumption in industrialized countries should also be implemented so as to offset improvements in Third World living standards.

This chapter provides a somewhat different view. We contend that, so long as markets function (or if policymakers respond appropriately when they do not) and if steps are taken to address poverty, then neither rising affluence nor growing population need result in reckless natural resource depletion and environmental degradation. This view is thus fundamentally — albeit cautiously — optimistic. We neither endorse nor contest the desirability of promoting family planning and the small-family norm in LDCs and encouraging re-examination of the values which underlie consumption patterns in MDCs. We do believe, however, that the rationale for either one is better sought elsewhere than in the area of global sustainability.

NATURAL RESOURCES AND THE ENVIRONMENT: AN ECONOMIC VIEW
Exhaustible resources, renewable resources, and pollution

Natural resources may be classified as *either exhaustible or renewable.* Exhaustible resources, including oil and minerals, consist of a fixed stock which can be replenished only over geologic time. Some exhaustible resources, such as copper, can be recycled; others, such as coal, can not.

Renewable (or regenerative) resources including topsoil, fisheries, or forests can be utilized or harvested without reducing the stock. If, however, the rate of utilization exceeds the "flow," or rate of regeneration, a renewable resource will be depleted and, in the extreme, can be destroyed forever.

The distinction between exhaustible and renewable resources is particularly important in assessing the scale at which population and living standards come into play. Most exhaustible resources are traded in world markets; thus, it is global trends in population and income which are relevant. Some renewable resources, such as farmland used to produce grain for the world market, deep-water commercial fisheries and commercially exploited forests, also respond mainly to global trends. Others, such as the ecologically fragile soil and forest resources on which rural households in LDCs depend for survival, are affected more directly by local trends in population, i.e. population distribution and living standards. *Production* consists of the combination of labor, man-made capital, and natural resources used to transform raw materials into usable goods. Unfortunately, all production gives rise to residuals which are at best useless and at worst noxious. For example, making a chair gives rise to woodchips; these can be burned as fuel, but this gives rise, in turn, to carbon dioxide and soot in the atmosphere. Residuals must be deposited somewhere in the environment. *Pollution*, from a strict economic point of view, is deposition of residuals in such a way as to reduce the asset value of the environment.

The damages from pollution are of three kinds: damage to health, damage to economic productivity, and reduction in the "amenity value" of the environment. The first two can be quantified with some degree of accuracy. The difficulty of quantifying the latter provides one of the strongest bases for involvement of government in the management of environmental assets.

From an economic point of view, managing an environmental asset — holdings of which may be drawn down by the deposition of residuals or augmented by environmental cleanup (assuming the damage is not irreversible) or by the adoption of less-polluting technology — is similar to managing a resource stock. In other words, socially optimal environmental management is not necessarily equivalent to environmental preservation. Though maintaining the environment in its pristine state may be the best policy in many cases, it may not be the best policy in every case.

Economic scarcity, markets, and price signals

Much of economics is concerned with how the underlying scarcity of goods, services, and factors of production is communicated to consumers and producers through price signals. *Economic scarcity* is an elusive concept, but can be understood basically as an insufficiency of supply relative

FIGURE 1: Energy Utilization Per Unit GDP

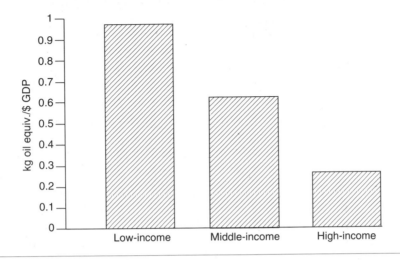

to demand. In the case of a natural resource such as petroleum, if demand goes up but supply is fixed, scarcity will result in a higher price, which will lead to a wide range of behavioral adaptations: natural resource firms will attempt to find and develop new resource deposits in order to take advantage of higher prices by expanding production, industrial producers will attempt to substitute other inputs for the resource in production, consumers will attempt to reduce their purchases of products whose fabrication requires the resource, and recycling and the development of resource-saving technologies will be encouraged. In other words, scarcity sets in motion forces to reduce scarcity. Dramatic examples of this phenomenon include the expansion of world oil exploration and supplies after the OPEC embargo of 1973-74, and the expansion of cropland and, eventually, food production after the poor harvests of the early 1970s.

Obviously, such responses will be more effective in some areas than in others. In the case of petroleum, minerals, and many agricultural commodities, markets are highly developed and function smoothly. In other areas, markets are nonexistent or do not function well. For example, say that a government has short-circuited market forces by setting "stumpage fees" — the fees charged to timber concessionaires for logging on public lands — at a fraction of the actual scarcity value of the trees. The subsidy will lead timber companies to cut too many trees too fast. Worse yet, as timber becomes increasingly scarce, the unrealistic price ceiling prevents the market from sending a "wake-up call," in the form of rising prices, to either producers or consumers.

Some environmental damage is taking place beyond the sphere of existing markets. In these cases, which include many forms of pollution, loss of tropical biodiversity, and global climate change, authorities must intervene. If producers are not charged for the residuals which they deposit in the environment, costs of production do not reflect environmental damage. As a result, the producer charges too little for, and consumers demand too much of, the good in question. Similarly, if producers (farmers, timber companies, and agribusiness) bear no costs for the extinction of species in tropical forests, the result will be sub-optimal conservation.

Do markets generally work in the area of natural resources and the environment, or do they generally not work? Most economists are confident that market failure, while not uncommon, is the exception rather than the rule; that where it exists, it is not terribly difficult to recognize, and that the means to counteract it are usually both effective and socially practical.[3] For example, a government could impose emissions taxes which reflect the social costs of pollution, therefore encouraging producers to pollute less in their own economic self-interest. Alternatively, a government could determine an acceptable level of emissions and impose a limit accordingly. Economists tend to favor approaches which imitate the operation of a market, such as the former, rather than direct regulatory approaches, such as the latter. In this, they are sometimes at odds with environmental advocates, who point out that market measures neither guarantee that the goal will be attained nor impose legal sanctions when it is not.

LIVING STANDARDS, NATURAL RESOURCES, AND THE ENVIRONMENT

Environmentalists tend to believe that affluence magnifies the environmental impact of a given population. Thus, they contend, as living standards rise in the Third World and as more people aspire to contemporary western levels of consumption, per capita environmental impact will increase. Most economists, on the other hand, are more sanguine. Part of the difference arises from economists' faith that, in most cases, demand pressures against the resource base will result in price signals to producers and consumers and that, if not, the problem will become apparent and can be rectified before irremediable harm results.

Economists also believe that, even holding prices constant, the relationship between income, natural resources, and the environment is not a straightforward one in which an increase in income necessarily results in a parallel increase in natural resource utilization and pollution.[4] Economic growth results in changes in the composition of *output* — the mix of goods and services produced — and improvements in the efficiency with which natural resources are used. Therefore, although natural resource consumption rises with income, it rises at a progressively slower rate, leaving an increasing share of income available to finance further development of substitutes for natural resources (largely man-made capital) and technical progress. Moreover, rising income results in greater demand for environmental quality. This manifests itself in consumers' insistence on policies to attain a higher level of environmental quality and, ultimately, leads to environmental investment which is financed by higher incomes.

In paragraphs that follow, we look at some of the empirical evidence in this area. This evidence is meant to be illustrative, rather than definitive, but the impression it leaves is a generally optimistic one.

Income and the demand for natural resources

Several factors moderate the environmental impact of rising income. First, as a country passes through the stages of development, the share of the "primary sector" (natural resources and agriculture) in output and consumption declines as resources are used more efficiently. For example, in 1990, commercial energy utilization per dollar of GDP was highest in low-income countries and lowest in high-income countries (see Figure 1).[5] The same pattern has been verified for agricultural goods, wood products, minerals, and so on.

Of course, energy use, consumption of agricultural goods, etc. rise on a per capita basis as per capita income increases (see Figure 2). In 1990, for example, per capita energy consumption in high-income countries was more than 15 times that in low-income countries. Over the long term, however, per capita consumption of primary products rises, not at the

FIGURE 2: Energy Utilization Per Capita

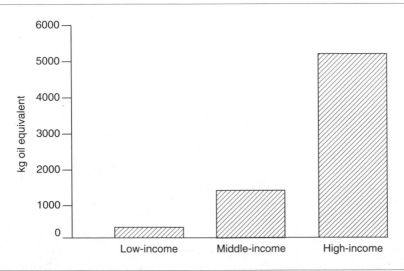

same rate as per capita GDP, but at a slower rate. The parameter which economists use to describe the relationship between changes in GDP and changes in the consumption of a given natural resource is the so-called elasticity of demand for the resource with respect to GDP; that is, the percent increase in resource consumption which will be caused by a one percent increase in GDP. For example, if the elasticity of demand is 0.5, a three percent rise in income will cause a 1.5 percent rise in resource consumption. If the elasticity is less than one, the share of natural resource in total spending (and output) declines when income grows; if it is one, it remains constant as income grows; if it is greater than one, it increases as income grows. Measured between 1965 and 1990, the elasticity of energy consumption with respect to GDP was 1.41 in low-income countries, 1.18 in middle-income countries, and 0.62 in high-income countries (see Figure 3). Extrapolation suggests that, at an income level sufficiently high, subsequent economic growth — occurring largely in the service and information sectors — will result in little further energy demand.[6]

But this is a long-run speculation, and in the long run, as Keynes put it so aptly, we are all dead. In the decades to come, rising world income will still spur increased demand for energy and other natural resources, and economic growth might lead to depletion of resource endowments long before the world reached the blissful stage in which output was "delinked" — or nearly so — from natural resource demand. In a scenario first described by the 19th-century English economist David Ricardo and

FIGURE 3: Energy-GDP Elasticity, 1965-90

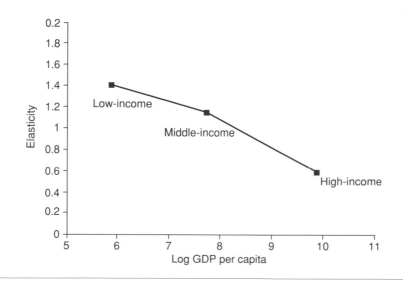

popularized by the *Limits to Growth* study,[7] natural resource scarcity can force a society to devote an ever-larger share of income to exploiting marginal energy supplies, cleaning up pollution, and eking out food and materials from ever more costly sources. Most modern economists argue, however, that operation of the price system as described above makes this possibility remote. They — and we — would agree with the remark attributed to environmental economist Anthony Fisher: The world is unlikely to go crashing full-speed into the wall of resource depletion or environmental collapse without warning from the price system that something is going terribly wrong.[8]

Income and the demand for environmental quality

The second factor that moderates the environmental impact of rising income is the fact that large, long-term increases in per capita income generate greater demand for environmental quality. As income rises, consumers insist on environmental policies and programs that meet this demand. Moreover, economic growth supplies the financial resources to implement such policies and programs.

In part because of data deficiencies, much less is known about the relationship between income and environmental quality than about the relationship between income and utilization of natural resources. Cross-country data from the 1980s presented in the 1992 *World Development Report* indicate that some environmental indicators, such as availability of safe drinking water and sanitation, improve with income growth whatever the level of development of the country in question. Others, such as urban air quality, worsen with income growth when a country is poor but improve with income growth once a country has reached a higher stage of development. Still others, such as carbon dioxide (CO_2) emissions per capita, worsen with income whatever the level of development.[9]

An admittedly unscientific way of illustrating the linkage between large, long-term increases in per capita income and demand for environmental quality is to compare air quality in the commercial centers of cities located in countries at various levels of development.[10] As illustrated in Figure 4, in 1983-86 the average ambient concentration of sulfur dioxide (SO_2) measured in commercial city centers was much higher in low-income countries than in high-income countries. More fine-grained data indicate that SO_2 concentration increases along with per capita income until a country attains a level of per capita GDP of roughly USD 1,000; after that point, there is an inverse relationship between per capita income and SO_2 concentration. Up to the turning point at USD 1,000 (about three times the level of income in India; roughly the level of income in Morocco or Colombia), economic growth worsens the SO_2 problem; after the turning point has been passed, economic growth

FIGURE 4: Ambient Concentration of SO_2, 1983-86

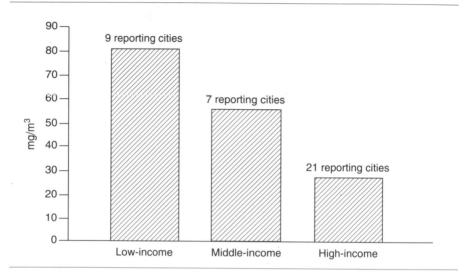

improves it. In the case of suspended particulates, the turning point appears to occur much earlier, at a level of per capita income of only a few hundred dollars.

These are merely statistical relationships estimated on the basis of a sample of countries; they capture no more than broad regularities. The effect of income growth in any given country at any given point in time depends on the policies and programs in place. Nonetheless, the data suggest that, while increases in per capita income might worsen pollution in some poor countries in the near term — and perhaps give rise to environmental disaster in countries where policymakers take no action in the face of worsening conditions — the effect of large, long-term increases in the standard of living is to alleviate some types of pollution, not worsen them.

Data on environmental quality trends in MDCs are somewhat more complete than data for LDCs. A 1991 OECD retrospective on environmental trends in member countries since the the 1970s found substantial improvement in most of the pollution areas surveyed.[11] The main exceptions were concentration of nitrogen oxides in the urban atmosphere, per capita production of municipal wastes, and per capita emissions of CO_2. GDP in member states grew by 80 percent over the period covered, so the overall improvement was achieved in the face of substantial increases in the standard of living. The direct cost to consumers was 1-1.5 percent of GDP invested in anti-pollution measures and policies.

However, although citizens of affluent countries demand better environmental quality at home, they still consume products that are produced in other countries at great environmental cost. Anti-pollution policies in the MDCs have led some polluting industries to relocate in middle-income LDCs, such as Mexico and Korea, where environmental policies are more permissive. Given the choice between extra income and a less polluted environment, the citizens of those countries have opted for the former. However, as incomes rise in these countries, so will demand for environmental quality. Polluting industries will then be forced to move again until, eventually, either the least-developed countries of the world play host to them or living standards are everywhere high enough that some industries and production technologies must be abandoned altogether.

Is this fair? Should poor countries be permitted to play host to polluting industries in exchange for economic growth, or should the international community prohibit the migration of such industries? These are rather deep waters. Law books are filled with prohibitions against actions which appear, on their face, to raise total social welfare; for example, laws that prohibit prostitution even though both consumer and supplier evidently believe that they would benefit from the transaction. Such laws are often justified on the basis of serving a more diffuse, higher social good, or on the grounds that the parties to the transaction do not understand the true costs and benefits. In reality, though, they usually reflect prevailing social values. In the case of poor countries and polluting industries, there is no reason to assume that international institutions will prove any better — or worse — at coping with equity issues, including the well-being of the least well-off, than have domestic institutions.

To summarize, then, it is clear that pressures against the natural resource base and the environment do not rise in lockstep with the standard of living. So long as private economic decisions reflect the underlying value of natural resources and environmental assets — which is to say that either markets function reasonably well or instances of market failure are addressed by public policy — economic growth need not be accompanied by worsening natural resource scarcity and environmental deterioration.

POPULATION, NATURAL RESOURCES, AND THE ENVIRONMENT

Economists also tend to be less alarmed than environmentalists about the effect of population increase on natural resources and the environment. Nonetheless, economists have identified conditions under which population growth may cause or exacerbate the effects of market failure. They are also devoting growing attention to the pernicious links between population, poverty, and the environment.

The tragedy of the commons

The market failure of most concern involves common-property, open-access resources. The "tragedy of the commons" arises not from common ownership per se, but from the fact that no person or legal entity controls access to a commonly held resource, so the price of using the resource is zero, and no one has an incentive to conserve it. Indeed, each person with access has an incentive to use up as much of the resource as possible before others get to it. Examples include deep-water fisheries, some coastal and inland fisheries, many areas in tropical forests, and semi-arid rangelands, etc.

Population size, and by extension, rapid population growth, exacerbates the tragedy of the commons by increasing the number of hands grabbing for a slice of the pie. It is important to observe, however, that population pressures do not, by themselves, give rise to common property, open-access problems; on the contrary, they are much more likely to encourage privatization and the evolution of rules limiting access to resources still held in common. The European enclosure movement of the 17th and 18th centuries provides an example of such a development over the very long term, but shorter-term responses of this type have been observed in the American west, Japan, South- and Southeast Asia, and Africa.

While slowing population growth may diminish pressures on commonly held resources, the problem itself can be eliminated only by establishing property rights. This need not involve rigid privatization along Western lines, which could weaken important sociocultural institutions or worsen the lot of the poor, for whom exploitation of common property, open-access resources has become the economic activity of last resort. In the case of rural Africa, researchers at the World Bank have identified many opportunities for alleviating pressures on the natural resource base while attacking poverty by strengthening traditional tenure systems and institutions.[12]

Public goods

Market failure also occurs in the realm of *public goods*. A natural resource is considered a public good when one person's consumption of that resource does not diminish another person's consumption. Examples are clean air, the global climate, and the existence of biological diversity. All of these can be "produced" like any other good, in the sense that investments may be made in pollution abatement, control of greenhouse gas emissions, and conservation. Unfortunately, since the benefits of a public good are available to all at no cost, no private entrepreneur has the incentive to produce it. Hence, in the absence of public intervention, there will be less than the economically optimal production of public goods.

It is cheaper on a per capita basis to produce a public good in a large rather than in a small population. This point is closely associated with economies of scale in supplying infrastructure[13] and is one of the benefits of large population size and rapid population growth. It is relevant to pollution control, provision of agricultural extension services, and a range of other topics of concern in the area of natural resources and the environment.

Externalities

Externalities are costs and benefits which arise in production or consumption and are borne by society as a whole, but not specifically by either the producer or the consumer. Many pollution problems arise from external costs; for example, when neither the owners of a coal-burning power plant nor the consumers who purchase the power are forced to bear the costs of acid rain which may fall hundreds of miles away. Of equal concern are external benefits, which arise, for example, when residents of Country A forgo burning coal in order to alleviate the atmospheric CO_2 problem, but receive no compensation for the benefit reaped by residents of Country B.

There is theoretical evidence that population growth gives rise to external costs[14] — that is, that private fertility decisions impose costs on others — on the assumption that there is an already existing market failure of the common-property type. In LDCs, high fertility gives rise to external costs which arise from the decreased per capita availability of common-property natural resources. Whether this conclusion can be generalized to cases of market failure other than those of the common-property type sounds plausible under the right assumptions, but is not known with certainty.

Under some circumstances, the pursuit of economic self-interest will spontaneously give rise to bargaining and negotiation processes that reduce externalities. While it would be reckless to claim that such processes will automatically eliminate externalities, evidence indicates that social and political institutions in free-market economies eventually come to grips with yawning gaps between the private and social costs of private decisions.[15] If such institutions for dialogue are better equipped to deal with moderate change rather than rapid change, then it is plausible that such processes might work better under conditions of slower population growth.

Poverty

Poverty is not a market failure per se, but gives rise to similar problems. Much rural environmental degradation in LDCs occurs in areas characterized by acute deprivation: inequitable access to resources of all kinds; insecurity; poor governance; and lack of opportunities for alternative economic activities or out-migration. Adjectives such as "rampant" are often used to describe poverty in LDCs, but these obscure the crucial fact that poverty is a problem which is both bounded and soluble. The World Bank estimates that the proportion of the total LDC population living in poverty remained at about 30 percent between 1985 and 1990.

The pernicious impact of poverty on the environment was a major theme of the 1992 World Bank *World Development Report*, which described poverty reduction as "essential for environmental stewardship." Policies to correct market failure will be of limited effectiveness and, from the standpoint of social justice, will be of limited benefit, if significant portions of the world's population continue to live in poverty.

Impoverished households living at the margin of survival are likely to deplete surrounding natural resources and environmental assets with little heed for future consequences. This is not because they are unaware of these consequences or are inherently reckless. Rather, it is because they are not able to undertake the types of substitution, adaptation, and behavioral change which are at the heart of the economic model. For example, when poor communities intensify agricultural production to cope with rapid population growth, it is often because they cannot afford the investments and inputs necessary to prevent soil degradation. As soil productivity falls, farmers who are unable to substitute other inputs for land and lack other sources of income will apply yet more labor to the same land, worsening the problem. Thus, the proximate cause of land degradation under these circumstances is population growth, but the ultimate cause is poverty.

Beginning with Adam Smith, economists have observed that poverty encourages high fertility and, by extension, rapid population growth. Yet there is little evidence running the other way — that high fertility and rapid population growth contribute to poverty.[16] There is, however, a strong prima facie case that, where poverty is already present, population growth and high fertility exacerbate its impact on natural resource depletion and environmental degradation.

We conclude that if markets function well, and if behavioral responses to market forces are not hampered by poverty, then as population grows — just as when income rises — pressure against natural resources will be translated into rising prices. Rising prices will, in turn, encourage adaptive behaviors. This is not grounds for complacence. When markets are structurally flawed, or when poverty inhibits the operation of market forces, population size and rate of growth usually magnify the adverse impact on natural resources and the environment. Under such circumstances, policies designed to reduce the rate of population growth may contribute to sustainability.

CURRENT SUSTAINABILITY ISSUES

Below, we examine the role of population and living standards in four environmental problems which have been identified by the U.N. as the greatest threats to global sustainability.[17]

Land degradation and soil erosion

Land degradation renders land unsuitable for cultivation, whereas soil erosion reduces its productivity. In the LDCs, land degradation arises from the combination of population pressure and poverty, which prevents farmers from adopting the methods necessary to conserve soil as they intensify cultivation.[18] It is not caused by rising living standards; on the contrary, it is largely due to poverty. Soil erosion in LDCs arises under similar circumstances. It is most pronounced in locales where population pressure and lack of alternative activities have forced cultivators onto marginal, usually steeply sloped, land. The direct cost of lost soil productivity in LDC agriculture is roughly 0.5-1.5 of total GDP, but the impact on the rural poor is far higher. Runoff of eroded soil gives rise to external costs, in the form of water pollution, siltation of waterways, harbors, and reservoirs, etc.; factoring in these social costs raises damages significantly.

Soil erosion in MDCs — especially the U.S., which is the world's largest agricultural producer and exporter — is related to both world population and economic growth, insofar as each of these affects global demand for food. Some studies of American agriculture have concluded that observed erosion rates may be economically rational from the farmers' point of view, since farmers can compensate for lost soil with other inputs, such as fertilizer. However, when external, off-farm costs are factored in, erosion is far higher than its social optimum.

Some experts have argued that MDC agriculture of the intensity and sophistication necessary to keep pace with global population increase — characterized by heavy reliance on trade and a handful of specialized producers, monoculture, permanent cultivation, intense irrigation, and application of pesticides, herbicides, fertilizers, etc. — is inherently unstable and environmentally costly.[19] Global agricultural output has generally kept pace with population and food prices have generally fallen over the long term,[20] but the agricultural sustainability debate will not soon be resolved.

Deforestation and loss of biodiversity

In global terms, most loss of tropical forest area is caused by *extensification* — the encroachment of impoverished, landless cultivators onto open, mostly secondary forest land.[21] This arises from the vicious combination of poverty and the common-property, open-access aspect of the lands involved. Although studies increasingly indicate that the rate of population growth is a significant factor in deforestation,[22] the household- and community-level dynamics of land-use change and, thus, the role of population at the micro-level are poorly understood. The role of commercial activities is greater than would be concluded based on the relatively modest forest area which they affect directly, because commercial development leads the way for cultivators. Aggressive timber exploitation is due less to growth of world demand for wood products, which has long been anemic, than to ill-conceived government policies designed to maximize near-term benefits in the face of weak world demand.[23] Loss of

FIGURE 3: Totally Protected/Protected Area (Percent)

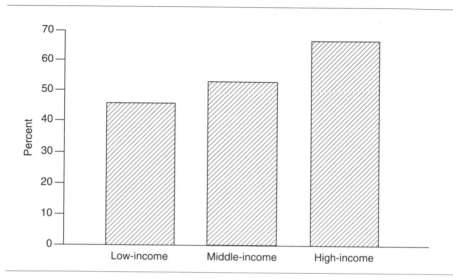

biodiversity is the result of tropical deforestation and is exacerbated by the public-good aspect of genetic diversity.

Water resources

Fresh water, while abundant at the global level, is unevenly distributed among regions. Some regions, such as Scandinavia, have an over-abundance of water, whereas others, such as the Sahelian belt in Africa, face severe and persistent drought. Scarcity of fresh water results from the interplay of supply and demand forces. On the supply side, water shortages can arise from natural hydrologic factors, such as inter-annual fluctuations in precipitation, in which case improvements in storage and distribution may be an effective response. Or, shortages can arise when available water resources are wasted in the form of runoff caused by deforestation, over-grazing of semi-arid lands, and erosion of hillsides; all of which are related, along lines developed above, to population and poverty.

On the demand side, agriculture plays the dominant role. According to data from the World Resources Institute, over 70 percent of global fresh water is used for irrigation, although this varies with local circumstances.[24] After agriculture, the largest water user is industry (23 percent), with domestic use coming in a distant third (8 percent). The proportion of total water use accounted for by irrigation declines with level of development: in low-income countries, it accounts for roughly 90 percent of total water use, in middle-income countries 70 percent, and in high-income countries, 40 percent. The proportion devoted to domestic consumption is 4, 13, and 14 percent, respectively; the remainder is used for industrial purposes. Regional, national, and local factors make for wide variation within broad categories, however. For example, in the U.S. as a whole, irrigation accounts for about 40 percent of total water use, but in California the figure is over 80 percent.

Clearly, the key to assuring future water supplies — including satisfying the needs of rapidly growing urban populations — lies in improving the efficiency with which the resource is used in agriculture. The threat to hydrologic sustainability posed by population growth and rising living standards pales beside that posed by continuing failure to charge prices which reflect the underlying economic scarcity of the resource. Seldom do farmers pay more than a fraction of the actual supply costs of the water they use; thus, it is not surprising that irrigation is the most inefficient user of water supplies, with waste often close to three quarters of total water diverted.[25] In some circumstances, policymakers have evidently concluded that it is socially impractical to charge realistic prices. For example, if true scarcity values were reflected in water prices, the California irrigated agriculture complex probably would not be sustainable over the long term. In other circumstances, action is socially practical but policymakers are evidently unwilling to come to grips with the problem. In yet other circumstances, it is hydrologically impossible to price water in conventional terms, and innovative policies and programs must be devised to imitate the operation, or least the outcome, of a market system.

Climate Change/CO_2 emissions

Climate change is a vastly complex topic. For our purposes, however, climate change can be considered as essentially the problem of CO_2 emissions caused by the burning of fossil fuels. In this area, more than any other, researchers have analyzed the impact of population growth and rising living standards on emission trends.[26]

Per capita CO_2 emissions are high in MDCs because of elevated levels of per capita production and consumption. Indeed, although they are home to only one quarter of the world's population, the MDCs generate nearly three quarters of all CO_2 emissions. On the other hand, CO_2 emissions per unit of GDP are high in LDCs because energy is used less efficiently and because LDCs rely more on "dirty" fuels, especially coal. In China and India, for example, governments heavily subsidize the exploitation of domestic coal resources. It has been estimated that, if subsidies on energy production were eliminated worldwide, CO_2 emissions would be reduced by 10 percent.

Between 1950 and 1985, world CO_2 emissions grew at three percent per year, world population grew at two percent per year, and emissions per capita grew at one percent per year.[27] As a simple matter of accounting, population growth was responsible for two-thirds of the increase in emissions, holding everything else constant. The relative contribution of population will be lower in the future because population growth is higher in LDCs, where per capita emissions are low, and because economic growth will — or at least we hope it will — be higher in LDCs, where emissions per unit of output are high. Energy use per unit output is likely to decline in MDCs and, eventually, in LDCs as well, along the lines predicted by Figure 3; in addition, LDCs will adopt cleaner energy sources. When these factors are all combined, analyses project that the contribution of population growth to CO_2 emissions will be something like one-half in 1985-2025 and one-fifth in 2025-2100. Projections of CO_2 emissions are relatively insensitive to changes in the assumed rate of LDC population growth.[28]

Such accounting exercises do little, however, to get to the heart of the matter, which lies in the public-good aspect of the global atmosphere, in the external costs countries impose on their neighbors by burning fossil fuels (especially coal) or clearing forest area[29] and in the external benefit which they confer by controlling emissions and conserving forests. The public-good aspect has led governments to agree, through the International Convention on Climate Change, on coordinated intervention to promote stewardship of the global atmosphere. It has been accepted that, having been responsible for most of the buildup of atmospheric CO_2 to date, MDCs should bear the burden of the costs of finding and implementing solutions. The externality aspects are being addressed — as economic theory would predict — in international negotiations. The outcome of negotiations will depend on how participants judge the costs and benefits of slowing climate change, the estimation of which is currently an active area of economic research.[30]

CONCLUSION

We have described the views expressed above, which are shared by most, although by no means all, economists, as "fundamentally optimistic," because they suggest that rising living standards, population growth, and wise stewardship of the ecosystem can go hand in hand. The economic perspective is not, however, Panglossian. Indeed, an economic analysis reveals that market failure and poverty foster patterns of resource utilization which are inefficient and inequitable, at best, and may endanger global sustainability, at worst. Nor is the economic view laissez-faire, to use a term frequently associated with economists, because it calls upon policymakers to take a range of far reaching, often painful, steps to address resource and environmental problems. Economists are no more or less prone than anyone else to worry about global trends — but we try to worry efficiently.

Population
Growth and Structure

....

CARL HAUB AND MARTHA FARNSWORTH RICHE

Population by the Numbers: Trends in Population Growth and Structure

....

"We are burdensome to the world, the resources are scarcely adequate to us; and our needs straiten us and complaints are everywhere while already nature does not sustain us."

"In many parts of the world, population is growing at rates that cannot be sustained by available environmental resources, at rates that are outstripping any reasonable expectations of improvements in housing, health care, food security, or energy supplies."

The first statement was written by Tertullian around 200 AD; the second by the World Commission on Environment and Development in 1987. Their similarity tells us that concern about population growth is not new. What *is* new is that the world's population is growing at an unprecedented rate.

"Recent demographic trends can be described without exaggeration as revolutionary, a virtual discontinuity with all human history," writes demographer Michael Teitelbaum.[1] Yet among the many myths about population, the one most widely held by Americans is that the "population bomb" has been defused. Paul Ehrlich's 1968 book of that name aroused concern among many Americans, but surveys show that, until this decade, concern diminished even as world population soared.

Many Americans find it hard to engage the issue of world population growth because they believe it is happening far away, because the United States is sparsely populated relative to most of the world, and because

Carl Haub is director of information and education and Martha Farnsworth Riche is director of policy studies at the Population Reference Bureau.

they are confused by the polarized arguments about the impact of population growth. Most of all, they find the numbers too abstract: like the budget deficit, they are so large that they are hard to grasp (see "Understanding the Numbers" page 108).

FERTILITY, MORTALITY, AND POPULATION GROWTH

Even among those who are concerned about population growth, misconceptions persist. For example, many assume that current population growth is a result of rising birth rates. This is not the case: over the last 40 years, birth rates have fallen in most parts of the world. However, because death rates have fallen even more quickly, more people are living long enough to have children, and the absolute number of births has gone up.

Declining mortality, then, is at the heart of the population "explosion." This is a relatively new phenomenon. Throughout much of human history, most people lived on the knife edge of survival, and high fertility was matched by staggering death rates. Goubert's classic study of French villagers relates that, in the town of Breteuil, the death rate tripled after a poor harvest in 1693.[2] During the 17th century, the populations of Denmark and Finland dropped by about one-fourth after a series of crop failures, and many Europeans were forced to resort to cannibalism during severe food shortages.[3]

FIGURE 1: Life Expectancy at Birth, MDCs and LDCs

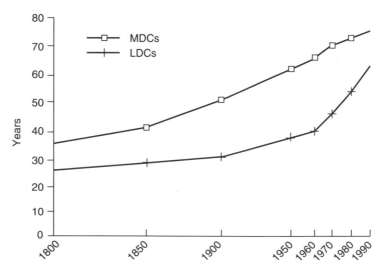

Source: PRB estimates before 1950, UN 1992 projections thereafter.

FIGURE 2: Average Annual Increase in Life Expectancy at Birth in MDCs and LDCs: 1800-1990

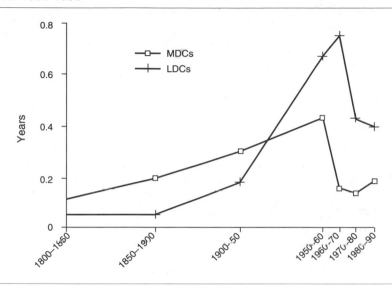

After centuries of struggle against the many causes of early death, humanity's numbers reached one billion by about 1800. This total had taken hundreds of thousands of years to amass. Population grew relatively slowly after 1800 — only about a half percent a year — but gathered momentum during the 19th century. By 1930, the world's population added its second billion.

After the Depression years and World War II, birth rates rose in the industrialized nations, led by a "baby boom" in the United States. But, more important for global population growth, death rates began dropping in the Third World, where two thirds of the world's people lived in 1950. This was made possible by a variety of health care advances, many of which were disseminated by United Nations agencies and private charitable organizations. Increased cooperation among nations in the postwar era hastened the spread of new vaccines, malaria-control methods, and techniques to reduce infant mortality. In a few short decades, these efforts succeeded in eradicating diseases that had plagued humanity for centuries.

As a result, life expectancy in the Third World climbed at a pace never before observed, as Figures 1 and 2 indicate. In 1950, average life expectancy in the developing countries was just under 40 years; by the early 1990s, it had shot up to 62, an increase of more than 50 percent. (Over the same period, life expectancy in the industrialized countries rose from 66 to 75 years.) The most dramatic gains have been made in

TABLE 1

POPULATION CHANGE 1900-1950

	1900	(millions) 1950	Number	Percent	Percent of World Growth
World	1,630	2,516	886	54.4	100.0
Less Developed:	1,070	1,682	612	57.2	69.0
Africa	133	222	89	66.9	10.0
Asia (less Japan)	867	1,294	427	49.2	48.1
Latin America	70	166	96	137.0	10.8
More Developed:	560	835	275	49.1	31.0
North America	82	166	84	102.6	9.5
Europe, Japan, former USSR, Oceania	478	669	191	39.9	21.5

POPULATION CHANGE 1950-1990

	(millions) 1990	Number	Percent	Percent of World Growth
World	5,292	2,776	110.3	100.0
Less Developed:	4,079	2,398	142.6	86.4
Africa	642	420	189.2	15.1
Asia (less Japan)	2,989	1,695	131.1	61.1
Latin America	448	282	170.1	10.2
More Developed:	1,213	378	45.3	13.6
North America	276	110	66.1	4.0
Europe, Japan, former USSR, Oceania	937	268	40.1	9.7

preventing the deaths of infants and children. Worldwide, infant and child mortality has been cut by nearly two-thirds since 1950.

Because children born in the 1960s and 1970s survived in greater numbers, there are now more young men and women of childbearing age than ever before. The age structure of developing countries is heavily weighted towards the young: in 1970, 42 percent of the population of the developing world was below age 15. The result is the phenomenon demographers call "population momentum." With such large *numbers* of people moving into the childbearing years, the total number of births continues to rise, even though individual couples are choosing to have fewer children than earlier generations did.

Mortality declines and population momentum have contributed to extraordinary demographic change in the Third World. After 1950, growth rates accelerated sharply, rising to 2.5 percent in the late 1960s. That rate may not appear especially high, but it will double the size of a population in only 28 years. Thus, after taking 200,000 years to reach 2.5 billion, the human population was growing at a pace that would double the global total with every generation. Today, about 90 million people — more than the current population of Mexico — are added to the world total each year. At this rate, about one billion people are added each *decade*.

As Table 1 indicates, the population surge that began in the 1950s is largely concentrated in the world's poorer countries — 95 percent of current population growth takes place in the developing world. This, too, is a departure from earlier trends. In the first half of the 20th century, population growth in the industrialized countries was proportional to those countries' share of world population — about one-third. At the same time, population growth in the Third World was surprisingly modest by today's standards. For example, between 1900 and 1950, Africa added less than 100 million people. Today, it adds that many in five years. However, in the industrialized countries, population growth fell sharply as birth rates dropped to historically low levels.

FERTILITY RATES AND THE DEMOGRAPHIC TRANSITION

In the late 1960s, when Ehrlich wrote *The Population Bomb*, Third World birth rates were uniformly high and it was unclear whether fertility could be substantially reduced. Women in developing countries averaged six children each, a level of fertility that would have led to runaway population growth had it been maintained. But birth rates have since dropped dramatically; women in developing countries now average four children apiece.

Demographers now expect world population to grow from the current 5.5 billion to about 10 billion in 2050 and to level off eventually at 11 to 12 billion. This expectation is based on the assumption that birth rates will continue their precipitous decline — an assumption which rests on the continuation of internationally supported family planning, health, and development activities. In the unlikely event that birth rates remain where they are today, world population would total 22 billion by 2050, 109 billion by 2100, and 694 billion by 2150. As we have seen, even small rates of increase can lead to explosive growth over time.

Explosive growth is halted only when a population achieves "replacement level fertility" — an average of about two children per couple. When that happens, couples just "replace" themselves, rather than increasing the size of successive generations. Eventually, the population will stop growing, thereby becoming stationary.

Reaching the two-child family represents a huge social and economic transformation in any society. This process is often referred to as the "demographic transition." Generally, the transition proceeds like this: the death rate falls due to improvements in public health and population increases rapidly, especially if the birth rate either does not fall or falls only slowly. In the last stage of the demographic transition, birth rates fall to replacement level, and population stabilization is achieved. The ultimate size of any population is determined by how quickly it reaches replacement level fertility.

The demographic transition is modeled on the experience of what we now call the developed world. In Europe, the transition took place gradually over the course of the 19th and 20th centuries. Between the end of the 18th century when Malthus wrote his famous *Essay on Population* and the start of World War I, the population of the United Kingdom grew from about 10 million to more than 42 million. Some analysts have estimated that, without emigration to America, Australia, India, Canada, and other nations, the population of Britain would have been larger by 30 million.

Today, the developing countries are experiencing an accelerated demographic transition. Thanks to the intervention of international organizations, the death rate dropped quickly. As in the now-developed countries, this spurred an enormous population increase. But there are significant differences between the demographic transition that occurred in the industrialized countries and the one that is now unfolding in the Third World.

First, the scale of the population increase in the Third World today is unprecedented in terms of absolute numbers. While earlier increases in Europe could be measured in millions or hundreds of millions, the developing world is now adding billions in a much shorter period of time. Second, in the now-developed countries, population grew concurrently with advances in technology and agriculture which helped provide for burgeoning numbers. This is not the case in much of the Third World, where rapid population growth is outpacing economic and social development. Finally, the European countries vented their population pressures by colonizing much of the rest of the world. Today the frontier is virtually closed.

PROGRESS IN LOWERING BIRTH RATES

Will developing countries complete the demographic transition by lowering birth rates to replacement level? As we have seen, substantial progress has been made, but there is a long way yet to go.

By the 1990s, 60 percent of developing country governments reported in a U.N. survey that their birth rates were "too high." Most of these countries maintain public or private population/family planning programs, and knowledge of family planning and availability of contraceptives has become more commonplace. While this is an important first step, it often takes years or even decades to affect the birth rate. The government of Kenya, for example, has sought to lower population growth since the 1960s, but its extremely high total fertility rate (TFR)[4] of approximately eight children per woman only began to fall in the late 1980s.

Asia and Latin America have made the most progress in this direction. In both regions, women now average about 3.2 children each, about half as many as before the 1960s. However, when the the large statistical weight of China's 1.2 billion people is excluded, the TFR for Asia is about 4.0. Fertility varies more in Asia than in any other region, ranging from a high of about seven children per woman in Arab and Moslem states to 1.6 children per woman in South Korea.

In China, a draconian "one-child family" population control program achieved a stunning decrease in fertility, at a considerable human cost. China's TFR now stands at 2.0 — quite low for a predominantly rural country. India, the world's second most populous nation, succeeded in lowering fertility from six children per woman around 1970 to a little less than four at present. However, the pace of decline in India slowed considerably around 1980, and little progress was made in the subsequent decade. Given that 38 percent of the world's population lives in China and India, fertility change in these two countries wields enormous influence on global population growth.

In Latin America, the speed and universality of fertility decline has surprised many demographers. Only a handful of countries have TFRs above 4.0, including Haiti and Guatemala, at 6.0 and 5.6 respectively. In the region's two largest countries, Brazil and Mexico, family planning spread rapidly after the early 1970s. Roughly 53 percent of Mexican women use contraceptives today, up from 13 percent in 1973,[5] and the country's TFR is down to 3.4. In Brazil, contraceptive usage of 66 percent has reduced fertility to 2.6 children per woman.

Still, replacement level fertility remains an elusive goal in Latin America. To date only Cuba, with its economy in shambles, is below this level. Elsewhere, fertility has reached a plateau *above* replacement level fertility. Costa Rica, for example, was widely hailed as a family planning success story in the 1970s, but its TFR stalled in the mid-1980s at about 3.5 children per woman. Recently, Costa Rica's birth rate began to decline again, but its TFR remains at 3.0. Similarly, Argentina, Chile, and Uruguay have all had TFRs in the 2.5 - 2.9 range for many years. Thus, fertility decline has been dramatic, but there is no evidence that it will, in fact, reach replacement level.

The 55 countries of Africa are in a far different situation. Throughout most of the continent, fertility remains at preindustrial levels and the use of family planning is only beginning. The most significant progress has been made in the Islamic countries of North Africa: the TFR has dropped to 3.9 in Egypt, 3.4 in Tunisia, and 4.2 in Morocco from preindustrial levels of six or seven children per woman. In sub-Saharan Africa, the use of family planning is just beginning. Sub-Saharan African women average 6.5 children each, the highest level in the world. Still, during the 1980s the number of family planning programs in the region increased sharply, and two out of three African governments responded to a U.N. survey that their birth rate was "too high." Significant birth rate declines have been observed in Kenya, Zimbabwe, and Rwanda. Nonetheless, fertility remains very high across Africa, even in countries where declines have taken place.

THE DEVELOPED COUNTRIES: LOW FERTILITY AND AGING POPULATIONS

While developing countries grapple with young and growing populations, industrialized countries face a different set of problems associated with populations that are declining and aging. In these countries, the process of demographic transition has been completed. Life expectancy has risen to record highs and fertility is at or below replacement level. In Western Europe, women average only 1.5 children each, with Italy the lowest at 1.3.

European governments are concerned about dwindling numbers of younger workers at one end of the age scale and rising numbers of pensioners at the other. In France, for instance, the number of 15- to 19-year-olds — the traditional ages for entry into the labor force — has been declining since the 1980s and is expected to fall from 4.2 million in 1990 to 3.5 million by 2025. As the native labor force declines, European countries increasingly depend on immigrant laborers, particularly for lower status and thus hard-to-fill jobs. The influx of migrant workers has prompted a backlash against foreigners in many European countries.

Clearly, most of the future world labor force will come from the developing world. By 2025, 84 percent of the world's working-age population may live in developing countries, up from 75 percent in 1990 and 65 percent in 1950. Only a relatively small share of these workers will emigrate[6] — most often to another developing country. Thus, the shift of the working (and consuming) population from the developed to the developing countries has strong implications for the economic organization of the world economy and the place of the industrialized countries within it.

Concerned about rising numbers of elderly relative to the working-age population, several European countries (France, Germany, Luxembourg, Switzerland, Bulgaria, and Hungary, among others) have reported to the U.N. that their birth rate is now "too low." Although these governments do not have explicitly pronatalist policies, many provide benefits for child care and maternity which are much more far-reaching than those of, say, the United States. Some observers believe that these benefits have kept European fertility from falling to still lower levels. For example, Sweden's TFR rose from its low point of 1.6 to 2.1 in 1991 — a rise that was generally attributed to generous maternity leave benefits and salary maintenance.

In the United States, the TFR has never dropped to the levels seen in Europe. The low point in U.S. fertility was reached in 1976 at 1.74. However, it soon rebounded to 1.8, then jumped again to 2.1 in 1991. Since 1976, most American women surveyed have said they expect to have two children. Therefore, demographers do not expect the U.S. fertility rate to increase by much, if any. The decline in fertility during the 1970s seems to have been the product of a final stage in the demographic transition — a shift to having children later in a woman's life. (This shift probably accounts, at least in part, for the present record-low fertility rates in some of the European countries, where women are only now being fully integrated into the labor force.) The major question in the U.S. is whether, when, and by how much the higher fertility rates of foreign-born women will converge with the lower rates of native-born women, especially non-minority women. Current U.S. projections assume a very small increase in the U.S. TFR — from 2.07 in 1993 to 2.15 in 2050 — mostly due to an increase in the proportion of women from racial and ethnic groups with higher fertility rates.

PREDICTING AN UNCERTAIN FUTURE

The future course of world population growth is largely unpredictable. Although much has been learned about fertility decline and its causes, population projections are still an inexact science. While we can logically expect that the TFR will fall in most, if not all, Third World countries, we cannot say when, where, or by how much.

How much will world population grow? To answer this question, it is important to keep in mind that 60 percent of the world's population lives in only ten countries. Three of these are industrialized nations — the United States, Russia, and Japan. Population is growing slowly in the U.S. and Japan, and it is currently declining in Russia. That means that what happens in the other seven countries — China, India, Indonesia, Brazil, Pakistan, Bangladesh, and Nigeria — will go a long way toward determining how much the world's population might rise.

Future world population size will be determined, as always, by the path of fertility and mortality. Mortality will play a lesser role in the future, as the developing countries complete the first stage of the demographic transition and death rates level off. An exception to this rule may be Africa, where mortality remains higher than in the rest of the developing world. Average life expectancy has risen slowly in Africa, from 38 years in the early 1950s to 52 years today. Mortality changes will have a greater effect on population size in Africa, but there, too, fertility changes will have the most impact.

Periodically, the United Nations issues a series of long-range population projections. The gap between the U.N.'s lowest projection for the year 2150 (4.3 billion) and its highest (694 billion) illustrates the uncertainty involved in foreseeing the demographic future. Truth obviously lies somewhere in this enormous range, but where?

The highest projected total, called the Constant Fertility scenario, assumes that all countries maintain their 1990 fertility level for all time. Few believe that such a scenario is likely. But if the human population has the potential to swell to such a fantastic size, how can the numbers used by most demographers today — 10, 11, or 12 billion — be so much lower? The latter figures come from the U.N.'s Medium projection, which foresees world population rising from 5.6 billion in 1994, 10 billion in 2050, and finally "stabilizing" at 11.5 billion by 2150. The very long time periods involved suggest that such projections are an ambitious statistical and prognostic endeavor. Moreover, the assumptions underlying this projection are highly uncertain.

The Medium projection assumes, reasonably, that fertility will continue to fall in developing countries, just as it did in the industrialized countries. However, it also assumes that all countries will achieve replacement level fertility — a far more questionable proposition. On one hand, European fertility has dropped to such low levels that population decline has already

FIGURE 3: Three General Profiles of Age Compositions

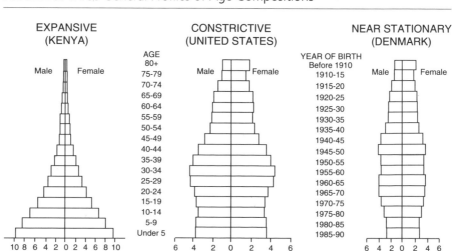

EXPANSIVE (KENYA)   CONSTRICTIVE (UNITED STATES)   NEAR STATIONARY (DENMARK)

Percent of Population   Percent of Population   Percent of Population

TABLE 2

POPULATION CHANGE 1990-2020

	1990	(millions) 2020	Number	Percent	Percent of World Growth
World	5,292	8,092	2,799	52.9	100.0
Less Developed:	4,079	6,739	2,660	65.2	95.0
Africa	642	1,452	810	126.1	28.9
Asia (less Japan)	2,989	4,571	1,582	52.9	56.5
Latin America	448	716	268	59.9	9.6
More Developed:	1,213	1,353	140	11.5	5.0
North America	276	326	51	18.3	1.8
Europe, Japan, former USSR, Oceania	937	1,026	89	9.5	3.2

POPULATION CHANGE 2020-2050

	(millions) 2050	Number	Percent	Percent of World Growth
World	10,019	1,927	23.8	100.0
Less Developed:	8,674	1,935	28.7	100.4
Africa	2,265	813	56.0	42.2
Asia (less Japan)	5,487	916	20.0	47.5
Latin America	922	206	28.7	10.7
More Developed:	1,345	-7	-0.6	-0.4
North America	326	-0	-0.1	-0.0
Europe, Japan, former USSR, Oceania	1,019	-7	-0.7	-0.4

begun to set in. But on the other, several developing countries have reached a plateau above replacement level fertility, even though their TFRs have been lower than 3.0 for quite some time.

What if fertility falls more slowly than the U.N. envisions, and settles at five percent above replacement level, as it has in New Zealand or Ireland? The world would then ultimately need to accommodate 21 billion people! What if the fertility drop is more rapid and settles at five percent below replacement level, as in the United Kingdom or France? In that case, world population would peak at just under 8 billion in 2050 and decline to 5.6 billion, just where it is today.

Future demographic trends will vary significantly by region, as they do now. Some countries will take a fairly direct route to the replacement

FIGURE 4: World Population: Past and Future

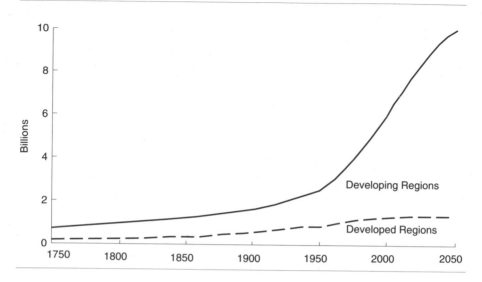

level, others will see their fertility plateau, still others may see their birth rate fall and rise again.

In Africa, the outlook for population growth ranges from 1.2 billion to 7.8 billion in 2150. In the former projection, population size would begin decreasing by 2150, in the latter it would continue to grow. The range between these two projections is even larger than the *total* current world population. In Latin America, projections for 2150 range from 519 million people (a little more than today) to 2.6 billion, an almost sixfold increase.

In Asia, much depends on China and India. As noted above, China's controversial population control program achieved a rapid decrease in fertility. One cannot help but wonder whether fertility would rise if an increase in democratic freedoms reduced some of the pressure on couples to have only one child. This began to happen in the mid-1980s, when there was some relaxation of family planning campaigns. U.N. projections for China now show a range from only 540 million in 2150 (less than half today's population) to 3.1 billion. In India, TFR in 1990 stood at 3.9, although provisional reports from the early 1990s suggest that it has declined below that level. The range of U.N. projections for India is unusually wide for a single country: 658 million to 4.6 billion.

The balance of Asia, which includes such demographically diverse countries as South Korea, Bangladesh, Pakistan, the Philippines, and Saudi Arabia, has a population growth potential about equal to Africa. The ultimate size of this region could range from 931 million to 7.1 billion, depending upon events in its largest countries: Pakistan,

Bangladesh, Indonesia, Iran, Vietnam, and the Philippines. In each of these instances, the gap between high and low projections depends on when and whether these regions attain replacement level fertility.

What is the demographic outlook for the U.S.? According to the Medium U.N. projections, shown in Table 2, the North American population will rise by 18 percent between 1990 and 2020 and decline very slightly between 2020 and 2050. As a result of more rapid growth elsewhere, the North American share of the world's population will drop from the present five percent to four percent in 2020 and to little more than three percent in 2050.

However, U.S. Census Bureau projections are considerably higher than those of the U.N., partly because the Census Bureau takes into account the recent increase in the U.S. TFR, and partly because the U.N. makes the unrealistic assumption that immigration will be zero after 2025. According to the middle series of Census projections, the U.S. will be home to 392 million people by 2050, up from 258 million today. In contrast, the U.N. projects a population of only 326 million for the U.S. and Canada combined. Along with Australia, the U.S. and Canada are the fastest-growing industrialized countries.

THE FUTURE IS IN THE HANDS OF THE PRESENT

Clearly, we cannot precisely predict how many human beings will inhabit the earth in 60 years, let alone beyond. That will be determined by the unpredictable: politics, social tastes and mores, and luck.

FIGURE 5: World Population: Number of Years to Add Each Billion

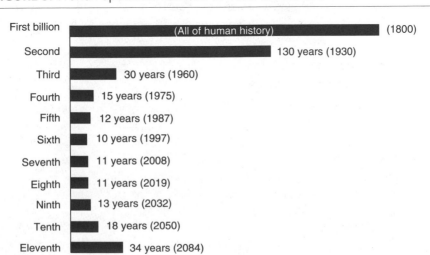

First billion	(All of human history) (1800)
Second	130 years (1930)
Third	30 years (1960)
Fourth	15 years (1975)
Fifth	12 years (1987)
Sixth	10 years (1997)
Seventh	11 years (2008)
Eighth	11 years (2019)
Ninth	13 years (2032)
Tenth	18 years (2050)
Eleventh	34 years (2084)

However, it is clear that the future size of the world's population is being decided now. Since the children born today are the parents of the future, the force of population momentum means that we are now writing the script for the world's population story 20 years hence. And that story will largely determine where, along the broad continuum of possible population sizes, the world will find itself in centuries to come.

Understanding the Numbers

Here's how one young demographer put world population growth in context for her grandmother:

Ronnie: Do you realize that the world's population has nearly tripled since you were born just 70 years ago?

Grandma: You mean there are three people now for every one person there was when I was born? I sure didn't. Well, if the population has grown so fast since then, what will it look like if I live to be 100?

Ronnie: When you turn 100, as I hope you will, the world is expected to have greater than a third more people than it does today. That means for every three people now, there'll be four then.

You might think this is nothing to be concerned about. But look at it from another perspective. It took billions of years of reproducing to reach our current size. But in only 25 more years, the population will be four times as large as it was when you were born! Unbelievable, isn't it? The world took billions of years, from the start of time until you were born, to reach its first 2 billion people. And in just 95 years, your lifetime I hope, almost 6 billion more people will be added to the world.

Grandma: What about you? If the world's population is growing this fast in my lifetime, what about yours?

Ronnie: Say that I live as long as the average American born when I was. In 1970, the world had fewer than 4 billion people in it. When I die it will have an estimated 10 billion, an increase of more than 250 percent just in my lifetime. If your life expectancy were the same as mine, the world population would have increased by about 4 billion in your lifetime and 6 billion in mine. So you can see that we are adding increasing numbers of people to the world each year—even though the rate of world population growth is decreasing.

— Ronica Rooks,
University of Maryland graduate student

History and Analysis of Population and Family Planning Programs

. . . .

PETER J. DONALDSON AND AMY ONG TSUI

The International Family Planning Movement

....

Dramatic changes have taken place in human society over the past 30 years. In Europe, the United States, and other industrialized countries, virtually all women have access to effective birth control to regulate the number and timing of children they have. Fertility levels have fallen to all-time lows in many of these countries. Even more remarkable, this revolution in childbearing has reached many poor agricultural societies in the Third World, where the concept of family planning was unknown a few decades ago. In some developing countries the use of modern contraceptives has grown from nearly zero to more than 25 percent of the women of childbearing age. Since the early 1960s, the average number of children born to women in the developing world has fallen from just over six to just under four children per woman.

What fueled the reproductive revolution in the Third World, where contraceptive use represented a radical shift from traditional ways? Among the most important influences were social changes, including increased education and employment for women and economic development. The active promotion of family planning by numerous individuals, organizations, and government agencies was another vital force behind the acceptance of contraceptive use. In this chapter, we trace the history of the international family planning movement from its turn-of-the-cen-

Peter J. Donaldson is president of the Population Reference Bureau. Amy Ong Tsui is deputy director of the Carolina Population Center and associate professor of Maternal and Child Health at the University of North Carolina at Chapel Hill. This chapter is adapted from "The International Family Planning Movement," Population Bulletin, *Vol. 45, No. 3, November 1990, published by the Population Reference Bureau.*

tury beginnings to the present, and briefly survey its impact on fertility in the developing world.

PRECURSORS AND PIONEERS

The history of the international family planning movement can be divided into four periods. During each period, the rationale for birth control and the public policies surrounding the provision of contraceptives changed in important ways.

The Earliest Advocates. The first period of international family planning movement history extends roughly from the late 19th century to the end of World War I. Birth control advocates spread the doctrine of voluntary motherhood, the recognition of female sexuality, and the right of couples—wives, in particular—to control fertility through the use of contraceptive devices. But this concern for women's health, sexual liberation, and family welfare was mixed with anxiety about the dangers of a rapidly growing population of poor people. During this period, the family planning movement in Europe was labeled neo-Malthusian because many people, evoking the 17th century writings of Thomas Malthus, believed that population growth, especially among the lower classes, was a threat to social welfare. Also during this time, a renewed interest in eugenics—the improvement of human genetic stock by controlling physical and mental traits—directed more attention to the neo-Malthusian movement.

The leaders of the family planning movement during this period were essentially reform-minded. The movement's pioneers, people such as Margaret Sanger and Charles Knowlton in the United States, Marie Stopes in England, and Paul Robin in France, promoted freer access to family planning information and services, including abortion in some cases, for rich and poor alike.

There was considerable opposition both in Europe and the United States to the work of these early family planning advocates, who were accused of destroying family life and promoting promiscuity. Opponents of the family planning movement fought to limit access to birth control devices and sought passage of the U.S. Comstock Law, enacted in 1873, prohibiting advertisements or prescriptions of contraceptives. Although the federal statute was repealed in 1916, in part because of Margaret Sanger's efforts, laws modeled on the Comstock Law stayed on the books in many U.S. states until the 1960s.

In the same year the Comstock Law was repealed, Margaret Sanger opened America's first birth control clinic in Brooklyn, New York. Five years later, in 1921, Marie Stopes opened England's first birth control clinic. Both clinics aimed at serving poor women, operating under their sponsors' belief that the poor deserved access to a vital health service already available to the more fortunate.[1] The work of Sanger, Stopes, and their allies helped advance the acceptability of family planning on both continents. These first family planning advocates were action-oriented, seeking, usually in the face of legal opposition, to disseminate information and services to married women to combat involuntary pregnancy.

The sixth of 11 children, Margaret Sanger was a trained nurse who advocated birth control access for over 50 years. She published numerous articles and pamphlets promoting birth control, such as "What Every Mother Should Know" and "Woman Rebel." Marie Stopes was a well-known botanist, the first woman to become a science lecturer at Manchester University, and a member of Britain's prestigious Royal Society of Literature. The involvement of other biological and medical scientists in birth control in the 1920s helped transform the family planning movement from a radical cause based on new and then-unpopular notions of social justice and women's rights to a broader-based movement that appealed to the middle class.

The surge of research on the biological factors of reproduction during this period laid the foundations for subsequent technological breakthroughs such as the birth control pill. Through their research some scientists were led into activism, focusing on the impact of birth control on differential population growth, eugenics, and the racial and economic composition of the population rather than its importance for conjugal welfare. Qualified endorsement of birth control by the influential American Medical Association in the mid-1930s symbolized the growing acceptance of family planning among the medical profession. At the same time there was growing public support for family planning.

The Interwar Years. The first phase of the international family planning movement ended abut the time of World War I, while the second phase dates from the end of World War II through the 1950s. The two world wars disrupted the expansion of international family planning activities and forced the public and policymakers to concentrate on other issues.

During the interwar years, however, a significant amount of birth control activity continued in the United States. Birth rates among the middle class were declining, aided only by basic contraceptive methods—such as the diaphragm and jellies.

The lower rates among more educated and more affluent Americans concerned some economists, politicians, and social scientists. Both population quality and industrial growth were thought to be threatened by these depopulation trends. Fear that the less desirable segments of the population were reproducing faster than the more desirable, and a belief that more consumers were needed to stimulate a stagnant economy placed birth control advocates on the defensive. Social welfare reforms were planned specifically to encourage couples to have large families again. The outbreak of World War ll interrupted these initiatives, and the post-war baby booms in the United States, Europe, and other countries removed the need.

Post-World War II Developments. The end of World War II is associated with two important changes in public attitudes toward population growth and family planning. First, confronted with the grisly consequences of Nazi Germany's attempts to promote racial purity, public support for the eugenics movement evaporated. Except among a few extremists, it was no longer popular to advance family planning and population control as means to establish a balance among different racial or economic groups. Economic reconstruction or development, and political stability became more frequently cited rationales for government involvement in birth control. Climbing postwar birth rates in much of the world gave credence to these issues and conferred more importance to the relationship between population growth and economic progress.

The second important change after World War II was an increase in the number and stature of private groups working to promote family planning services. Private family planning associations gained momentum in many countries in the postwar years and became important focal points for family planning activities. A conference in Stockholm in 1946 provided the initial forum for planning an international association of private family planning groups. At subsequent international conferences in England in 1948, and in Bombay in 1952, the constitution of what is now the International Planned Parenthood Federation (IPPF) was drafted and ratified. The original members came from India, the United Kingdom, the United States, the Netherlands, Sweden, West Germany, Singapore, and Hong Kong.

The IPPF is the most prominent and widely recognized private sector effort to support family planning internationally. Its objectives were (and are) to promote family planning and population education, to support family planning services, and to stimulate and disseminate research on fertility regulation. In 1990, IPPF had 107 family planning member affiliates representing 150 countries. Funding for IPPF's $70 million annual budget comes from governments, private foundations, and individual donors. Funding of local IPPF affiliates may involve grants from IPPF itself or other international groups as well as funds raised locally from government sources, service fees, and donations.

After World War II, the newly established United Nations helped direct attention to population issues and strengthen links between international groups working in the field. In 1946, the U.N. Economic and Social Council established the Population Commission, composed of representatives from member governments. The commission has met 25 times since to review demographic and family planning studies that it sponsored and to report to the general U.N. membership on population problems. At the same time, a Population Division was established within the U.N. Secretariat to provide member states with information on population change and a better understanding of its determinants and consequences.

The 1960s: A Surge of Activity. The third period in the evolution of the international family planning movement began around 1960 and lasted until the mid-1980s. This period was characterized by a significant increase in support for family planning among both industrialized and Third World countries.

The most influential manifestation of support by developed countries was the incorporation of family planning as a central component of the development assistance activities of the U. S. Agency for International Development (A.I.D.). A.I.D.'s population program exemplifies the view that large-scale, government-sponsored programs were the best way to increase contraceptive use and slow population growth. Many A.I.D. staff, along with experts outside government, actively promoted large-scale national family planning programs funded by international aid. Convinced that rapid population growth was an urgent problem in devel oping countries, they believed that contraceptive use should be widely promoted through a variety of channels, using health personnel with different backgrounds and levels of expertise.

In 1968, the U.S. Congress first earmarked foreign assistance appropriations to support family planning activities. Over the following two decades, A.I.D. contributed a total of $3.9 billion to population and family planning assistance throughout the Third World, making it the largest single donor of population aid.[2]

Another key event of this period was the establishment of the United Nations Fund for Population Activities (UNFPA) in 1967, later renamed the United Nations Population Fund. The goals of the UNFPA are to promote an awareness of population problems and their relationship to social and economic development, the implementation of population policies, and the spread of family planning to better the health and well-being of the individual, family, and community. The UNFPA was established as a trust fund and financed by voluntary contributions from U.N.members. Over the first 15 years of its existence, the UNFPA allocated over $1 billion to member states in support of population and family planning programs.

Another landmark event of this period was the issuance of the Teheran proclamation following the 1968 United Nations Conference on Human Rights, held in Iran. Article 16 of this proclamation identified the ability to control one's fertility and therefore access to the means of fertility control as a basic human right.

The publication of Paul Ehrlich's *The Population Bomb* in 1968 was another influential event. Ehrlich's appearances on popular television talk shows and surge of media attention galvanized public awareness of the consequences of rapid population growth.

In the United States, a number of organizations were founded to publicize the perceived dangers of rapid population growth. Among these groups were Zero Population Growth, the Population Crisis Committee, and the Population Institute. The Population Reference Bureau, which had been founded in 1929 as part of the first burst of concern over population growth, expanded its dissemination efforts to respond to the growing public demand for information on population trends.

Population studies gained greater stature in academia as well. Population programs were started or expanded at the Universities of Chicago, Michigan, North Carolina, and Pennsylvania, and at Brown, Johns Hopkins, and Princeton Universities. Because fertility levels are generally the major determinant of population growth rates, many programs concentrated their research and curricula on fertility. Much of the applied research and training, for foreign as well as American students, was related to family planning.[3]

In 1974, the United Nations sponsored the World Population Conference. This conference, held in Bucharest, marked the transformation of the concern about high birth rates in the Third World from an interest primarily of the United States and a few Asian countries to a legitimate political issue in a much wider circle of countries. The conference, attended by delegations from U.N. member countries, succeeded in drafting the first World Population Plan of Action. This achievement occurred despite the fact that population and family planning were politically sensitive issues in many of these countries.

The U.S. delegation to the 1974 conference suggested that developing countries establish targets for lowering population growth rates, and work to achieve them. In response, a large block of Third World nations argued for the establishment of a new international economic order that would enable developing countries to gain a competitive position in the world market. They believed that this new order would not only advance economic development but also slow population growth. These nations saw a greater long-term benefit in attacking the underlying barriers to economic development than in the expansion of family planning programs by foreign donors. "Development is the best contraceptive" became the often-repeated motto for many representatives from the developing countries. The 1974 World Population Conference marks the first significant conflict between the United States and Third World countries regarding population growth and family planning.

The 1980s to the Present. The 1980s marked the beginning of the fourth phase in the history of the international family planning movement. This period was characterized by the increasing commitment of the public and policymakers in the Third World to family planning, both to slow population growth and to improve the health of women and children. The period also witnessed the waning of U.S. government concern about the effects of rapid population growth and of its support for family planning.

In the United States, a new political conservatism brought the election of a president and many representatives opposed to government support of abortion and family planning. Politically appointed policymakers and managers within the U.S. government worked to carry out their wishes. Efforts to dismantle the A.I.D. programs that helped support developing country family planning programs were launched by influential conservative leaders of the New Right.

The confrontation between the old and the new policies was apparent at the second World Conference on Population held in Mexico City in 1984.[1] The official American position at the conference expressed two important revisions to the United States' previous support of international population policies. First, population growth was now said to have a neutral, not detrimental, effect on the economic development of poor countries. Second, the U.S. government announced that it would cease to support nongovernmental organizations that performed or promoted abortion, even if they did so with their own resources. This expanded the ban on the use of U.S. foreign assistance funds for abortion in effect since 1974.

At the 1974 Bucharest Conference many developing countries were reluctant to concede that population and family planning might be important. But by the time of the Mexico City Conference, representatives of developing nations agreed that too rapid population growth could have adverse consequences and that family planning programs were important for both health and economic development.

Ironically, by 1984 the United States delegation no longer supported this position. Following the Mexico City Conference, the United States withheld its contributions to the UNFPA and the IPPF, the two largest international family planning organizations. In 1985 the U.S. withheld $10 million of the $46 million that had been granted UNFPA by the U.S. Congress on the basis that $10 million was equivalent to the amount UNFPA contributed to support projects in China where coercive abortions were said to be occurring. The following year the United States discontinued all support to UNFPA. [More recently, the Clinton administration reaffirmed U.S. support for UNFPA and population assistance generally. See "The Politics of U.S. Population Assistance," by Sharon L. Camp, page 122.]

IMPACT OF FAMILY PLANNING PROGRAMS ON FERTILITY

Do family planning programs lead to lower birth rates? This question has raised considerable controversy among those who study fertility and family planning behavior in the Third World. On the one hand, many social scientists argue that economic development is a necessary precondition for fertility decline. The social and economic transformations brought by development and rising standards of living force couples to reevaluate the desirability of large families and motivate them to use family planning. This view has been called the "demand" argument. In contrast, family planning advocates put forth a "supply side" argument—if birth control information and services are widely available, access to these services can prompt couples to limit their childbearing, leading to significant fertility declines without substantial economic development.

In a 1978 study, demographer John Bongaarts identified four key factors that determine fertility levels in a population: (1) the age at marriage, (2) contraceptive use rates, (3) the number of months of postpartum sterility (approximately length of time a mother breastfeeds her newborn baby), and (4) the rate of induced abortion.[5] Of the four, contraception has the strongest effect, with age at marriage second, in most developing countries today. The main contribution of family planning programs has been improving the availability and accessibility of contraceptive services. The international family planning movement and the expansion of service delivery throughout the developing world have exposed millions of couples to these services. It is likely that these organized efforts exerted some impact of their own on fertility levels.

Social and economic development, in contrast, affects a broader range of variables: the age at which women marry, their motivation to use contraception, their ability to spend time breastfeeding their children, and on improvements in infant survival. Primary schooling, women's labor force participation, and rising household income all work to shift individuals' preferences toward smaller families. For example, Singapore's rapid industrialization, strong educational programs, and rise as a modern city state have created many work opportunities for women as well as men. Many young adults are currently opting either not to marry or to marry at older ages. Singapore's family planning program was so successful in reducing birth rates that, in the late 1980s, the government, fearing population decline in the future, enacted tax incentives and other benefits to encourage individuals to marry and have children.

Family planning programs and socioeconomic development are not the only impetus for fertility decline. Stringent government policies to increase or reduce births can have an effect. When the Romanian government severely limited access to legal abortions in November 1966, the birth rate in the country tripled in nine months, going from 12.8 to 39.9 per 1,000 between December 1966 and September 1967.[6] China's vigorous pursuit of its "one child for one couple" policy has yielded rapid fertility decline in that country.

Religious or cultural forces also may work against the forces of development, keeping fertility high. In spite of early support for family planning policy, many Middle Eastern countries with strong Islamic traditions have had little fertility change. There, as in many sub-Saharan countries, the idea of limiting the number of births was so culturally unacceptable that family planning programs were introduced as a means for promoting better maternal and child health by helping women space their births.

Disease, especially sexually transmitted diseases (STDs) like gonorrhea, can also lower fertility independently because they can cause sterility. Fertility levels in some sub-Saharan African nations have remained below the regional average because of the high incidence of STDs in their populations.

Can family planning programs increase contraceptive use and reduce fertility without the effects of development? A recent study by Robert Lapham and Parker Mauldin sheds new light on these questions.[7] Following the design of an earlier landmark study,[8] they tracked indicators of economic, social, and demographic conditions between 1965 and 1980 to evaluate family planning program impact. They measured the quality of national family planning programs and national commitment based on a set of 30 different factors, grouped into four components: (1) policy and stage-setting activities, (2) service and service-related activities, (3) record-keeping and evaluation, and (4) availability and accessibility.

The Lapham and Mauldin study confirmed that high-quality family planning programs can help lower birth rates and increase contraceptive use, independent of social and economic change. Perhaps more important, their research showed that family planning programs are most effective in countries that are experiencing steady socioeconomic development. A strong family planning effort in a modern socioeconomic setting was associated with higher levels of contraceptive use and greater and more rapid birth rate declines

A number of other studies have outlined the impact of family planning on fertility levels,[9] while others have addressed issues of measuring program impact.[10] Evaluation of the Matlab, Bangladesh, demonstration project has provided a prime example of the measured impact of a family planning project in a local rather than national or multinational setting.[11]

More recently, a major study of successful family planning programs has helped identify the interrelationships between program inputs, management and performance, and demographic change.[12] Unfortunately, this aspect of program evaluation has, to a large extent, taken a secondary role to assessing the reasons for program impact. Future research will likely direct more attention to the ways that specific characteristics within a program's environment affect the quality and efficiency of family planning services.

The family planning versus socioeconomic development debate has lost much of its fervor over the past decade. The issue preoccupied Western population scholars and some policymakers, but it was less relevant to those actively involved in the delivery of family planning services. International donors and host country professionals were more concerned with improving the availability and accessibility of birth control in Asia, Latin America, and Africa. These actors may have recognized early on that family planning is one of many important development strategies available to resolve the negative consequences of high fertility in a developing country.

CHANGING ORIENTATIONS

There is a general consensus today that family planning should be judged as a human right and health measure and not just for its demographic or economic impact. This new emphasis is likely to become more important as existing family planning programs mature, newer ones evolve in African, Asian, and Latin American countries, and public concerns abut rapid population growth lessens.

In part, this shift in emphasis reflects the changing role of the United States and other developed countries in the international family planning movement and as supporters of family planning programs in the developing world. During the earliest days of the movement when people and institutions from the developed countries were guiding and financing developing country family planning programs, more emphasis was put on what specialists call the "aggregate effects" of changes in population and fertility. Aggregate effects include such things as the impact that a particular rate of population growth or numerical increase has on savings rates, school enrollment, or infant mortality. With the institutionalization of family planning programs in different countries and increase in numbers of couples practicing contraception, attention in the movement shifted from aggregate effects to a concern for the impact of family planning on individual users and their families. Today, specialists are no longer exclusively interested in the relationship between population growth and economic development, but also in such topics as the health consequences of contraceptive use or the impact of lower fertility and slower growth on family welfare and the quality of life.

The international family planning movement has been an important part of the transformation of the developing world that has occurred since the end of World War II. Most people think of this transformation in terms of higher incomes, increased life expectancy, more industrial employment, and other elements of the transition from traditional agricultural societies to modern industrial ones. Family planning has been a crucial part of these changes. In a sense, it has been a liberating influence. By enabling individuals in developing countries to regulate their fertility, the international family planning movement has allowed them to take advantage of the opportunities of the modern age.

SHARON L. CAMP

The Politics of
U.S. Population Assistance

....

ew global issues generate debates with the emotional intensity of
those that surround international population and family planning
programs. This is true in the U.S. today, despite substantial public
support for family planning, despite nearly three decades of generous U.S. assistance to world population efforts, and despite the growing
legitimacy of population programs in the world community.

In the U.S., the most powerful opposition to population assistance has
come from conservative anti-abortion groups — many of whom oppose
all forms of artificial birth control. These groups held U.S. population
assistance hostage throughout the Reagan and Bush administrations.[1]
But opposition is also occasionally voiced by liberals, including feminist
health advocates and anti-poverty groups worried about coercive population control. Criticisms from both ends of the political spectrum can
sound surprisingly similar.[2] Charges of "genocide against the poor," "cultural imperialism," and "contraceptive dumping," as well as arguments
that "development is the best contraceptive" have been made by both
conservative and liberal critics of U.S. population aid.

The resulting cacophony of criticism tends to divert attention from
the consensus which has emerged on both the need to slow population
growth and the importance to women and their families of control over
the timing of pregnancy and childbearing. Political debates have also
diverted attention from increasing scientific evidence of a large unmet
demand for family planning and of a powerful synergism at work

Sharon L. Camp is the former vice president of Population Action International.

between family planning programs and other development initiatives, especially those which benefit women.

Since the change of presidential administrations in 1993, some of the battles over U.S. population assistance have subsided. But it will take major new initiatives from the Clinton administration to reestablish the U.S. as a leader in world population efforts.

POLITICAL SUPPORT FOR FAMILY PLANNING EFFORTS IN THE U.S. AND ABROAD

About 130 national governments currently subsidize family planning services, including about 65 developing-country governments which specifically seek to slow population growth. Virtually all of the world's largest developing countries have national population policies, including: China, India, Indonesia, Brazil, Nigeria, Pakistan, Bangladesh, Mexico, Iran, Vietnam, the Philippines, Thailand, Turkey, Egypt, and Ethiopia. Of the 25 large countries which account for over 80 percent of world population growth, only the U.S. and Russia have no national population policies.[3]

This represents a considerable change from the mid-1960s, when Congress first earmarked foreign aid funds for population assistance. At that time, only 12 governments had national population programs, which were funded primarily by the U.S. and other foreign aid donors.[4] Today, developing countries provide about 75 percent of all funding for their family planning programs — over $3.5 billion a year.[5]

While the U.S. remains the largest bilateral donor to international family planning programs ($431 million in fiscal year 1993),[6] on a per-capita contribution basis the U.S. ranks only sixth, behind Norway, Finland, Sweden, Denmark, and the Netherlands. Great Britain, Germany, and Japan still provide little support for family planning efforts.[7] After a decade in which the inflation-adjusted value of international donor assistance remained essentially flat, total assistance has taken a significant jump in recent years, from $629 million in 1987 to roughly $1 billion in 1990.[8] Major donor contributions to the United Nations Population Fund more than doubled between 1981 and 1991, despite the U.S. withdrawal in 1986.[9] Unfortunately these increases still fall far short of developing country requests for assistance. Between now and the end of the decade, donor assistance needs to be quadrupled if we are to meet the growing level of demand for fertility control among an ever-larger number of reproductive-age couples.

Despite the controversies which developed during the Reagan and Bush administrations, Americans continue to express strong support for family planning. Although world population problems are not "front of the mind" issues for most Americans, public opinion is strongly in favor of both domestic family planning programs and global population efforts. By margins of 85 to 90 percent, Americans surveyed in recent years say they are "seriously concerned" about population growth, believing it contributes to world hunger and poverty, environmental degradation, political instability, illegal immigration, and other problems. A near two-thirds majority supports U.S. government funding of family planning programs overseas even if those programs involve abortion. A large majority of Americans opposed Reagan-era policies, maintained by President Bush, which resulted in the withdrawal of U.S. support from the United Nations Population Fund and the International Planned Parenthood Federation.[10]

Public attitudes about family planning and population issues probably account for strong, bipartisan congressional support of domestic and international family planning programs. Congress first earmarked foreign aid funds for international family planning programs in 1965 and has continued to prod reluctant administrations through much of the programs' history.[11] Until the summer of 1984, when the Reagan White House announced its controversial new policies, there had been few open debates in Congress over U.S. population assistance.[12] In fact, until 1984 there had never been a recorded floor vote in either chamber of Congress on international family planning issues. Even in the midst of the partisan debates over policy issues in the 1980s, a bipartisan majority provided moderate to large increases for U.S. population assistance in every year that Congress managed to pass a new foreign aid appropriation.

Congressional support for population programs has become even stronger in recent years. For fiscal year 1993, the 102nd Congress provided a record $100 million, 40 percent increase, in population aid (not including money from the Development Fund for Africa) despite deep cuts to other foreign aid accounts.[13] For fiscal year 1994, the 103rd Congress approved an increase of $80 million, despite overall cuts to foreign aid of nearly $1 billion below fiscal year 1993 levels and $1.4 billion below President Clinton's budget request.[14]

FAMILY PLANNING AND THE POLITICS OF ABORTION

If population and family planning programs enjoy such broad support, how were the Reagan and Bush administrations able to slash funding for those programs in the 1980s? Some analysts believe that family planning programs got into trouble largely because they became enmeshed in the domestic U.S. abortion debate. It is therefore natural for people outside the reproductive rights community to argue that the two issues — family planning and abortion — ought to be separated in order to build a broader political consensus for family planning and population efforts. Most politicians and many grassroots environmental and development groups make a strong plea for such a separation. But for political, technological, and programmatic reasons, it is actually quite difficult to separate the two issues.

The political reality is that many U.S. anti-abortion groups — including the American Life League, Human Life International, Pro-Life Action League, Operation Rescue, and other smaller groups — are openly opposed to *all* methods of birth control, except periodic sexual abstinence, which is the only method presently approved by the Vatican.[15] This helps explains why it has proved so difficult for supporters of family planning to reach any lasting political compromise with the U.S. anti-abortion movement on, for example, federal funding for family planning services, contraceptive research, or adolescent sexuality programs — all programs that could reduce the incidence of abortion.

Many anti-abortion groups and religious conservatives believe the mere availability of contraceptives creates (in their words) a promiscuous "contraceptive mentality" favoring "recreational sex." While most Americans believe family planning reduces the need for abortion, the religious right believes sex education and contraceptive availability *cause* abortion by undermining traditional family values.[16] Anti-abortion groups, therefore, do not seek merely to reform family planning programs by cleansing them of abortion. They want the U.S. government out of the family planning business entirely. Their tactic — highly effective under the Reagan and Bush administrations — has been to tar family planning with the pro-abortion brush and to label every critical family planning vote in Congress a "scorecard" pro-life vote. The fact that most family planning proponents do support abortion rights as part of comprehensive reproductive health care has helped make the "pro-abortion" label stick to the family planning movement.

Because there is not a bright clear line between many modern contraceptives and abortion, the anti-abortion groups are also able to argue that most contraceptives are "abortifacients." Most highly effective contraceptive methods do, in fact, make changes in the uterine lining which would, if all other methods of action failed, prevent a fertilized egg from implanting. Some older IUDs and most low-dose oral contraceptives in use around the world may work this way some of the time.[17] Although most people think they know the difference between abortion and family planning, the scientific distinction is blurred and may get fuzzier with new birth-control drugs, like RU-486, which could work differently depending on how and when they are used.

It is also difficult to separate abortion and family planning activities programmatically, particularly in settings where lack of education and inadequate public information about modern contraception contribute to high levels of unintended pregnancy even among contraceptive users. In these settings, clients often seek family planning help for the first time only after a missed menstrual period creates fear of unwanted pregnancy. If safe abortion services exist, most family planning providers feel obligated to counsel or refer clients who definitely want an abortion to facilities where they will get competent care, including follow-up contraceptive services. In a growing number of countries around the world, abortion — whether legal or not — plays an important role in birth control, as it has through most of human history. In many developing countries today governments have intentionally allowed restrictive abortion laws to lapse and have encouraged private groups to establish safe abortion services to treat contraceptive failure. Although good quality family planning programs will eventually reduce the need for abortion, *both* abortion and contraception are presently on the rise in most developing countries, reflecting a rapid increase in the demand for fertility control.[18]

POLICY CHANGES UNDER REAGAN AND BUSH

The role of abortion in birth control enabled conservatives to undermine political support for U.S. population assistance programs, beginning in the early 1980s. The categorical federal family planning programs, established by Congress in the late 1960s under both the domestic public health program and the U.S. foreign aid program, came under attack almost as soon as President Reagan took office.[19] In 1980, President Reagan proposed to eliminate the $160 million domestic Title X family planning program by rolling it into a health block grant to the states and then substantially reducing federal health funds available for such grants. It was assumed, with historical justification, that without federal funding family planning programs would shrink dramatically in many states. In 1983, president Reagan's Office of Management and Budget sought to eliminate the entire $235 million line-item budget for population assistance under the Agency for International Development (AID). Both of these proposals, made in one form or another in most subsequent years of the Reagan and Bush administrations, failed in part because of strong congressional and public support for family planning programs.[20]

But other attacks on the programs succeeded, specifically those designed to: (1) cripple family planning programs with restrictions, especially on abortion services; (2) divert federal funding to so-called "pro-family" groups opposing all forms of contraception and promoting "natural family planning;" and (3) defund family planning organizations specifically targeted by the anti-abortion lobby.[21]

This lobbying effort was singularly successful on the international front. At the 1984 United Nations International Conference on Population held in Mexico City, the U.S. delegation to the conference announced, in essence, that the U.S. government no longer considered population growth a serious world problem. The delegation also declared that organizations that provided abortion information or services, even with their own funds, would henceforth become ineligible for any kind of U.S. family planning assistance.[22]

This so-called Mexico City policy led almost immediately to U.S. withdrawal, after 17 years of support, from the International Planned Parenthood Federation (IPPF).[23] IPPF, which has been a major force in international family planning efforts for forty years, now represents voluntary family planning associations in 125 countries. In 1984, IPPF spent roughly $400,000 on abortion-related services (none with U.S. funds) out of a budget of about $52 million. Since the U.S. government represented a quarter of its budget, U.S. withdrawal forced IPPF to cut its worldwide technical assistance staff sharply, cancel plans to support 13 new national family planning associations in Africa, and generally retrench around the world.[24]

After a long court battle, the Reagan administration also succeeded in defunding the international programs of Planned Parenthood Federation of America, the U.S. affiliate of IPPF and for many years the largest intermediary for U.S. population assistance, with approximately $20 million in projects in up to 70 countries.[25] The Mexico City policy eventually denied U.S. foreign aid support to a number of highly effective family planning efforts around the world, including those in India, where early abortion is legal for health and socioeconomic reasons.[26] The loss of aid may have a significant demographic impact: India adds more people to the world's population each year than all the countries of sub-Saharan Africa combined.

Congress first attempted to override the Mexico City policy in 1984 and tried many times thereafter. The last attempt was made late in the Bush administration. The fiscal year 1992 foreign aid authorization bill contained a provision reversing the Mexico City policy that passed the Senate by voice vote and the House by a 222 to 200 vote margin. The conference report on the bill failed to pass the House, however, partly as a result of the President's threat to veto the entire foreign aid bill over its population provisions (see below).[27] In both chambers a clear majority favored reversal of the Mexico City policy, but lacked sufficient votes to override a presidential veto. Anti-abortion groups also made repeated, unsuccessful attempts to write the Mexico City policy into law, resulting in a stalemate which lasted from 1984 until President Clinton's inauguration in 1993.

On the domestic front, Planned Parenthood was clearly the target of parallel Reagan-era changes in Title X family planning regulations, which followed within a year of the announcement of the Mexico City policy. Widely known as the "gag rule," these regulations denied Title X monies to any family planning program that provides abortion services, counseling or referral. Reagan's Title X regulations, subsequently supported by President Bush, were upheld by the Supreme Court in 1991, in *Rust v. Sullivan.*[28] The 102nd Congress narrowly failed to override presidential vetoes of legislative provisions blocking implementation of the gag rule. However, President Clinton abolished the gag rule during his first week in office.[29]

Perhaps the most important victim of Reagan-era policy changes was the United Nations Population Fund — seen by most other donor countries as the lead agency in international cooperation on population issues. (The fund is known by its former acronym, UNFPA.) UNFPA supports programs in about 140 countries and territories, including a number of programs that receive no U.S. foreign assistance. It does not provide support for abortion services or counseling, and was therefore not affected by the Mexico City policy. It does, however, support family planning projects in countries which include abortion as part of reproductive health care. One UNFPA-supported program, in the People's Republic of China, was accused of employing widespread forced abortion. In 1985, under pressure from U.S. anti-abortion groups opposed to the Chinese program, the Reagan administration withheld $10 million of the congressionally mandated $46 million U.S. contribution to UNFPA.[30] The U.S. made no contribution to UNFPA for seven years after 1985. In 1989, President Bush vetoed the entire fiscal year 1991 appropriation of $14 billion for foreign assistance, principally in order to block a $15 million congressional earmark for UNFPA.[31]

LOST OPPORTUNITIES

The Reagan-Bush policies have left a legacy of lost opportunities. Most seriously, those policies prevented population aid from keeping pace with growing developing-country requests for assistance. Before Reagan took office, President Carter asked Congress to double international population assistance in response to the recommendations of the *Global 2000 Report.*[32] While Congress might well have balked at such a large single-year increase, presidential leadership would surely have produced substantially increased funds over time, given already strong congressional support for the program. Instead, following a pattern established in several earlier administrations, presidents Reagan and Bush routinely asked for less in population assistance than Congress had appropriated in the prior year. For fiscal year 1993, the U.S. government spent an estimated $431 million for population assistance. That amount, when adjusted for inflation, is substantially *lower* than President Carter's request for fiscal year 1982.[33] Since 1980, nearly 30 additional countries have launched national family planning efforts and the number of couples of childbearing age worldwide has increased by almost 50 percent.

The U.S. government's retreat from leadership of world population efforts during the 1980s may bear at least part of the blame for the slower progress made during the last decade. During the 1970s, world contraceptive use grew by 53 percent and average family size declined 22 percent, with the greatest gains occurring in East Asia. During the 1980s, although many additional countries adopted population policies, contraceptive use grew by less than 20 percent and family size declined less than eight percent. Among the factors contributing to slower progress was the shortage of funds for family planning efforts relative to demand. Although Japan and many European countries increased their contributions to UNFPA, IPPF, and other international family planning efforts in the 1980s, population assistance from all countries remained essentially stagnant in inflation-adjusted dollars throughout the 1980s.[34] The World Bank and many European donors complain that developing country governments don't ask for population assistance unless funds are earmarked for that purpose. However, the fact is that donor country assistance has fallen at least 60 percent below the level of requests for assistance from developing countries. In 1992, for example, the United Nations Population Fund estimated that its program allocations were $180 million compared to its total requests of $550 million for population and family planning assistance.[35]

One can argue whether substantially increased funds for family planning or stronger U.S. political leadership during the 1980s would have, by themselves, produced significantly higher levels of contraceptive practice worldwide. But it is clear that official U.S. ambivalence about population problems and its preoccupation with abortion helped delay a global commitment to early population stabilization and derailed progress in many countries on expanded reproductive rights for women, especially increased access to safe, legal abortion.

Conservative religious opposition to family planning has increased since the early 1980s, fueled in part by the export of anti-birth control activism from the U.S. By pandering to ideological and religious opponents of family planning programs, the Reagan administration legitimized a higher level of opposition to world population efforts from the Vatican and a variety of religious fundamentalists, thus undermining political support for reproductive rights wherever they were vulnerable around the world.

NEW INITIATIVES AND NEW CONTROVERSIES

After a decade of delay, world population problems in the 1990s are both more urgent and more expensive to solve. There is still time to avoid another population doubling, but only if the world community acts very quickly to make family planning universally available and to invest in other social programs, like education for girls, which help accelerate fertility declines. During the decade of the 1990s the number of couples of childbearing age in the world will increase by about 18 million a year. Added to the growing proportion of married women worldwide who already say they want to stop or postpone childbearing, the record increase in fertile-age couples is fueling an extraordinary growth in unmet demand for family planning services.[36] The world did not have a decade to dither away on debates about abortion and "supply-side demographics." In the 1990s, the U.S. and other governments must make up for lost time by making a quantum jump in their financial commitments to family planning and related development programs.

The support which the Clinton administration gives to American population aid, and the quality of American leadership at the 1994 International Conference on Population and Development in Cairo, will therefore be key to determining when and at what level world population is stabilized. Happily, the Clinton administration got off to a good start. On January 22, the 20th anniversary of *Roe v. Wade*, a newly inaugurated President Clinton reversed a long list of Reagan- and Bush-era restrictions on domestic and international family planning programs. In his first presidential budget request, Clinton also proposed a substantial $100 million increase for U.S. population assistance, $50 million of which was earmarked for the United Nations Population Fund.[37]

Unfortunately, administration initiatives on population are still politically vulnerable. Although the anti-abortion lobby no longer threatens continued U.S. population assistance, their opposition remains in force here and overseas. Moreover, international family planning programs face renewed criticism from the liberal end of the political spectrum. Many of these critics believe that overconsumption and uneven distribution of resources, not population growth, are at the root of most environmental problems.[38] At the same time, they question whether organized family planning programs are the most effective means of lowering birthrates. The debate is often dominated by extremes. On one side are those who see population stabilization as an international imperative so important to economic development or the global environment that governments (like China) are justified in taking whatever measures are needed to bring human fertility under control. On the other side are those who believe that governments have *no* legitimate interest in individual reproductive behavior and, therefore, no right to promulgate national or international population policies. Some on this latter side would go so far as to argue that any government-sponsored family planning service, in the absence of women's empowerment, is inherently coercive.[39]

In recent years, as environmental activists in this country and overseas have become increasingly concerned about the environmental effects of population growth, feminist health groups and many women of color have expressed real alarm about the possibility of a "population control juggernaut," driven by a powerful environmental lobby, which would legitimize a whole range of family planning abuses against women. These concerns reached a fever pitch in the women's tent at the Earth Summit in Rio in 1992, and have created a somewhat treacherous political climate for population initiatives in the new Congress and Clinton administration.[40]

At the heart of the debate, and perhaps at the heart of its resolution, lie several assumptions about why couples in developing countries have large families. Despite the substantial progress made by family planning programs, it is still widely perceived among U.S. policymakers and opinion leaders that there are major cultural, religious, social, and economic barriers to widespread family planning use. In particular, poverty and the low status of women are believed to diminish motivation to practice family planning. Many people still believe that large families reflect the needs of parents in poor countries — for old age security, for labor around the house or the farm, and as insurance against high levels of child mortality.

The prevalence of these assumptions has allowed other international development constituencies to argue that their own agendas — be they women's development, child survival, or poverty alleviation — are prerequisites to success in family planning. They are half right: development can help reduce birthrates. But they are also half wrong. Development measures cannot match the demographic impact of improved access to high quality family planning services. And ideas that are only half right, especially politically safe or popular ones, are potentially dangerous because they can lead to bad policy choices. Policymakers could make no greater mistake than to divert family planning funds into other development sectors on the mistaken assumption that "development is the best contraceptive."

WHAT FAMILY PLANNING CAN ACCOMPLISH

After 30 years of organized family planning efforts, we have increasing evidence that well-designed and managed family planning programs have a powerful, independent effect on fertility — greater than that of any other single intervention in virtually all settings around the world. In some 70 developing countries where there have been measurable declines in family size over the last several decades, organized family planning programs have accounted for about 50 percent of the declines.[41] This is a remarkable record, given that population assistance represents only 1.5 percent of total official development assistance, and government family planning programs generally represent less than one percent of national budgets. In very poor countries, like Bangladesh, family planning programs have played an even bigger role, accounting for up to 90 percent of fertility declines.

No part of the world is completely resistant to the idea that couples can and should make conscious decisions about childbearing. The greatest progress has occurred in East and Southeast Asia, especially in countries like South Korea and Taiwan, where fertility has actually dropped below replacement level through a mutually reinforcing combination of strong family planning programs and broad-based economic growth. Family planning advocates can also point to progress on parts of the Indian sub-continent, in the Middle East and North Africa, and most recently in sub-Saharan Africa. In Kenya, over the last decade, desired family size has dropped from eight children to four — family-size norms halved in the demographic equivalent of a nanosecond.[42]

The Kenyan experience suggests that the pronatalist values which characterize most traditional societies have proven quite malleable in response to public education campaigns, community outreach programs, and — more generally — social and economic changes. Profound changes are underway in virtually every developing country. For example, most citizens of the developing world are already part of a cash economy. Many communities have at least a primary school and rudimentary health services. By the end of the decade, half the world's population will live in cities. Women are entering the paid labor force in record numbers, usually out of necessity and often in occupations incompatible with unplanned pregnancies. All of these factors change the opportunity costs of childbearing and child rearing, and they are contributing to a revolution in reproductive behavior. This revolution in attitudes and behaviors is reinforced by an increasingly international mass media.

PRIORITIES FOR U.S. POLICYMAKERS

Governments can clearly accelerate these trends, *both* by expanding access to information and to safe and effective birth control methods and by making strategic investments in those sectors — like education for girls — which have the greatest indirect effect on contraceptive use, the timing of first childbirth, and family-size norms.[43]

Today there are roughly 350 million married couples in developing countries, and a substantial but unenumerated number of sexually active unmarried women and men, who are not using any method of family planning.[44] Most of these couples are not actively trying to get pregnant and have heard of at least one modern contraceptive method. Why aren't they practicing family planning? For hundreds of millions of people, the answer is the exorbitant cost of modern contraception. A recent study of contraceptive costs in 125 countries showed that the $35 cost of a year's worth of 100 condoms in Ethiopia amounted to one-third of per capita annual income of $135. In Kenya, the private-sector cost of contraceptive sterilization was almost 90 percent of annual per capita income. The same study showed that free or subsidized public-sector services reach less than 50 percent of the people in many countries, and less than 20 percent in many African countries.[45] Even when services are nominally available, many people are discouraged from using them because of distance, inconvenient hours of service, the arrogant attitudes of providers, limited contraceptive choices, misinformation about side effects, and inadequate counseling and follow-up.[46]

Still, it will take more than universal availability of high-quality family planning programs to bring about early population stabilization. Fortunately, there are other measures the world community could take — quickly and inexpensively — to improve the prospects for population stabilization. Each of these measures demonstrates a powerful synergism with organized family planning programs.

First, governments need to take specific measures to get and keep girls in school through at least the seventh or eighth grade — a level of education which correlates consistently with later childbearing and lower fertility. Partly as a result of population growth and economic recession, there are now more children out of school than ever before, and most of these children are girls. The World Bank estimates that it would cost about $3.2 billion a year to close the gender gap in education at the primary and secondary school level.[47] More detailed recent estimates indicate the real costs are somewhat higher.[48] But no other investment would provide higher returns in social and economic development, and no other intervention (other than family planning services) would have a more powerful effect on fertility.

Second, governments need to ensure that parents who choose to have only one, two, or three children can be confident those children will survive to adulthood. While declines in child mortality are not a precondition for declines in fertility, low rates of child mortality are necessary over time if replacement-level fertility is to become widespread. UNICEF estimates that additional expenditures of $3 billion a year on safe motherhood and child survival programs could reduce child mortality by a third and maternal mortality by half.[49] These investments need no demographic rationale, but their long-term effect on prospects for population stabilization is not insignificant.

Other women's development initiatives could also contribute to early population stabilization.[50] Women's paid employment outside the home has a tremendous impact on family size in every culture and at all income levels. In some settings, so does women's access to credit, new labor-saving technology, or control over land and other productive resources.

Greater economic independence may help give individual women the autonomy and security they need to make their own decisions about childbearing.

Finally, the U.S. and other governments need to do more to expand educational and economic opportunity for the poorest of the poor (a category in which women-headed households are everywhere overrepresented).[51] Whether headed by women or men, families with no hope of a better life have little incentive to plan the number or timing of their children. Where there is no hope for the future, women will, in their own words, "have all the children God sends."[52]

U.S. policymakers must become much more sensitive to the realities of life for such women. We must also ensure that the world's population is not stabilized at the cost of women's reproductive freedom. But, by the same token, the world community cannot allow the human population to triple to 15 or 16 billion — which could result if our current demographic trajectory is not modified. In this critical decade, we must choose between two possible long-term demographic scenarios: a population less than double or at least triple our current numbers. The choice will have extraordinary impact on world efforts to alleviate the absolute poverty in which one-fifth of the world's people now live, to protect the global environment, and to promote peaceful democratic change.

If our generation fails to bring human numbers and human needs into balance with the earth's resources, we risk leaving our children — American children included — a poisoned inheritance, a world much more polluted, crowded, unjust, and dangerous than the one our parents left to us.

ADRIENNE GERMAIN AND JANE ORDWAY

Population Policy and Women's Health: Balancing the Scales

....

To most people, the "population problem" means "overpopulation" — primarily in the Third World, where three-quarters of the world's people live. "Overpopulation" conjures up images of malnourished and dying children, burgeoning slums, deforestation and desertification, an unending cycle of poverty, disease, illiteracy, and social and political chaos. Population growth, along with poorly planned industrialization and environmental destruction, are seen as threats to sustaining life at acceptable levels in the future.

Hoping to change this devastating prospect, family planning and related programs have supplied millions of women in the Third World with contraceptives which would otherwise be unavailable to them. However, a tendency to neglect other aspects of women's reproductive health has often undermined or negated the achievement of effective and widespread contraceptive use. For example, inappropriate contraceptive use due to poor counseling and high discontinuation rates due to side effects or infection, among other causes, are common in the Third World.

The "population problem" and possible solutions need careful review and redefinition. A "reproductive health" approach, with women at its center, could considerably strengthen the achievements of existing family planning and health programs, while helping women to attain health, dignity, and basic rights.

Adrienne Germain and Jane Ordway are vice president and associate director for public education, respectively, of the International Women's Health Coalition.

STRATEGIES TO DATE

Population growth in the Third World was first recognized as a severe problem by the United States and by some Third World countries in the early 1960s. The primary strategy launched to reduce birth rates was development and distribution of modern means of contraception through family planning programs. Initially financed primarily by the United States and other Western governments, and more recently by Third World governments and major international development organizations, family planning programs have sought to facilitate contraception by the largest possible number of fertile couples. Generally, these programs have viewed women as producers of too many babies, and measured their accomplishments in numbers of contraceptive or sterilization "acceptors" and statistical estimates of "births averted."

Additional related programs have included maternal and child health (MCH), Child Survival, and the recent Safe Motherhood Initiative, all of which, at least in theory, recognize family planning as a key to achieving their own goals. But MCH programs have tended to emphasize child health, attending to women's health as a means to that end rather than an end in itself. Similarly, the newer "Child Survival" programs focus on children's immunization, growth monitoring, and oral rehydration (a treatment to prevent death due to diarrhea). These programs provide few if any services for the woman herself and tend to place heavy demands on the mothers' time, financial resources, and skills.

Since 1987, attention has been drawn to the mother herself through the Safe Motherhood Initiative (SMI), which seeks to reduce pregnancy-related mortality through prenatal care, emergency obstetrical services, and postnatal care. Contraception is recognized as an important means to prevent the highest risk pregnancies. Although SMI recognizes that botched abortions are a major cause of maternal mortality, provision of safe abortion services is not an explicit component of SMI.

Today, most Third World countries have national population or family planning programs as a population control measure or as a health measure. Several new contraceptives and improved sterilization techniques have been developed in the last two decades. The accomplishments of population and other related health programs have been substantial. However, their effectiveness has varied considerably both within and across countries. Much remains to be done to enhance their achievements and extend their services.

Few, if any, of these programs yet recognize or provide services for major reproductive health problems that affect fertility decisions directly and indirectly, such as infertility, sexually transmitted diseases (STDs), or violence. Nor do they attend to the health of girls in their formative years (ages 5-15) when health, nutrition, and sex education could prepare them for safe sexual relations and childbearing as adults. Few family planning and related programs effectively include men.

MORE THAN A PROBLEM OF CONTRACEPTION

Solving the "population problem" requires more than simply provision of contraceptives. Fertility control involves the most intimate of human relations, complex behaviors, and substantial risks. To control their own reproduction, therefore, women must also be able to achieve social status and dignity, to manage their own health and sexuality, and to exercise their basic rights in society and in partnerships with men.

Early sexual relations and pregnancy, however, curtail education, employment, and other social and political opportunities for millions of young women in the Third World, just as they do for one million teenage women in the United States every year. Prevention of adolescent pregnancy will require social acceptance of sex education and contraceptive services for teens, wide-ranging support for development of young women's self esteem, and other interventions that are politically or other wise challenging.

Third World women who become pregnant face a risk of death due to pregnancy that is 50 to 200 times higher than women in industrial countries. Pregnant adolescents frequently face obstructed labor that culminates in death or serious physical damage. Sixty percent of pregnant women in the Third World are anemic, which makes them especially vulnerable to problems in pregnancy and labor that result in death. Over half, in some countries 80-90 percent, of pregnant women give birth without trained assistance or emergency care. As many as 250,000-375,000 women are estimated to die annually when giving birth. This tragedy is intensified manyfold by its impacts on the families left behind.

Fears about the safety of modern contraceptives are strong deterrents to contraceptive use. Women must bear most of the social and health risks of modern contraception, partly because contraceptive methods available to men are extremely limited in number and appeal. Condoms have no side effects and can be very useful in preventing spread of disease, but men are often reluctant to use them and women are not in a position to persuade them to. Similarly, vasectomy, safer and simpler than female sterilization, is practiced far less in the Third World. Thus, population stabilization requires development of new and improved contraceptive methods.

Increasingly, women in the Third World who do not want to be pregnant avoid pregnancy by using contraception effectively. But millions of women have unwanted pregnancies. Many of these carry their pregnancies to term and end up with one to three more children than they want. Every year an estimated 30-45 million pregnant Third World women who cannot accept a birth resort to abortion. And every year, at least 125,000 of them — and quite possibly at least twice that many — die in the process. Uncounted others are rendered sterile or suffer severe chronic health consequences. Those who survive abortion often face greatly increased risk of death in subsequent pregnancies.

Sexuality and sexual relationships are fraught with other dangers for girls and women that also affect their views about fertility and contraception. First, millions suffer from STDs, including AIDS, transmitted by men. As a result of STDs, botched abortions, harmful surgical practices, or their partners' infertility, among other causes, millions of women are subfertile or infertile. They live in dread of divorce and social ostracism because they cannot bear children. Second, millions are subject to violence due to their gender — rape, incest, and emotional and physical battering by husbands or relatives.

WHAT DO THIRD WORLD WOMEN WANT?

Women want to be healthy and to have as many children as they want, when they want them, without risk to their own or their children's health. They want services for their own health, for safe delivery of healthy infants, and for child health. They want means to space or to limit their childbearing that are easily available.

Many women now excluded as a matter of policy from contraceptive services want access. In many countries, these include the young, the unmarried, and those who do not yet have a child. Many women need easier access to services. Those who are employed or have multiple demands on their time often face major obstacles in obtaining services, including inconvenient clinic hours; travel over long distances just to reach a clinic or contraceptive dispenser; hours of waiting in the clinic or repeated visits for service because of shortage of supplies or personnel, among other reasons.

Women determined to avoid birth want safe services for terminating an unwanted pregnancy. Unlike women in most Western countries and in China, most Third World women face highly restricted access to such services due to legal restrictions or the failure to make services available, even where legal, as in India. One third of Third World women live in countries where abortion is prohibited altogether or is permitted only to save a woman's life. Even women who qualify for safe abortion under the law may often not be able to reach a hospital, or obtain enough money to pay for the procedure, or persuade a medical committee that they are eligible for pregnancy termination.

Women also want services to meet their multiple reproductive health needs. Millions of Third World women face the discomforts and consequences of reproductive tract infections, including the personal and social trauma of infertility. Little or no counseling or treatment is available to women suffering from infertility or STDs, often a consequence of their partners' sexual behavior rather than of their own. Where the services do exist, they frequently are not all available from the same source, but must be sought (more conspicuously, expensively, and inconveniently) at diverse locations.

Sex education for boys and girls and support for girls and women who are the victims of violence are very rarely available. Women themselves may be reluctant to demand such services because of social and other taboos.

Faced with the time and costs of seeking health care, women frequently give priority to their children's health care over their own. To avoid the death, sterility, chronic health problems, and unwanted children that are the consequences of unwanted or mistimed pregnancies, family planning services, including safe abortion, must be made convenient, their quality improved, and their scope expanded. More and better contraceptives, for both men and women, must be developed, along with programs to enable women (and men) to undertake sexual relations safely.

"Population" is a fundamentally human problem. The solutions must be both humane and responsive to the complexities of people's behavior. For both humanitarian and political reasons, those concerned about population growth need also to reaffirm their commitment to individual well-being. That commitment can be enacted by making reproductive choices possible, by modifying program approaches to emphasize quality of care, and by recognizing and seeking to meet women's multiple reproductive health needs. The potential scope for innovation is broad. In setting program priorities, it is essential to recognize that the woman is important in her own right, as well as the key actor in fertility regulation and in infant and child health. Her needs, not just those of her children, family, and society, must be central. Alliances for this purpose will be to the benefit of all.

Population Policy, Reproductive Health, and Reproductive Rights

....

MAHMOUD F. FATHALLA

From Family Planning to Reproductive Health

....

"All couples and individuals have the basic right to decide freely and responsibly the number and spacing of their children and to have the information, education and means to do so."
— International Conference on Population,
Mexico (United Nations, 1984)

The concept of birth planning is not new, nor is it a human invention. In nature, births are timed to enhance the survival prospects of the young. Among wild animals living in temperate latitudes where seasonal climate changes dictate changes in food availability, breeding is carefully timed so that the metabolic demands of lactation coincide with an increased food supply for the mother. Mammalian oestrus cycles, in which the female is receptive and attractive to the male only during limited periods, also help ensure that births take place at an optimal time.

Humans, too, are equipped with biological mechanisms to facilitate birth planning.[1] Puberty occurs later in humans than in any other mammal, to ensure physical and mental maturity in the mother (and father) and to allow enough time for the child to benefit from the transmission of intergenerational knowledge and skills before becoming a parent. Even after the onset of menarche, there is a "grace period" during which young women are usually infertile. After a child is born, the mechanism

---

Mahmoud F. Fathalla is a senior advisor in biomedical and reproductive health research and training at The Rockefeller Foundation.

of lactational amenorrhea provides for child spacing. Finally, menopause helps ensure that any child born will have access to a reasonable period of care from a healthy mother. As when our species first set foot on earth, family planning remains an effective strategy for child survival. Delaying the first birth, adequate spacing of births, and putting a voluntary limit to childbearing remain cornerstones of any child survival strategy.

Of course, men and women have always sought greater control of their fertility than that afforded by physiology. Since early days in human history, men have used withdrawal, or coitus interruptus, as a means of birth control. Women, on the other hand, had neither the power nor the safe and effective means to control fertility until the modern era. The lack of tools did not prevent them from trying, often risking their health, future fertility, and even their lives in the process. In almost every culture historians have found ancient, traditional contraceptive recipes used by women. For example, Egyptian papyri dating from 1850 B.C. refer to plugs of honey, gum acacia, and crocodile dung, which were used as a contraceptive vaginal paste. Whatever the effectiveness of these and other methods, their universality throughout history demonstrates the serious intent with which women have pursued control of procreation.

SOCIETY AND FAMILY PLANNING

No society, culture, religion, or legal code has ever been neutral about reproductive life. The reproductive rights of individuals have always been framed within certain socially accepted norms. Although human reproductive capacity is great — according to the *Guinness Book of World Records,* one Brazilian couple had 32 children — there is no record of any society in which average fertility approached even half the level of maximal female fertility. Society has always imposed brakes on childbearing, most notably through the custom of marriage, which legitimizes reproduction only within a socially sanctioned union.

Most societies have regulated the use of contraceptive practices and technologies. Where contraception was outright prohibited, the predominant objection was often to contraceptive control by women. Patriarchal societies reasoned that if women had control over their reproduction, they would also have the unthinkable: control over their own sexuality.

In the last few decades, traditional fertility controls have been joined by increasing government involvement in reproductive life. According to the United Nations Policy Data Bank, in 1988, 102 of the world's 170 governments maintained policies intended to affect fertility: 21 countries had policies to increase fertility; 20 sought to maintain current fertility levels; and 61 had policies to decrease fertility.

Many analysts believe that governments have a legitimate interest in influencing the birth rate. As former French minister Simone Veil has written:

...the basic objective of the State is to promote the economic and social development of the country and to ensure the maximum well-being of its citizens. The objective applies to the future as well as to the present. ... Avoiding a long-term weakening of a country brought about by a dangerously low birth rate and, inversely, avoiding excessive population growth when it becomes an obstacle to economic development and to the well-being of the population must certainly be among the basic goals of all governments.[2]

Although their interest may be well-intentioned, government interventions to affect fertility have a checkered history. While some have been applauded for their beneficial effect on health and human welfare, others have been excoriated for heavy-handed tactics and abuses of human rights.

Measures taken by governments to influence fertility can be classified into two broad categories: indirect and direct.[3] Indirect measures, for example, may seek to lower fertility by improving the status of women so that women have life options in addition to motherhood. Others may aim to enhance child survival so that couples will not need to have large numbers of children in anticipation of child losses. Still others provide care and protection for the aged, in order to reduce the need for children to safeguard old age security. Each of these measures may be seen as beneficial in their own right, apart from any intended effect on fertility. From an ethical perspective, then, these indirect interventions are widely accepted.

Direct interventions fall across a broader ethical spectrum. Some direct interventions — such as provision of family planning services to lower fertility, or parental benefits and family allowances aimed at increasing fertility — are usually seen as ethically benign. Others, such as incentives and disincentives to have children, are more ambiguous. (These are explored more fully in "Ethical Issues in Population and Reproductive Health," by Ruth Macklin, page 191.)

Some direct interventions, however, are clearly ethically objectionable. For example, in 1976, the national population policy of India permitted state legislatures to enact laws mandating compulsory sterilization. During the "national emergency" period which followed, several million forced sterilizations were performed. On the opposite side of the same coin, the Romanian despot Nicolae Ceausescu declared that "the fetus is the socialist property of the whole society. Giving birth is a patriotic duty, determining the fate of our country. Those who refuse to have children are deserters, escaping the law of natural continuity." Ceausescu's aggressive pronatalist policy required all Romanian women up to age 45 to undergo monthly gynecological examinations in their workplaces. Factory physicians received their full monthly salaries only if plant employees achieved a state-stipulated monthly quota of pregnancies.[4]

As far as the health and well-being of women are concerned, there is little difference between coerced motherhood and coercive population control. Both interventions deny women the dignity of making a choice in their reproductive life. I would argue that fertility by choice, not by chance or by force, is a basic requirement for women's health and well-being. A woman who does not have the means or power to control her fertility cannot have the joy of a pregnancy that is wanted, or avoid the distress of a pregnancy that is unwanted. She cannot freely plan her life, pursue her education, or undertake a career. Nor can she plan her births to ensure more safety for herself and better chances for her child's survival and healthy development.

THE CONCEPT OF REPRODUCTIVE HEALTH

The concept of reproductive health evolved to articulate the totality of health needs in reproduction. It recognizes that family planning is an important component of reproductive health, but emphasizes a more comprehensive approach to the the needs of women and men. As an alternative to narrowly focused population programs, a reproductive health approach encompasses, among other things, family planning, prevention and treatment of sexually transmitted disease and other reproductive tract infections, and prevention and management of infertility. It recognizes unsafe abortion as a major health problem and emphasizes the special needs of adolescents. More broadly, reproductive health implies that people should have the ability to safely regulate their fertility and to practice and enjoy sexual relationships; that women should go safely through pregnancy and childbirth; and that reproduction should be carried to a successful outcome through infant and child survival, growth, and healthy development. Evidence suggests that the various elements of reproductive health are strongly interrelated, and that improvements in one element may result in gains in another.[5]

In these days of competitive demands on scarce resources, there is a need to make priorities in the allocation of investments. Criteria for setting priorities in health include the magnitude of the problem and its impact as well the availability of cost-effective interventions. By these criteria, are reproductive health programs a worthwhile investment?

First, let's assess the magnitude of the problem. The World Health Organization (WHO) and the World Bank have made quantitative assessments of the global burden of different diseases, with the results expressed in terms of disability-adjusted life years (DALYs) lost as a result of the disease.[6] Although the overall burden of disease was not found to be different in women than in men, the relative importance of different causes was strikingly different. Maternal causes, which are of course exclusive to women, were estimated to account for almost 30 million

DALYs lost. The global burden of sexually transmitted diseases is estimated to be 17.2 million DALYs for women, as compared to 3.8 million DALYs for men. Reproductive health problems thus constitute the major burden of disease in women.

We turn, then, to the question of whether there are cost-effective remedies for reproductive health problems. The WHO/World Bank study cited above found that, for the 15-44 age group, the percentage of disease that can be substantially controlled with cost-effective interventions was estimated to be 43.9 percent for women, compared to 17.5 percent for men. Clearly, such an investment would have a profoundly beneficial effect on the world's women.

The impact of reproductive health problems extends well beyond the level of individual lives. This is clear in the case of sexually transmitted diseases — especially AIDS — where failure to control the spread of disease exacts an enormous social toll. It is, however, most striking in the area of fertility regulation. Some have argued that high birth rates in the developing countries have disturbing implications for economic development, global stability, and the balance between population and natural resources.

Investments in reproductive health are also necessary to correct striking inequities. WHO's Alma Ata Declaration of 1978 stated that:

> The existing gross inequality in the health status of the world's people, particularly between developed and developing countries, as well as within countries, is politically, socially and economically unacceptable, and is, therefore, of common concern to all countries.[7]

In no area of health is this inequity as great as in reproductive health. If we look at mortality differentials around the world, we find that while the crude death rate in the developing countries is about ten percent higher than in the more developed countries, the infant mortality rate is almost six times higher, the child mortality rate is seven times higher, and the maternal mortality rate is 15 times higher.[8] Almost 99 percent of maternal deaths take place in developing countries. In some rural areas in Africa, the lifetime risk of maternal death is as high as one in 20, compared to one in 4,000 for a woman in North America.[9] Indeed, the rates of maternal mortality in rich and poor countries show a greater disparity than any other public health indicator. Maternal mortality in developing countries is a tragedy that demands the attention not only of the health profession but of the entire international community. I submit that maternal mortality should not be considered simply a health problem. It is a human rights issue on which governments should be held accountable.[10]

For these reasons — the crushing burden of disease, the availability of cost-effective interventions, an impact that transcends national boundaries, and a deep concern about social inequity — a special national and international effort is justified in the field of reproductive health.

REPRODUCTIVE HEALTH AND FERTILITY REDUCTION: THE SEARCH FOR COMMON GROUND

Clearly, reproductive health programs merit increased investment for all the reasons noted above. Some analysts also believe that reproductive health programs can be equally, if not more effective than "population control" programs in reducing fertility. Others are not convinced, fearing that promising experiments in reproductive health cannot be "scaled up" to work on a national level, or that funding for broad reproductive health initiatives will dilute the strength of scant family planning funds. Are concerns about population growth and reproductive health at opposite poles, or is there a common ground?

There are extremists in both camps. On one end of the spectrum, there are those who believe that programs and policies based on voluntary family planning and respect for reproductive rights will not bring about substantial reductions in fertility. They maintain that demand for family planning will not stabilize population growth in time to avert catastrophic outcomes, including famine and ecological collapse. On the other end, there are those in the reproductive health/reproductive rights camp who maintain that there is no such a thing as a population problem, or that the real issue is the inequitable distribution of power between the haves and have-nots, or that population programs are simply an attempt to maintain patriarchal control over women and their bodies.

Beyond these extreme positions, there is a lot of mutual understanding as well as some genuine and valid concerns. Many reproductive health advocates acknowledge the need to achieve a balance between human numbers and needs and the natural resource base essential for any development. They would, however, maintain that we should not confuse the means and the ends. Contraceptive use and bringing down fertility are not ends in themselves; they are means to improve the quality of life. It is people — not numbers — who should be at the center of the process. If the ultimate objective is to improve the quality of people's lives, it does not make sense to present them with only one component of the reproductive health package.

Reproductive health advocates understand that women need to protect themselves from unwanted pregnancy. But, they contend, women also need to protect themselves from potentially deadly sexually transmitted diseases and from losing their life or health in the process of pregnancy and childbirth. And they make the case that the solution to the population problem will not be made in government boardrooms but in people's bedrooms. People are more likely to use family planning when they find that their other reproductive health needs are addressed.

Among those concerned with population growth, most acknowledge the importance of reproductive health care as a complement to family planning. They are concerned, however, about limited resources. If resources are spread too thin, they argue, there will be little impact. Adherents of this view highlight the urgency of the population problem, and the heavy penalty for delayed action. Moreover, they note that family planning *per se* has a significant impact on other elements of reproductive health. In the absence of resources to provide the entire reproductive health package, does it not make sense to provide at least one important component? Should the best be the enemy of the good?

A QUESTION OF RESOURCES

There is indeed a common ground between population and reproductive health concerns, but there remains a question of resources.

Reproductive health should be seen for what it is: a determinant of our future, with an impact that transcends the individual, the family, the community, and the country. Empowering people, and particularly women, to regulate and control their fertility is a central component of the reproductive health/reproductive rights package, but there are other legitimate and closely related needs which must be met. Reproductive health should be raised to a higher level of political commitment, at the national and international level, with mobilization of the necessary resources.

Can we afford to do it? More appropriately, can we afford not to do it? We have the resources, if we would spend on human welfare a minute fraction of our investment in warfare. For a change, let human wisdom prevail.

JUDITH BRUCE

Population Policy Must Encompass More Than Family Planning Services

·····················

The consequences of rapid population growth can be viewed in light of the carrying capacity of the globe, but also, and more relevantly, in relation to human capacities and interdependence; that is to say, the *caring capacity* of human beings for each other. To be effective, our policies must comprehend the impact of population growth on global resources, on prospects for governance, and on the realization of national development goals, while concomitantly seeking to improve the welfare of families and individuals alive today.

Population policy, though potentially encompassing a range of subjects and instrumentalities, is equated in the mind of the public, and that of most policy makers, with fertility reduction implemented solely through family planning programs.

The tendency in the population community, in both our policy and our rhetoric, has been to issue vigorous claims for increased investment in family planning programs, followed by tepid prescriptions for the supportive social and economic measures governments might take. We have concentrated our advocacy, our energy, and our far too-limited resources on the provision of services alone. In fact, the discourse typically begins with a description of the demographic disasters ahead, or current, and the inexorable movement from the image of crisis to a seemingly simple and complete solution: the provision of family planning services. Even

Judith Bruce is a senior associate at the Population Council. This chapter is adapted from the text of testimony given before the U.S. House of Representatives Foreign Affairs Committee, September 22, 1993.

governments gravely concerned about rapid population growth have unnecessarily limited the scope of their activities:

- They have made family planning services a lonely centerpiece of their efforts.
- They have focused their messages and attentions solely on women, as if women alone could bring about fertility decline, and as if their sexual and parental roles were determined autonomously.
- They have treated familial negotiations over fertility regulation and the time and monetary costs of children as a private matter out of the domain of public policy.

I am the last person to say that the provision of voluntary family planning services of adequate quality is not vital or that the resources for such programs should not be greatly expanded. However, I also believe we must not oversell the potential of these programs. For example, the Population Council estimates that if unwanted fertility were eliminated, we would move the developing world only one-third of the way towards population stabilization.[1] Addressing the remaining two-thirds of the expected growth in population will require measures which address two other components of population growth; people's desire for more than two children and population momentum.

People's desire for more than two children. The average desired family size in sub-Saharan Africa remains between five and six, according to the latest data available. Although the desired number of children in countries of Latin America, Asia, and North Africa is lower, it still exceeds three in most cases.

Population momentum. Replacement-level fertility is reached when each man and woman becomes the parent of just two surviving children. After decades of replacement-level fertility, population stabilizes. But even if fertility were immediately brought to replacement level, the population size would increase for some time, from the momentum caused by a young population age structure. Population momentum alone would account for nearly half (49 percent) of the projected population growth over the next century, even if replacement-level fertility is achieved in 1995.

Thus, for population policy to be successful, it must be expanded to encompass more than family planning services. We must pursue the neglected two-thirds of the population growth equation — reducing demand for children and slowing population momentum — through selective, creative, and morally-sound social and economic investments. We must make sure that these investments affirm this generation's women, men, and children and not simply lay the foundation for future generations. Only then will these policies find political constituencies and be accepted by people in societies with the highest fertility levels.

Our task, however daunting, is ethically correct and politically attractive. It is the basis of good government. It calls for more active sharing from the adult generation to its children. It calls for eliminating gender inequalities in schooling and the marketplace. And it calls for a far more equitable distribution of rights and responsibilities between men and women in terms of sexuality, family planning, and the time and monetary burden of childrearing.

BEYOND SERVICES: DEALING WITH THE NEGLECTED 67 PERCENT OF POPULATION GROWTH

A major goal for all countries in the coming years is to move beyond the current population "strategy" of simply offering family planning services, by encouraging fertility reduction through explicit social and economic policy measures which are consistent with human rights and sound development practices. Responsibility for encouraging demographic change should be vested at the highest levels of government planning rather than confined to 'ministries of health' which have little policy voice in broader governmental councils and only claim leftovers in the budget process. Development plans and national budgetary guidelines across all sectors should be reviewed and weighed in light of their population implications.

There are three closely-related measures that must form part of a more comprehensive population policy that reduces the demand for children and slows population momentum:

Increasing Women's Access to and Control of Valued Resources

To moderate the risks of non-marriage, late marriage, and low fertility, women need special social and economic supports. Throughout the reproductive years and beyond, women need assurance of adequate livelihoods under their own control if they are to successfully regulate their fertility. Research shows that women who are earning and can control those earnings negotiate more effectively with partners over sexuality and fertility matters. Women who earn rely more on themselves and less on their children for security now and in old age. As a result, women with steady livelihoods are likely to desire fewer children, as well as invest more effectively in those they do have.

We cannot reduce population momentum — approximately half of future population growth — without fundamentally altering the behaviors and choices of girls and women aged 15-24. It is between the ages of 15-24 that most girls become sexually active (whether voluntarily or not), marry (whether chosen or not), and have their first children (whether planned or not). Our approach to social and economic planning must support young women so that they are not pressured by partners or family into early sexual relations or unchosen and oppressive marriages.

We must also pursue policies to help ensure young women are not denied making choices about their fertility, left unsupported in pregnancy and delivery, or I believe most grievously, do not find themselves left with the dramatic responsibility of raising children with their own limited resources. The difficulties young women face in effectively "planning" their personal lives are underscored by a recent series of studies in Latin America and the Caribbean. These studies found that 22 percent to 63 percent of ever-married women aged 15-24 with one live-born infant had premarital — presumably unplanned — conceptions.[2] Among unmarried women, from one-half to two-thirds of first pregnancies were reported as unintended. There is increasing evidence of coercion in sexual relations both inside and outside of marriage, particularly in the adolescent age group, and high rates of abandonment by fathers of children conceived premaritally.[3,4]

What does this mean for population policy? An expanded population policy would give attention to the development of social choices and livelihoods for women across the reproductive ages. But this will require more than simply mounting new efforts expressly directed at women. We also must review the internal contradictions and missed opportunities in overall development portfolios. In the same country, for example, one may find aid programs that support family planning services and urge women to have smaller families, as well as investments in large-scale agricultural intensification projects which may create more uncompensated demands on women's time, increase their reliance on child labor (and hence their demand for children), and undermine their bargaining powers in marriage.

Any mechanisms which demote women's livelihoods, pressure their already scarce time, or implicitly or explicitly increase male control over female labor are inherently unjust. However, they also have major implications for fertility. The perception that women lack economically viable options reinforces the stereotype in young girls and their families that security can only be guaranteed by early marriage, high fertility, and dependence upon males (husbands, fathers, and sons). More immediately, a large body of research finds that as women's income and control over that income increases, so does the health and education levels of their children, especially those of girls. Disinvestment in women results in long-term disinvestment in children, which carries negative fertility implications to a second generation.

Thus, we must promote socially acceptable and economically valued roles for young women — apart from early and rapid childbearing. Young women should be included in formal and non-formal education,

community development programs, and participation in income generation and small-scale credit programs. They must also receive honest sex
education and access to fertility regulation/reproductive health services.
Without these explicit policy and program measures, the youngest repro-
ductive-age females will be less able to delay the age of onset of sexual
activity, find protection from unwanted pregnancy, postpone first births
into the early twenties, and lengthen the space between subsequent
births. These are the central factors determining population momentum.

Intensify Investments in Children

Although there is much we must do to improve the lives of future generations, it is efforts in two areas — reductions in childhood mortality and
increases in educational levels — that are most likely to reduce fertility.

Uncertainty among parents about the survival of children encourages
high fertility in several ways. The possibility that children will die discourages parental investment in their children — a perverse, but real, effect.
Second, parents who feel 'at risk' require excess births to ensure the survival of at least some children into adulthood. Both of these effects can
be counteracted by undertaking programs and policies to improve the
nutrition, home environments, and treatment of disease in young children. While there can be no doubt about the value of such measures in
their own right, they are also a critical component of population policy.
No country in the developing world has experienced a sustained fertility
reduction while child mortality rates remained high.

Our current program approaches to improving child health have
underestimated the time costs of these strategies to women and overestimated mothers' power in families. These programs have often failed to
draw on other family members, especially fathers. Furthermore, we cannot overlook how parental preference for boy children undermines
parental willingness to invest time, food, and health resources in all children, male and female.

Increasing school attendance and grade completion rates have the
effect of raising the cost of children to parents, but also of preparing children for the emerging economies of their countries. Of all forms of
investment which are thought to affect fertility behavior, the education of
girls stands out as the most consistent. Arguably, educating a girl means
liberating the least empowered person in the family — the youngest
females — from the heavy burdens assigned to them in the domestic
hierarchy. Educating girls means empowering them as they move from
the age of six to puberty; the onset of their reproductive capacity. If we
want girls to have nontraditional fertility choices, we must begin early.

Women who have completed seven years of school have on average
three children less than their unschooled counterparts, according to

recent U.N. studies.[5] The observed reduction in fertility among these women is linked to delayed marriage, effective fertility regulation, and more modern forms of investment in children. Educated mothers have higher expectations for the education of their children, especially their daughters, thereby passing on the fertility advantages of improved education to a second generation. The children of women with even a few years of education are less likely to die than the children of uneducated women. Overall, primary school education, especially for girls, is a crucial element in setting the stage for sustained fertility decline.

Include Men in the Fertility Transition

Third and finally, population policy needs to concern itself with men's sexuality, fertility, and their share of the costs of children. Whereas men reap a disproportionate benefit from children, women carry a vastly disproportionate share of the penalties and responsibilities associated with unprotected sexual activity, contraception, childbearing and child rearing. Even where governments fully subsidize the costs of fertility regulation, women carry an unfair burden. For example, in some places the ratio of women to men having surgical sterilization is four to one, even though the male procedures involve considerably less monetary and time costs as well as health risks. And though the act of childbearing cannot be shared, child rearing surely can. It is not uncommon for mothers to spend seven times more hours caring for young children than fathers.

The increasing numbers of female-supported families — between 25 percent and 40 percent of all households in some settings — indicate that even men's economic share of the cost of children is declining. This inequality in parental responsibility is always unfair, but where resources are scarce, the already-skewed dependency burden created by high fertility is greatly intensified. When men and women have markedly different time and financial obligations to children, women may have a greater stake in limiting and spacing their fertility, which men may not share.

In some countries, there is evidence that groups of men are having many, many more children than the national average. In Mali, for example, of women aged 45-49, the median number of surviving children was 4.3. However, the men to whom these women were married had typically fathered 8.3 children.[6] In Kenya, married men are fathering nearly 10 children over their reproductive lifetimes, while their wives report 6-7.[7] Much of this 'excess' male fertility takes place after age 45.

There is no reason to suppose that fertility decline can occur without male participation or even leadership. Men were key actors in the fertility transition of 19th century Europe. Though the day-to-day burden of child care was not equally shared between mothers and fathers, history shows that fathers were expected to support their children.

Fertility fell under such circumstances despite a lack of any modern methods of contraceptives.

Population and development policies must examine how the costs of children are shared between the state and the family, as well as between parents. There are countries in which fathers' names appear on birth certificates only if the parents are married. There are countries in which children have claims on fathers' resources only if their father is living with their mother. Child-maintenance procedures, as difficult as they are to enforce, set norms about parents' responsibility for their children. In some countries, over 50 percent of the children may be potentially denied access to their fathers' economic resources, either because of the circumstances under which they were conceived or born, the current marital and sexual relationships of their biological parents, or their living arrangements.[8]

We must include in population policy all of the normative, judicial, program, and economic measures at our disposal to encourage male responsibility in family planning and parenting. We must draw young men into discussions about sexual responsibility, contraception and cooperation between men and women to avoid unwanted pregnancies and the transmission of disease. All too often our programs create perverse incentives for men to absent themselves: many family planning messages assume that women are uniquely interested in and responsible for contraception. Women who have uninformed or uncooperative partners are not given the opportunity to bring partners in for consultation on fertility regulation. Child health programs rely on the mother/child link almost exclusively and give scant attention to the father/child link. Programs and policies must promote fathers' roles as carers, nurturers, and reliable supporters of their children, regardless of their relation to the child's mother. The link between sexual, reproductive responsibility and better parenting, and vastly reduced levels of unplanned and unwanted children is unquestionable.

IMPLEMENTATION: A SIMPLE PROPOSAL

As for implementation, the overriding change proposed is to elaborate population policy far beyond its current family-planning emphasis. An elevated concern with population policy should be spread throughout all sectors of our bilateral programs and in our collaboration with other donors. We must promote initiatives to realign social and economic investments which favor fertility decline.

If 'population' is to be an important subject in our bilateral assistance and international policy efforts, it would seem logical to assign policy planning and budgetary oversight for this at the highest levels of The State Department and USAID. The proposed reorganization includes the establishment of a Bureau for Global Programs, and within it, an Office of Program and Policy Coordination. This unit could coordinate technical support to ambassadors and USAID directors and staff in reviewing and potentially revising major country portfolios. By integrating a population perspective across sectors, we can simultaneously foster productivity among the poorest and most vulnerable members of society, enhance parent to child investments, promote social justice, and accelerate voluntary fertility decline.

STEVEN W. SINDING, JOHN ROSS, AND ALLAN ROSENFIELD

Seeking Common Ground: Unmet Need and Demographic Goals

······················

The international family planning movement began forty years ago largely out of concern for human well being. A central premise and rationale of this movement was, and is, to enable individuals — women in particular — to exercise control over their own reproduction.

However, beginning in the 1960s, as governments became gradually more concerned about rapid population growth, national interests sometimes eclipsed concern for individual welfare. Particularly in Asia, the explosive rate of population growth caused governments to worry about their ability to provide adequate levels of health, education and other social services, and to keep food production on a par with demographic increase. Donor countries began to push policies and aid programs designed to reduce population growth rates. Many governments established demographic "targets" for fertility decrease, and family planning programs served largely as instruments to meet those targets.

At the same time, policymakers worried that individual motivation to limit family size was not sufficient to bring about fertility decline on the scale they deemed necessary. Some demographers and other social scientists concurred and held that voluntary family planning programs alone

Steven W. Sinding is director of the Population Sciences Division of The Rockefeller Foundation. John Ross is a senior research associate with the Futures Group. Allan Rosenfield is the dean of the School of Public Health at Columbia University.

would not stem high rates of population growth.[1] Thus, the great debates of the 1960s and 1970s: development versus family planning, and voluntarism versus direct interventions to influence reproductive behavior. These debates were partly the result of inadequate understanding — lack of information about how much fertility was unwanted and how rapidly reproductive desires would change if family planning information and services were provided.

The pessimists, some of them famous names in demography and economics, developed strong and persuasive theories about why the poor and the uneducated, living on the margin of subsistence, *should* want large families. Family planners, on the other hand, saw substantial demand wherever good services were available, but their position was less successful in the struggle for policy preeminence. The economic determinists carried the day at the Bucharest World Population Conference of 1974. Their position, that structural social and economic changes are a necessary prerequisite for fertility decline, prevailed in the form of such phrases as "development is the best contraceptive" and "take care of the people and population will take care of itself."

This perspective encouraged ambivalence about family planning on the part of some governments. Partly as a result of this ambivalence, direct donor funding for family planning stagnated on a per capita basis after 1974, although several developing countries did substantially increase their funding for family planning services in the years immediately following the Bucharest Conference. Still, as the result of donor agency cutbacks, family planning services probably did not spread as rapidly as they might have and the quality of those services was not as high as it could have been.

A few governments, pessimistic about the efficacy of voluntary family planning and continuing to fear the consequences of population growth, imposed policies that were to a greater or lesser degree coercive. These policies, often implemented in clumsy and heavy-handed ways, produced severe backlashes in a few places. Citizens' organizations reacted angrily to efforts to compel adoption of particular methods of family planning. A revulsion emerged against administratively set demographic targets and field-worker quotas, which required employees of state family planning programs to enlist given numbers of contraceptive "acceptors."

Both approaches to slowing population growth — social and economic development programs intended to reduce desired family size and direct interventions to change reproductive behavior — went forward without key information. Planners lacked accurate data about family size preferences, and they could not determine whether birth rates would decline if people were given the option of full and convenient access to family planning services.[2]

The World Fertility Survey and the Demographic and Health Surveys have changed this picture. The internationally comparable data produced by these research programs have shown a high level of unwanted fertility in almost all countries — a level too high and too consistent over time to be due to unreliable expressions of fertility preferences. Furthermore, the preferences revealed by these surveys have proved to be remarkably strong predictors of subsequent contraceptive and fertility behavior in all the countries in which they have been conducted. As a result of these survey programs, it is no longer necessary for governments to operate in ignorance. They now have data on unwanted fertility and on people's desire to control it. There is conclusive evidence of unmet need and unmet demand for family planning.[3, 4, 5]

At the same time that news about demand for family planning spread within the development community, the environmental movement, building up to the 1992 Earth Summit, renewed global calls for action on population growth. Many environmentalists believe that population growth must be curbed in order to achieve a sustainable balance between people and resources. But the environmentalists' concern reawakened fears among women's groups and others about coercive, target-driven demographic policies. In response to earlier abuses, women's health advocates worked throughout the 1970s and 1980s to reorient family programs toward respect for individual choice and rights. However, despite gains made in some countries and agencies, many anticipated that environmental concerns would produce a wave of aggressive population policies and massive low-quality family planning program efforts. They feared that people would be forced to act against their perceived self-interest, both in terms of the number of children they could bear and in terms of the quality of services that would be available to them.[6]

In the face of this concern, and armed with so much recent data on individual fertility preferences and desires, we decided to compare the projected demographic results of meeting unmet need for family planning services with those of meeting the demographic targets set by governments. While the analysis set forth below is approximate, it suggests that very significant demographic impact would result from family planning and reproductive health programs that attempted no more than to satisfy the stated reproductive wishes of individual women, particularly the women of the developing world. The analysis strongly suggests that such a rationale — helping women achieve their goals — would decrease fertility by as much, or more, than is called for in most countries' demographic targets.

UNMET NEED

Demand for family planning is measured by the now familiar concept of *unmet need,* identified with the work of demographer Charles Westoff and colleagues.[7] Various measures of unmet need have been used, but the basic sense is that many women experience unwanted pregnancies, yet do not obtain contraceptive protection. Women's health advocates Ruth Dixon-Mueller and Adrienne Germain[8] have recently enlarged the concept of unmet need, stressing that needs also exist among the unmarried, among users whose method is unsafe, ineffective, or unsuitable, and among those with unwanted pregnancies who lack access to safe and accessible abortion.

These further elaborations are well taken, and all would produce a higher estimate of unmet need than the one used in this paper. We employ a narrower definition here in order to use a consistent series of data on need, targets, and contraceptive prevalence. Besides being comparable across numerous countries, the measure used is in a general sense representative of the size of the potential clientele for family planning services, although it does not address the issue of the *content* or the quality of services. A broader definition of unmet need that tried to incorporate these issues would certainly expand our estimates and reinforce our conclusions.

The paper attempts to calculate what the demographic effect would be in each country if unmet need were satisfied, in addition to the prevailing level of contraceptive use. Where the information is available, this estimate is then compared with the targets stipulated by the government, converted where necessary to prevalence of contraceptive use.*

PRESENT TRENDS IN FERTILITY AND CONTRACEPTIVE USE

From Figure 1 it can be seen that overall contraceptive use in the developing countries has expanded dramatically since 1960 but that the

---

* *Targets are stated in various forms by the relatively small number of countries having them (20 by U.N. count). If the target is stated as a fertility measure or a population growth rate, it is converted to the corresponding level of prevalence, i.e. the proportion of couples who would need to be practicing family planning. The equations used for the relation of prevalence to the crude birth rate and the total fertility rate are based upon international correlations:*

$$TFR + 7.03 - (.0662) \text{ (Prevalence)}. \qquad CBR + 46 - (.42) \text{ (Prevalence)}.$$

In a few cases the target was stated as either the NRR or GRR (net or gross reproduction rate). We referred to the UN projections for those countries, which show the relation of those measures to the TFR for the same future dates. We then related the TFR to prevalence by the above formula.

FIGURE 1: Married Women in Reproductive Ages 1960-2010

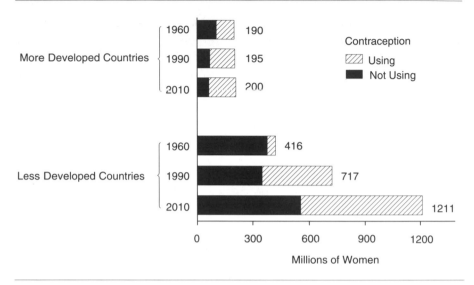

Figure 2: Total Demand for Contraception by Region and Fertility Level

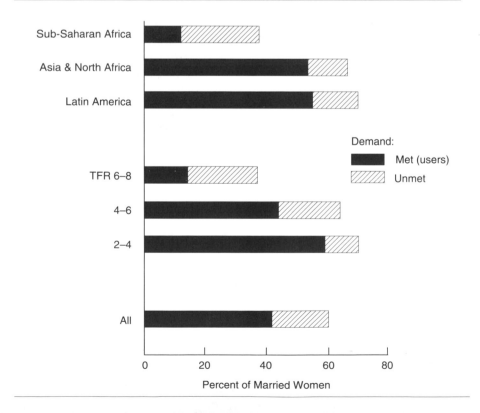

FIGURE 3: Current Total Fertility Rate and Predicted Fertility Rate After Five Years — Demographic and Health Surveys, 1986-1989

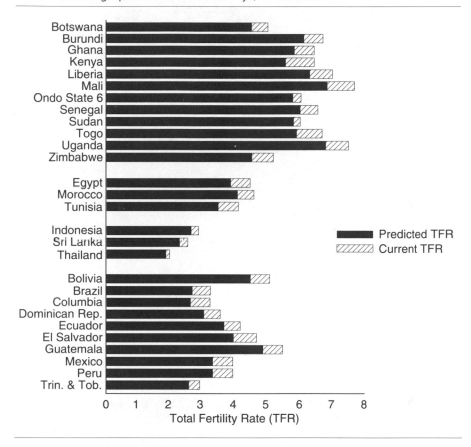

total number of women not using any form of contraception has declined only slightly because of the enormous expansion in the number of women in the reproductive age groups. Projections to the year 2010 show that, even though the number of women estimated to be using contraception again expands quite dramatically, the number of women not using contraception also increases substantially. This illustrates the enormous need for contraceptive coverage if replacement level fertility is even to be approached in the first quarter of the next century.

Figure 2 shows the proportions of all women who want no more children or wish to space the next birth and are not currently using contraception. As can be seen, sub-Saharan Africa lags far behind Asia, North Africa, and Latin America in the proportion of need that is unmet. It is also interesting to note that the proportion of need that is unmet tends to be highest in countries with the highest fertility, which also tend to be

FIGURE 4: Wanted and Observed TFR by Level of Development
(18 DHS Countries)

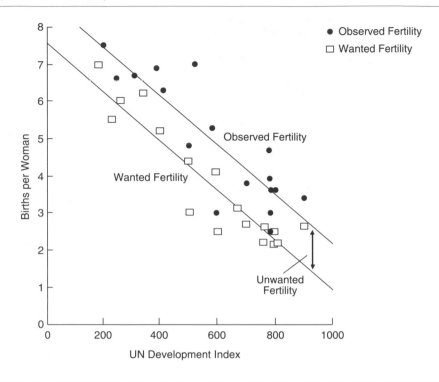

FIGURE 5: Demand for Family Planning and Donor Resources Available

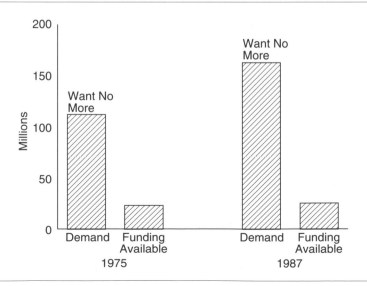

the least developed. This indicates that, even though overall demand for family planning tends to be lower in such countries, they have nonetheless done proportionally less to address unmet need than better-off countries in which demand is higher.

Figure 3 shows the decline in fertility that can be expected five years after the Demographic and Health Survey was carried out in the countries listed, based on respondents' statements about future fertility intentions. Despite the relatively high fertility rates in Africa, a number of countries show significant prospective declines in fertility on the basis of women's statements about desired future fertility. Indeed, some of the African countries (e.g. Kenya, Mali, Togo, and Uganda) show prospective declines that equal or exceed those of lower-fertility countries in which family planning services and information are already much better established, like those in North Africa and Latin America. (The prospective declines in Asia are quite low because of the comparatively high contraceptive use and low fertility rates in those countries.)

Figure 4 shows the relationship between fertility and development, and demonstrates that the gap between actual and desired fertility is constant at all levels of development. That is, desired and observed fertility both drop sharply as development occurs, but the extent of unwanted fertility remains relatively constant across the entire spectrum. Thus some level of unsatisfied demand for family planning exists at all levels of development.

Finally, Figure 5 shows the remarkable increase, from 1975 to 1989, in the number of women worldwide who are estimated to want no more children. While the number of women who want no more children is only a partial measure of the demand for family planning (it excludes those who use family planning for spacing purposes, a particularly important factor in Africa), the increase of 50 million women over the 12-year period demonstrates a strong increase in the need for family planning. However, over the same time period the funding available for family planning has remained virtually stagnant in 1975 dollars; nominal increases in funding have been offset by inflation.

SATISFYING UNMET NEED VERSUS ACHIEVING TARGETS: THE EVIDENCE

Table 1 lists 17 of the 21 countries of the developing world which, according to U.N. and other sources, have quantitative demographic targets. These 17 have conducted surveys recently, so the estimated unmet need for family planning can be calculated. The last column compares the target with total demand (i.e. prevalence already attained plus unmet need) for these countries. A positive number indicates that the satisfaction of unmet need would result in a contraceptive prevalence rate (CPR) higher than that which would be achieved by meeting the demographic target. For example, in Tunisia, with a target CPR of 51 percent and an actual CPR of 50 percent, meeting the target would have resulted in an increase in prevalence of one percent, whereas satisfying unmet need would have resulted in a 15.7 percent increase over the 1988 CPR. Thus, the figure in column four is 14.5. In 13 of the 17 countries, satisfying unmet need would exceed the government-established targets by amounts ranging from one to 31 percent. In India, the unmet need figure is based only on desire to terminate childbearing; if spacing had been included, satisfying unmet need would exceed target prevalence by a larger margin. Nigeria and Ghana have targets whose realization would require more than simply satisfying expressed need for family planning, and in both of these countries service availability is still severely con-

TABLE 1: Comparison of Targets and Unmet Need

Country	Date of Prevalence and Unmet Need	Prevalence (1)	Unmet Need (2)	(a) Date of Prevalence Target	Prevalence Target (3)	Excess over Target (4)
LATIN AMERICA						
Haiti *	1989	10	27	2000	60	-23
Jamaica *	1989	55	20	2000	74.5	0.5
Mexico	1987	52.7	24.1	2000	71.3	5.5
Peru	1991-92	59	16	1995-2000	61	14
SUB-SAHARAN AFRICA						
Ghana	1979-80	12.9	26.6	2000	46	-6.5
Kenya	1989	26.9	28.9	2000	40	15.8
Mauritius *	1985	75	3	1987	71.7	6.3
Nigeria	1990	6	20.8	2000	46	-19.2
Tanzania	1991/92	10	55	2000	31	24
MIDDLE EAST AND NORTHERN AFRICA						
Egypt	1988-89	37.8	25.2	2000	60	3
Tunisia	1988	49.8	15.7	1991	51	14.5
Turkey *	1988	60	19	1995-2000	62.5	16.5
ASIA						
Bangladesh	1991	39.9	41	1995	50	30.9
India	1988-89	43	18.3	2000	60	1.3
Indonesia	1987	47.8	13		55	5.8
Pakistan	1990-91	11.9	28	1998	29	10.9
Thailand	1987	65.5	11.1	1991	79.4	-2.8

Columns : (1) + (2) -(3) = (4)

strained. In Thailand, prevalence is already high and this, plus the relatively small amount of unmet need (by the definition used here), does not exceed the target of 79 percent prevalence.

THE GLOBAL PICTURE

Next, we look at the issue on a global basis. What would the demographic effects be if the family planning programs worldwide focused upon unmet need? Demographer John Bongaarts estimates that about 100 million women, or approximately 17 percent of the women in the developing world outside of China, presently have an unmet need for contraception[9] (compared with a figure of 21 percent using Westoff's more liberal procedure[10]). What would be the global effect of addressing this

---

NOTES ON TABLE 1

Where unmet need estimates differ between Bongaarts and Westoff and Ochoa, the more conservative (lower) Bongaarts figure is used.

*   Taken from "Family Planning Surveys" in Robey et al., which base unmet need on fecund, sexually active women regardless of marital status, whereas DHS surveys base unmet need only on women who are married or cohabitating.

India: unmet need includes only those wishing to limit childbearing; it omits those wishing to space.

Nigeria: Amenorrhoeic women (nearly 30 percent of currently married women) are excluded from calculations of current unmet need.

Tanzania: unmet need is estimated approximately as follows: of married women of reproductive age aged 15-49, 42 percent wish to space (over two years) and 23 percent wish to limit, for a total of 65 percent. Of these, ten percent are using (any method), leaving 55 percent in need. Infecund and don't know responses excluded. See Tanzania DHS Survey 1991/92, tables 5 and 7.

Sources of Prevalence and Unmet Need:

Bongaarts, John. 1991. "The KAP-Gap and the Unmet Need for Contraception," *Population and Development Review 17*, no. 2: 293-313. (For actual figures, see: Bongaarts, John. 1991 *The KAP-Gap and the Unmet Need for Contraception.* Working Paper No. 23, The Population Council, New York.)

"Family Planning Practices in India," Third All-India Survey (1988-1989) Operations Research Group.

Robey, Bryant et al. 1992. "The Reproductive Revolution: New Survey Findings," *Population Reports,* Series M, No. 11.

Westoff, Charles and Luis Ochoa. 1991. "Unmet Need and the Demand for Family Planning," *DHS Comparative Studies No. 5.*

SOURCES OF TARGETS:

Taken chiefly from United Nations sources:

United Nations. 1989. *Review of Recent National Demographic Target-Setting.* New York.

United Nations. 1987. *World Population Policies* (3 Volumes). New York.

need? International data show that a 15 percent increase in contraceptive prevalence is associated with about a one point decline in the total fertility rate (TFR), so that satisfying the 17 percent unmet need should reduce the TFR by 1.13 points. If one included China, where near-universal contraceptive use means there is essentially no unmet demand (by the narrower definition used here), there would be an overall decline of .79 points, or a reduction in the TFR for the entire developing world from 3.90 to 3.11. That is well below the target of 3.34 which has been set by the U.N. for the year 2000. The fall, from just under four to just above three births per woman, represents a fertility decline of 44 percent. The result would be made even more dramatic by satisfying the higher estimate of unmet need by Westoff, as it would be by recognition of the additional categories of unmet need identified by Dixon-Mueller and Germain.†

A focus on unmet need cannot guarantee either the early satisfaction of all need or a rapid reduction in fertility. Of course the same must be said of some targets: They too have not been met, or not met on schedule, and neither approach can promise an arrival at replacement level fertility in the near term. Both approaches require a continued enlargement of service outreach toward the underserved, continued efforts to improve contraceptive technology, and continued efforts to raise programmatic standards. They differ however in their assumptions and in their approaches to the public. While the full 17 percent of unmet need is unlikely to be satisfied in the near future, much of it can be, and that is all that can be reliably said about most demographic targets as well.

HEALTH BENEFITS OF ADDRESSING UNMET NEED

A softening of target emphases should help alleviate another prevailing tension, that between population and health programs. Advances in contraceptive protection and advances in child survival are mutually supportive. Reduced infant mortality encourages birth planning and fertility decline, and increased contraception improves child survival. With more contraception and safe abortion, there are fewer pregnancies and births and therefore fewer unsafe abortions and infant and maternal deaths. In addition, health gains can be expected from avoidance of closely spaced and other high-risk births. Availability of contraceptives to facilitate birth spacing is particularly important in African countries, where much

---

† *Our colleague W. Parker Maudlin has pointed out that if 85 percent of current unmet need (100 million women) were satisfied immediately the developing world's TFR would be about 2.89, matching the UN low variant projection for 2000. Further, if only one-third of unmet need were satisfied the TFR would be about 3.34, matching the UN medium projection for 2000.*

unmet need comes from the desire to space births more widely. There are also nutritional and other gains that raise the quality of life for the child, not to mention the advantages to the mother and family from the lessened burdens of excessive fertility.

According to one estimate, close to 14 million infants and children under age five die each year, most from preventable causes.[11] A number of health initiatives, including oral rehydration therapy (ORT) and the expanded program of immunization (EPI), are having an impact on infant and childhood mortality. These programs can be substantially reinforced by addressing couples with unmet need, who are often the same couples sought out by the ORT and EPI programs.

Maternal mortality is dramatically higher in developing countries than in the industrialized West; the World Health Organization estimates that maternal deaths in developing countries range from 100 to 1000 per 100,000 live births, ratios that are 10 to 100 times higher than in the U.S. and Europe.[12] More maternal deaths occur in India in one week than in all of Europe in an entire year. Many of these deaths occur in village settings from causes that are relatively immune to local measures, causes that cannot easily be relieved except through formal health infrastructures that are unfortunately absent. Thus pregnancy prevention in the first place emerges as a vital service. Using WFS and DHS data, one demographer[13] has estimated that nearly 50 percent of maternal deaths might be prevented if all women who wish no further children had reliable contraceptive protection.

Rarely can one program initiative have a significant impact on two major policy objectives. Satisfying unmet need is that rare initiative. The voluntary use of contraception by all women and couples who wish to avoid pregnancy would yield very large health benefits to both mothers and children and would go far toward meeting the societal need for lowered population growth rates.

POLICY IMPLICATIONS

The policy implication is clear: Family planning programs should abandon the use of demographic targets and instead express objectives in terms of meeting the stated desires of their clientele. This cannot be done without extensive organizational effort; large populations will continue to lack access to high quality contraceptive services unless program structures are substantially strengthened and extended. Furthermore, in some cases modifications must occur in the organizational culture of programs, in supervisor-worker relationships, and in provider attitudes toward their clients.

Finally, this does not by any means suggest that the provision of family planning services is the only thing that needs to be done to achieve replacement level fertility and eventual population stabilization. Clearly, continued intensive efforts to enhance the conditions of women, alleviate poverty, raise educational levels, and improve living standards in the developing countries will be required if desired family size is to eventually coincide with replacement level fertility. But, it is equally clear that improving access to, and the quality of, reproductive health services for all who have stated a need for those services will carry the world a very long way toward replacement level fertility.

In short, while it is quite appropriate for countries to set long term demographic objectives in the context of a comprehensive development strategy (one that includes raising literacy, enhancing educational opportunities for women, lowering infant mortality, and improving maternal and child nutrition), it is neither desirable nor necessary for those objectives to be applied as targets for the provision of family planning and reproductive health services. This analysis strongly points to a reconciliation between satisfying individual needs and achieving societal objectives. That should come as very good news to those who fear that targets encourage coercive family planning programs and to those who favor user oriented, demand-driven service delivery programs.

A N R U D H   J A I N   A N D   J U D I T H   B R U C E

Quality:
The Key to Success

....

In many developing countries the provision of family planning services to meet the needs of individuals and couples to have the number of children they want coexists with growing interest in reducing the birth rate. In others, however, the demographic rationale is the overriding policy concern. Placing the burden of societal fertility reduction on a family planning program is always inappropriate and its role should be reaffirmed in terms of making services of adequate quality, and information on various contraceptives freely available.

We define quality in terms of the way individuals (or clients) are treated by the system providing services. The following six elements of care (see box on page 173 for definition) should be considered in judging the quality of family planning services:

- Choice of contraceptive methods
- Information given to clients
- Technical competence
- Interpersonal relations
- Mechanisms to encourage continuity
- Appropriate constellation of services.

The improvement in quality of services decreases fertility mainly by increasing acceptance, continuation, and thus prevalence of contraceptive methods. The magnitude of this impact, however, depends upon the intensity of couples' motivation to regulate their fertility. Couples or indi-

Anrudh Jain and Judith Bruce are deputy director of programs and senior associate, respectively, at the Population Council. A version of this chapter originally appeared in People, *Volume 16, Number 4, 1989.*

viduals strongly motivated to control their fertility will accept hardship, risk, and poor services to attain their goals. On the other hand, couples or individuals with no interest in spacing or limiting will not change their objectives because attractive services are available. It is those individuals and couples in the middle range who are most likely to have their behavior influenced by the quality of the service offered.

The empirical information to link the six elements of quality to fertility reduction is very limited. A review of small-scale diagnostic studies suggests that positive adjustments in any of the elements (more choice of methods or better information supported by written materials, for example) has beneficial effects on individuals' knowledge, behavior, and satisfaction. We illustrate the importance of quality by presenting selected results for two elements: choice and follow-up.

There are powerful arguments and evidence that providing a choice of methods improves program performance and individual satisfaction. The proposition that providing a choice of methods increases the effectiveness of family planning programs is based on the following three main ideas:

- that individuals and couples pass through different stages in their reproductive life and therefore over time their needs and values will change;
- that multiple methods provide for switching for individuals who find their initial choice unacceptable or unhealthful;
- that the availability of a variety of methods makes it more likely that — given erratic contraceptive supplies — at least services for some methods will be available.

One recent analysis has taken a particularly hardheaded look at the impact of lack of choice on individuals' contraceptive use. One researcher followed a group of 2,500 new acceptors of various methods in East Java, Indonesia, and found that the continuation rate after one year depended upon whether a client's choice was 'granted' or 'denied.' Eighty-five percent of the women who did not receive the method they had originally requested had discontinued within a year. The comparable discontinuation rate among those who had received the method they requested was 25 percent. Whether a client's choice of method was 'granted' or 'denied' turned out to be more powerful than all other independent variables in explaining the tendency to continue or discontinue the use of contraception.

The literature on the impact of choice on contraceptive prevalence provides support to the following four assertions:

- The addition of a method results in a net addition to contraceptive prevalence. This assertion is based on the experience in Hong Kong, India, South Korea, Taiwan, and Thailand.
- One-method family planning programs are inadequate to meet individual fertility goals. This assertion is based on simulation models for

Elements of Quality

Choice of methods refers both to the number of contraceptive methods offered on a reliable basis and their intrinsic variability. Which methods are offered to serve significant sub-groups as defined by age, gender, contraceptive intention, lactation status, health profile, and — where cost of method is a factor — income groups? To what degree will these methods meet current or emerging need (for example, adolescents)? Are there satisfactory choices for those men and women who wish to space, those who wish to limit, those who cannot tolerate hormonal contraceptives, and so forth?

Information given to clients refers to the information imparted during service contact that enables clients to choose and employ contraception with satisfaction and technical competence. It includes: information about the range of methods available, their scientifically documented contra-indications, advantages, and disadvantages; screening out unsafe choices for the specific client and providing details on how to use the method *selected*, its possible impacts on sexual practice, and its potential side-effects; and finally, an often neglected element, explicit information about what clients can expect from service providers in the future regarding sustained advice, support, supply, and referral to other methods and related services, if needed.

Technical competence involves, principally, factors such as the competence of the clinical technique of providers, the observance of protocols, and meticulous asepsis required to provide clinical methods such as IUDs, implants, and sterilization.

Interpersonal relations are the personal dimensions of service. Relations between providers and clients are strongly influenced by a program's mission and ideology, management style, and resource allocation (for example, patient flow in clinical settings), the ratio of workers to clients, and supervisory structure.

Mechanisms to encourage continuity can involve well-informed users managing continuity on their own or formal mechanisms within the program. It can rely upon community media, or on specific follow-up mechanisms, such as forward appointments, or home visits by workers.

Appropriate constellation of services refers to situating family planning services so that they are convenient and acceptable to clients, responding to their natural health concepts, and meeting pressing preexisting health needs. Services can be appropriately delivered through a vertical infrastructure, or in the context of maternal and child health (MCH) initiatives, postpartum services, comprehensive reproductive health services, employee health programs or others.

Quantifying Quality: Examples from Africa

Barbara S. Mensch

Until recently, family planning programs in developing countries have focused more on quantity — the volume of contraceptives dispensed — than on the quality of services offered. Faced with a demographic mandate, many donors and providers have perceived quality of care as a luxury that resource-poor countries could not afford. But that perception has begun to change. As the accompanying article by Judith Bruce and Anrudh Jain makes clear, there is a new emphasis on improving the quality of family planning services and reproductive health care. This emphasis makes sense on many levels: it is unquestionably beneficial in its own right; and, from a demographic perspective, couples are presumably more likely to accept and continue using contraception when they have access to high quality services.

To draw attention to the quality of family planning services in developing countries, Bruce and Jain developed a quality of care framework. Researchers at the Population Council have used the framework to guide them in their examination of service delivery sites in Kenya, Burkina Faso, Côte D'Ivoire, Ghana, Nigeria, Tanzania, Zaire, and Zimbabwe. The data collected in these "Situation Analysis" surveys demonstrate that, although the very word "quality" implies a certain arbitrariness, a systematic description of quality of care can be achieved. And, if quality can be described, it can be monitored and improved.

The surveys revealed serious problems with some aspects of care. The lowest

IUD experience in Taiwan and sterilization in India.

- Availability of multiple methods improves continuation of contraceptive use. This assertion is based on follow-up studies in Taiwan and Matlab, Bangladesh.
- Contraceptive prevalence depends upon the number of methods available at multiple service points in a country.

The fourth assertion is based on the work of demographers Lapham, Mauldin and Jain. The relationship between contraceptive prevalence (CPR) and the availability score (AVAIL) for 72 developing countries implies that an increase of one point on the availability score, on average, is associated with an increase of about three percentage points in the use of contraception. In other words, the widespread addition of one method (four points on the availability score) to the choice of methods available in a country would be associated with an increase of about 12 percentage points in the use of contraception. Controlling for the level of a country's socioeconomic development reduces the effect of

scores were consistently given to indicators measuring communication with clients. Providers typically fail to discuss contraceptive side effects with clients, although the lack of such information may lead women to discontinue use of the method if a problem develops. In Tanzania, for example, only one quarter of clients were told about potential side effects; and fewer than ten percent were told how to deal with side effects. Frequently, providers do not ask which contraceptive method clients prefer or about physical conditions which may contraindicate the use of certain methods. For example, fewer than one-third of clients in Ghana, Nigeria, and Tanzania were asked whether they were breastfeeding, yet may have been given estrogen-containing birth control pills, which undermine lactation. Providers rarely inquire if new clients are experiencing pelvic pain or unusual vaginal bleeding or discharge, which could indicate that serious reproductive health problems are not being detected and treated.

Communication between provider and client is an important indicator of the quality of family planning services. The potential benefits of providing and eliciting adequate information are large, both in terms of reproductive health and in terms of attitudes towards and use of contraceptive services.

There is a longstanding debate in the population field about whether and how "supply-side factors," including quality of services, contribute to a decline in fertility. However, in the many countries in which a basic service system is in place, the marginal cost of improving its effectiveness would seem to be in the client's as well as the program's interest.

Barbara S. Mensch is an associate at the Population Council.

availability on contraceptive use from 12 to six percentage points; but it still remains highly significant.

This statistical relationship does not take into account the extent to which it would be feasible to add methods to the delivery system in poor countries. It also does not imply that the addition of each successive method will continue to increase the use of contraception under all circumstances.

Whereas the proven positive effects of follow-up mechanisms are fairly unimpressive because the literature is thin, the negative consequences to program performance of a failure to promote continuity of care can be convincingly modeled. One such model contrasted results of a high acceptance-low continuation program with a low acceptance-high continuation program. The latter scenario had a markedly greater impact on contraceptive prevalence, which suggested that programs would have a much greater impact if, rather than putting undue emphasis on recruitment, they took better care of the users they already had. It is important to point out that programs should not try to improve continuation only through long-acting contraceptive or irreversible methods, because dependence on any one such method is unlikely to be adequate to meet individual fertility goals or to achieve societal goals of fertility reduction.

In summary, the empirical information that links quality of care elements with clients' knowledge, satisfaction, contraceptive use, fertility and health, though limited, strongly suggests that improvements in the quality of services will result in a larger, more committed clientele of satisfied contraceptive users. Over the long term, this expanded base of well-served individuals will translate into higher contraceptive prevalence and, ultimately, reduction in fertility. Within private and commercial programs, where clients provide all or partial cost-recovery, the laws of the marketplace suggest that better services at the right price will attract more patrons. Within publicly supported programs, both clinic and community-based, it is likely that improvements in the quality of services will result in greater initial acceptance and more sustained use.

JODI L. JACOBSON

Abortion and
the Global Crisis in
Women's Health

·····················

Induced abortion is the most widely practiced form of family planning in the world. Each year on average, an estimated 50 million women worldwide resort to abortion to prevent unwanted births[1] More than 20 million of those abortions are illegal procedures, the majority of which are either self-induced or performed by untrained providers under unsanitary conditions.[2] Complications from unsafe abortions are leading contributors to illness and death from reproductive causes worldwide.

Access to safe abortion services, therefore, is a critical determinant of women's human right to control the number and spacing of births,[3] and of women's ability to safeguard their health and their lives.

The majority of women throughout the world, however, remain without access to safe, affordable abortion services. As a result, they increasingly find themselves caught between the forces pushing down fertility rates — attempts by government and international agencies to reduce birthrates, economic and social pressures for smaller families, and women's own fertility desires — and their inability to regulate fertility safely and effectively.

THE CONTRACEPTION CONNECTION

Based on available data about legal procedures, abortion appears to rank fourth in terms of birth control methods used, behind female sterilization, intrauterine devices (IUDs), and oral contraceptives. Use of

Jodi L. Jacobson is the director of the Health and Development Policy Project.

these other methods, however, is heavily concentrated in China, India, Western Europe, and the U.S., whereas induced abortion is practiced in every country.[4]

The prevalence of abortion in a given country is partly a reflection of women's access to contraception, as measured by, among other things, the degree of difficulty in obtaining the necessary information and supplies, the availability of methods to meet their personal and health needs, and their ability to negotiate contraceptive use with their partners.

In Poland, for example, many women wish to limit their childbearing, but only about 12 percent of the population at risk of pregnancy is using reliable birth control. Fertility regulation is severely hampered by lack of sexuality education and information, high prices and inadequate supplies of contraceptives, and the censure of contraceptive use by the Catholic church. Yet even with a very restrictive abortion law, Poland has one of the highest rates of induced abortion in all of Europe.[5]

But not even diligent use of contraceptives can eliminate the need for safe abortion. Assuming they wanted no more than two children, seven out of 10 women using a 95 percent effective method of birth control would still have to undergo at least one abortion in their lifetimes to achieve their desired family size.[6]

LIMITED ACCESS

On the basis of an analysis of legal status alone, the conclusion could be drawn that access to safe abortion services has increased rapidly over the past two decades. Looking past the laws at policies and practices, however, reveals that the costs and consequences of unsafe abortion actually are on the rise.

Abortion laws usually are grouped according to "indications," or circumstances under which abortions can be performed. These categories are broad, representing a diverse set of statutes. Countries with the narrowest laws either completely ban abortions or, with indications known as "life endangerment," restrict them to cases where pregnancy poses an immediate risk to the mother's life.[7]

Some laws allow abortions in case of rape or incest. Others consider risks to physical and mental health. Still others permit abortion of a severely impaired fetus. Some societies allow abortion for what are known as "social" reasons, as when parents determine an additional child will bring undue burdens to an existing family. The broadest category is recognizing contraceptive failure as a sound basis for abortion, or allowing procedures on request (usually within the first trimester).

Liberalization of abortion laws began in the 1950s, as recognition of the need to reduce maternal mortality and increase reproductive choices became widespread. Liberalized abortion laws have achieved many of these goals. France, Tunisia, the United Kingdom, and the U.S. are a few examples of countries where the relative number of births due to unintended pregnancies and deaths due to illegal abortion fell following liberalization. For instance, abortion-related mortality among U.S. women fell from 30 per 100,000 live births in 1970 to to five per 100,000 in 1976.[8]

In theory, about 40 percent of the world's women now have access to induced abortion on request.[9] The key phrase here, however, is "in theory." Changes in laws are a necessary but far from sufficient condition to create universal access to safe, affordable abortion services. Because social ambivalence about abortion is widespread, what happens in practice often does not reflect the laws on the books.

In many countries women find it difficult to exercise their legal rights to obtain abortion. The reasons for this difficulty include stringent medical regulations, burdensome administrative requirements, lack of information or referral networks, lack of trained providers, extreme centralization of services, and local opposition or reluctance to enforce national laws. Other variables include the barriers posed by lack of personal resources, particularly money. These conditions exert an equal and sometimes more important influence on women's access to abortion than do laws.

In the U.S., for example, access to abortion services through the first trimester is guaranteed by law. However, states are now able to impose restrictions, such as 24-hour waiting periods. Because they may require two or more visits to a distant clinic, such restrictions make it more difficult to obtain an abortion, especially for working women. Access is also determined in large part by income level. The majority of states deny low-income women the public funding necessary for them to obtain abortions. And more than three-fourths of U.S. counties have no abortion provider, further complicating access to abortion for women without the means to travel.[10]

Access may also be limited by medical regulations governing how, where, and by whom abortion services can be provided. In most countries with liberal laws, abortions must generally be performed by licensed providers (though not necessarily physicians), a regulation both the intent and effect of which is to protect public health. Some countries take this one step further, by requiring that abortions be carried out only in designated hospitals or centers, or by highly trained specialists.

Often, such restrictions succeed only in delaying abortion until later stages of pregnancy — when the procedure poses a greater risk to a woman's life and health and is more expensive and technically complicated to perform. "First trimester abortions carried out by well-trained practitioners carry a very low risk of complications for the woman," states the International Medical Advisory Board of the International Planned Parenthood Federation. But "beyond 10 weeks of gestation the health risks of abortion rise with each week of pregnancy, the risks of late trimester abortion being three to four times those of the first trimester."[11]

Such contradictions in law and practice are universal. Few legal abortions are performed under the health exception in the countries of Latin America and Africa, for example, even where these indications technically apply under the law. Legal abortion rates under the health indication in Israel, New Zealand, and South Korea, by contrast, are comparable to those countries allowing abortion on request.[12] Fear of imprisonment under Poland's new and more restrictive abortion law is leading doctors to refuse to examine women even when they meet the law's criteria, reports Wanda Nowicka, President of the Federation for Women and Planned Parenthood in Poland.[13]

And legal barriers remain a major hurdle for a large share of the world's women. One in four of the world's women lives in a country where abortion is against the law except to save the mother's life, or where it is totally banned. Yet as the mounting toll of unsafe abortion makes plain, it is the number of maternal deaths and illnesses, not the number of abortions, that is most affected by legal codes.[14]

SAFE ABORTION SAVES WOMEN'S LIVES

Throughout the countries of the South, the lifetime risk of maternal death is between 80 and 600 times higher than in Western Europe and the U.S. Each year, according to the World Health Organization (WHO), at least a half-million women die from pregnancy-related causes, 99 percent of them in the developing world. For every woman who dies, several others suffer serious, often lifelong health problems — among them hemorrhaging, infection, abdominal or intestinal perforations, kidney failure, and permanent infertility. Complications of unsafe abortion are leading contributors to these illnesses and deaths.[15]

Based on available data, several researchers have attempted to calculate the number of abortions that take place worldwide each year. Estimates indicate that from one third to half of all women of reproductive age undergo at least one induced abortion. According to calculations by Stanley Henshaw, deputy director of research at the Alan Guttmacher Institute, some 36 to 51 million abortions were performed worldwide in 1987. He estimates the annual number of illegal procedures at over 20 million annually. Other estimates put the total number at between 40 million and 60 million. Either set of figures implies that there is close to one induced abortion for every two to three births worldwide.[16]

Comparative regional estimates of the number of unsafe abortions and maternal deaths are difficult to come by, mainly due to lack of resources allocated toward this end. The available data are generally drawn from hospital or community-based studies that offer but a fragmented picture of the real situation. In Brazil, for example, studies indicate that 20 to 35 percent of all pregnancy-related deaths are due to complications of unsafe abortion. Yet because of legal restrictions, bureaucratic indifference, and social disapproval, abortion in Brazil is largely an undocumented and clandestine activity.[17]

The same holds true across Latin America. "Data on the extent of induced abortion in Latin America are inconsistent," writes John Paxman, adjunct professor of Health Services at Boston University, and his colleagues, "...Substantial underreporting is a major problem. Surveys based solely on hospital data or small, but focused, surveys tend to produce gross underestimates of induced abortions, because accurate reporting is discouraged by the nature of the subject and the general illegality of [unsafe abortions] leading to hospitalization."[18] Data for most countries of Africa and Asia are also difficult to obtain.

WHO studies in various settings indicate that the share of maternal deaths caused by induced abortion ranges from seven percent to more than 50 percent. On average, between 20 and 25 percent of maternal mortality is attributable to illegal or clandestine abortion. In Latin America, complications of illegal abortion are thought to be the main cause of death in women between the ages of 15 and 39.[19]

Abortion-related deaths are estimated to reach 1,000 per 100,000 illegal abortions in some parts of Africa, as opposed to less than one death per 100,000 legal procedures in the U.S. Hospital admissions in African cities, virtually the only available indicator of abortion trends, are rising in tandem with reliance on abortion as a method of birth control. Khama Rogo, a medical doctor and researcher studying abortion, notes that in East and Central Africa at least 20 percent of all maternal deaths are due to complications of induced abortion, and that the share has reached 54 percent in Ethiopia.[20]

Hospitals in many developing countries are literally inundated with women seeking treatment from complications of unsafe abortion. Over 30 percent of the beds in the gynecological and obstetric wards of most urban hospitals in Latin America are filled with women suffering abortion complications.[21] At Mama Yemo Hospital in Nairobi, Kenya, some 60 percent of all gynecological cases fall into this category. And at a hospital in Accra, Ghana, between 60 and 80 percent of all minor surgery performed related to the aftereffects of illegal abortions in 1977; half that hospital's blood supply was allocated to related transfusions.[22]

Complications from illegal abortions drain scant health resources. A study of 617 women suffering abortion complications in Zaire, for example, found that 95 percent required antibiotics, 62 percent required anesthetics, and 17 percent required transfusions.[23] In a similar study in Latin America, demographer Judith Fortney concluded that illegal abortions require, on average, "two or three days in the hospital, 15 or 20 minutes in the operating room, antibiotics, anesthesia, and quite often a blood transfusion.[24] In many hospitals, each of these resources is relatively scarce and their use for abortion patients may mean that other patients are deprived." In 1987, the estimated cost of treating 9,440 spontaneous abortions in Chile's National Health System, based on the average stay of 5.9 days per patient, was $1.4 million.[25]

Restricting access to safe abortion services also increases the financial burdens on low-income women and their families. Procuring a clandestine abortion can be expensive, but for those women who suffer complications, the costs are higher still. In Thailand's rural Chayapoom province, women suffering complications of illegal abortion severe enough to require hospitalization lost an average of 12 days of time from their normal activities; those whose complications did not require a hospital stay still lost six days.[26]

WHY WOMEN CHOOSE ABORTION

Broadly speaking, abortion rates are governed by the cultural and economic pressures on family size in a given society in concert with the mix of laws and policies that determine access to family planning. The numbers of illegal versus legal procedures in a given country, the degree to which pregnancy termination is used to regulate fertility, and the demographic makeup of the groups relying most heavily on this method are all shaped by the circumstances of individual women, social and economic pressures to limit or delay childbearing, the availability and reliability of contraceptives, and by the legal, cultural, and political climate that surrounds abortion services.

In Latin America, abortion rates have been consistently high for over two decades, despite restrictive laws and the firm opposition of the Catholic church. Indeed, there is evidence of a long tradition of induced abortion in the region. In 1551, the King of Spain was notified that the indigenous population in his Venezuelan colony practiced induced abortion, through the use of medicinal herbs, to prevent their children from being born into slavery.[27]

During the 1970s, the International Planned Parenthood Federation (IPPF) estimated that abortion rates in Latin America and the Caribbean were higher than in any other developing region: an estimated one-fourth of all pregnancies in Latin America were intentionally aborted in that period, compared with less than 10 percent in Africa and 15 to 20 percent in South and Southeast Asia.[28]

Fertility rates have fallen since the 1960s, but there is evidence of desire for even smaller families throughout the region. Data from the World Fertility Survey in the 1970s showed that while the average family had at least four children, over half the women interviewed wanted only two to four. In every Latin American country except Paraguay, over half the women with three children wanted no more children.[29] Because political and religious opposition has kept contraceptive availability to a minimum, the number of illegal abortions remains high and shows no sign of diminishing. Paxman and his colleagues cite estimates of the number of induced abortions in Latin America each year ranging from 2.7 million to 7.4 million, which represents roughly 10 to 27 percent of all abortions performed in the developing world.[30]

Today, induced abortion accounts for about one-fourth of total fertility control in Latin America. Although the use of contraceptives has increased steadily since the 1960s, it is still at a relatively low level and supplies are unevenly distributed throughout the region. Access to services is unequal: those most at risk of unwanted pregnancy — teens, single women, and low-income women — are those for whom contraceptives and safe abortion services are most out of reach.

A study of 602 women seeking treatment for incomplete abortion at a clinic in Bogota, Columbia reveals some of the issues and conditions confronted by women in preventing unwanted pregnancy and unwanted births. The study, conducted by Margoth Mora and Jorge Villarreal of Orientame, a non-profit organization providing high-quality reproductive health services to women in Columbia, revealed many roadblocks to higher levels of contraceptive use among women.[31] For example, women's knowledge of their own reproductive physiology — a precursor to successful contraceptive use — was low. Seventy-one percent of the women knew how a pregnancy starts, but only half knew the period when there was a greater likelihood of conception, and only 10 percent understood the process of menstruation. Partner attitudes were also problematic. Of the women's partners, 39 percent thought that women should be responsible for contraception; 17 percent thought women's use of contraception was synonymous with infidelity; and eight percent thought women should not use contraceptives at all. "These deeply rooted male attitudes can lead to contradictory behavior that is harmful to women," state the authors. "On one side, women are made to feel responsible and guilty for becoming pregnant, and on the other they are made to feel they should not use contraceptives to avoid pregnancy."[32]

A picture similar to that of Latin America is developing in Africa, where the number of induced abortions and the related health and social costs of illegal or clandestine procedures are likely to continue to rise throughout this decade. The predominantly young population is characterized by high fertility and low rates of contraceptive use. Access to both contraceptives and safe abortion services is limited geographically and by income. Although desired family size in Africa remains high relative to other regions, the desire to limit family size is growing.

Social and cultural limitations on women in Africa are an equally important factor in the abortion equation and may be much harder to change than laws. Nolwandle Nozipo Mashalaba, a private family practitioner in Botswana, sees the lack of communication between African couples on matters of sexuality and the desire to maintain male dominance as the primary roadblocks to reducing pregnancies that women themselves may not want. Mashalaba notes that where "men migrate...for work, they keep the wife in a continuous state of pregnancy and lactation as a way to keep her (possible) infidelity to a minimum."[33]

WOMEN'S STATUS AND UNWANTED FERTILITY

There are many reasons why women suffer unwanted and personally untenable pregnancies. Contraceptive failure and lack of access to or knowledge of suitable methods are among them. Gender inequity and discrimination are others. Millions of women lack the power to freely determine at what stage they become sexually active, whom they are bonded to, when sex will take place, or when and how to bear children.

Women may be forced into unwanted sexual contact and unwanted pregnancies due to fear of domestic violence, under sexual coercion, or through arranged and often forced marriage. Even where the means to prevent pregnancy are available to women, lack of spousal support often leads to high rates of contraceptive failure. Unwanted pregnancy not only limits personal autonomy, but also impedes the struggle for women to become equal partners in society and efforts to improve health among women and children.

Access to safe abortion plays an integral role in allowing individuals to meet their fertility goals without compromising their health and enables countries to more rapidly make the transition from high to low fertility. The tremendous social gains to be reaped from eliminating illegal abortions cannot be ignored. Reducing unsafe abortions and unwanted pregnancies would save billions in social and health care costs, freeing these resources for other uses. Reforming restrictive laws and committing resources to safe services may stir opposition. But failure to do so exacts an enormous emotional and economic toll on society — and sentences countless women around the world to suffering and death.

CHRISTOPHER J. ELIAS

AIDS: An Agenda for Population Policy

....

The World Health Organization estimates that, in 1994, approximately two million people will become infected with the Human Immunodeficiency Virus (HIV) — the organism that causes the Acquired Immune Deficiency Syndrome (AIDS).[1] The vast majority of these infections will occur in less developed countries, where the means of transmission is primarily through heterosexual intercourse and roughly equal numbers of men and women will be affected. Women are the fastest-growing population of persons living with AIDS and the HIV infection. In some U.S. cities, AIDS has already become the leading cause of death among vulnerable groups of women.[2] In highly endemic regions, such as sub-Saharan Africa, it is common to find infection rates of greater than 20 percent among women of reproductive age.[3]

The global epidemic of HIV — with its rapid spread and stunning fatality — has attracted the attention of policymakers from virtually all sectors and all parts of the world. The impact of AIDS on development has been assessed by economists and national security advisors, as well as health and disease control officials. Their consensus is that, in addition to causing an incalculable amount of human suffering, AIDS poses a serious threat to sustainable development and to the global economy.[4] Clearly, this epidemic demands attention in the deliberations in Cairo at the IVth International Conference on Population and Development. However, in order to identify an effective and timely

Christopher J. Elias is an associate at the Population Council.

Priorities in Contraceptive Research: The Case for Barrier Methods

Judy Norsigian

Current epidemics of sexually transmitted diseases (STDs), especially AIDS/HIV infection, should change prevailing views of what constitutes the "ideal" birth control method. Because many couples find it burdensome to use two separate methods for STD prevention and birth control, the best method would offer both kinds of protection. To a great extent, barrier methods — condoms, in particular — have that advantage.

Women's health activists and consumer groups have long emphasized the need for improved barrier methods. In contrast, the major organizations conducting contraceptive research, as well as their funding agencies, have given highest priority to long-acting methods that usually involve a minimum of user control. Examples of the latter include implantable and injectable progestins (Norplant and DMPA) and several types of contraceptive vaccines. Although there has been some increase in funding for barrier-method research in recent years, vaccines and other long-acting methods still command the lion's share of research funding.

Why do women's health advocates urge greater emphasis on barrier-method research? First, as noted above, some barrier methods provide protection against STDs, as well as against pelvic inflammatory disease (PID). Unlike the intrauterine device (IUD) and hormonal methods, barrier methods cause very few side effects or serious health problems. Moreover, use of barrier methods usually requires a woman to learn some basic information about her body, which is often an important first step towards better sexual relation-

response, it is essential to consider the context in which this deadly epidemic has emerged.

HIV infection is just one of several dozen sexually transmitted diseases (STDs) that cause significant morbidity and mortality among men and women throughout the world. Recent estimates indicate that approximately 250 million new sexually transmitted infections are acquired annually.[5] For a host of anatomical and biological reasons, women are particularly susceptible to the ill effects of sexually transmitted infection. These infections are often asymptomatic or, at best, vaguely symptomatic among women. This fact, combined with the social stigma associated with "venereal disease" and the general inadequacy or inaccessibility of health services for women in many settings, leads to poor recognition and treatment. As a result, women with untreated STDs often suffer from infertili-

ships and more effective fertility control. And, unlike the IUD or other methods that can only be removed by a medical provider, barrier methods are readily reversible.

Some women's health advocates are concerned about the paucity of data on long-term health effects of implants and the potential for irreversible problems with vaccines. Moreover, long-acting, provider-controlled methods such as vaccines and implants may present opportunities for abuse in settings where women's rights to decent treatment and choice are not recognized. On the other hand, these methods offer several advantages, such as long-term efficacy without repeated administration, ease of distribution and storage, and cost-effectiveness. Also, because their use is separated from the act of sexual intercourse, vaccines and similar methods may be used with secrecy — an advantage for women who have little control over when and how they will have sex. However, these methods do nothing to change the conditions which limit women's control over their sexual relationships.

Thirty years ago, researchers hoped for the development of a "magic bullet" contraceptive — one that was perfectly safe, effective, and easy to use. Such expectations now seem unrealistic. Given the complexity of sexual relationships and the numerous factors that determine why, when, and how people use contraceptives, even the existence of a near-perfect method would not guarantee that women will be able to control whether, when, and with whom they become pregnant. However, better methods will certainly help. And in the context of the growing STD epidemic, barrier methods clearly deserve greater attention and a larger share of both research and service delivery dollars.

Judy Norsigian is co-director of the Boston Women's Health Book Collective

ty, ectopic pregnancy, chronic pain, genital tract cancer, and complications of pregnancy and delivery. Sexually transmitted diseases are also a frequent cause of congenital infection and neonatal morbidity and mortality. Among newborns, for example, eye infections resulting from gonorrhea acquired during delivery remain an important cause of preventable blindness.

Unfortunately, prior to the emergence of AIDS, efforts to prevent and control STDs were among the most poorly financed of all health sector activities. As a consequence, efforts to combat the spread of HIV began with a serious disadvantage. There were few well developed and proven primary prevention programs and clinical services for diagnosing and treating infection, and those were often exclusively targeted to high-risk groups, such as prostitutes.

There are several reasons why it is important to address the full spectrum of sexually transmitted infections. First, primary prevention strategies, such as those which encourage people to reduce their number of sexual partners and use condoms, are broadly effective in lowering the risk of all STDs, including HIV. Secondly, infection with other STD organisms at the time of exposure to HIV has been shown to significantly augment the likelihood of the virus' successful transmission. Ulcerative genital infections, such as chancroid and herpes, in particular, greatly increase the risk of HIV transmission. Recent information also implicates the more common, non-ulcerative STDs, such as trichomoniasis, as cofactors that amplify the spread of HIV. Consequently, efforts to prevent and treat other STDs have become an important strategy for the primary prevention of AIDS.

Just as AIDS must be addressed as one of many sexually transmitted diseases, STDs are best understood in the larger context of reproductive tract infections (RTIs). RTIs are of three general types: sexually transmitted diseases — including gonorrhea, chlamydia, and HIV infection; endogenous infections caused by overgrowth of organisms normally present in the vagina, such as yeast infections and bacterial vaginosis; and iatrogenic infections that are associated with medical procedures, such as abortion or insertion of intrauterine devices (IUDs). Through the advocacy of the International Women's Health Coalition and other concerned agencies, greater attention has been brought recently to the issue of RTIs.[6]

A broader concern for RTIs, as opposed to a more narrow focus on STDs, is preferable because it more accurately reflects women's need for reproductive health services. Numerous studies conducted in both clinics and communities indicate that RTIs are extremely common and most often go unrecognized. For example, in a recent household survey in two rural villages in Egypt, investigators found that approximately half of the women had one or more RTI.[7] These infections result in a tremendous burden of morbidity and mortality for the world's women.

POPULATION POLICY: MEETING THE CHALLENGE

How can population policy best respond to these challenges? The departure point for a more effective response begins with recognition of the need to focus our attention more broadly on the reproductive health of women and men. Highly categorical programs aimed exclusively at averting unwanted births have left a legacy of family planning programs that are narrowly focused on contraceptive delivery. This singular emphasis has impeded the adoption of a broader reproductive health agenda that would include attention to sexually transmitted disease, including AIDS.

We must remember that, for many women throughout the world, family planning clinics present one of the few opportunities to obtain medical services. The future challenge for family planning programs, therefore, will be to expand their constellation of services to meet a broader range of women's reproductive health needs without compromising the quality of existing contraceptive delivery. This will require a significant investment of additional resources, as well as careful program design and evaluation to identify the most effective service models.

In addition to providing a broader range of reproductive health services, population policy must re-examine the priorities for contraceptive technology development in light of the global epidemic of STDs and AIDS. We are currently in an unfortunate situation where our most effective contraceptive technologies provide no protection against sexually transmitted infection. Some technologies, such as the IUD, may actually increase the complications of such infection by facilitating the spread of infection to the upper reproductive tract. On the other hand, the most effective means for avoiding sexually transmitted infection — condoms and spermicides — have a relatively poor track record as contraceptives. Urgently needed are improved means of barrier contraception that provide enhanced protection against both infection and unwanted pregnancy. Optimal use of barrier methods will require not only the development of better technology, but also the reorientation of provider attitudes toward these methods — attitudes which have historically characterized these methods as "less effective" based solely on their contraceptive efficacy. For some women the need for protection from STD infection may be as great or greater than the need for effective contraception. Recently, there has been considerable interest expressed in the development of vaginal microbicides as a method to improve women's ability to reduce their risk of infection by HIV and other STDs.[8,9]

A final implication of the AIDS epidemic for population programs regards the need to extend the reach of our efforts to better include men and youth. The impetus for this action obviously comes not just from an enhanced awareness of the threat of STDs. As indicated in other chapters of this book, thoughtful reflection on the dynamics of unmet demand, abortion, and gender violence all suggest the need for greater male involvement in education efforts and service delivery. Given that the most effective means of preventing the further spread of STDs and AIDS currently available (e.g. condoms)requires the full and dedicated participation of the male partner, a broader reach may be essential to the ultimate success of prevention programs. Similarly, young people throughout the world are among those at greatest risk of acquiring a sexually transmitted infection, and yet, they are often entirely left out of traditional family planning services.

In summary, the rapidly expanding epidemic of the HIV infection and the associated morbidity, mortality, and economic devastation of

AIDS have brought renewed attention to the serious threats posed by sexually transmitted diseases. An adequate response to STDs and AIDS requires that population policymakers expand both the reach and the scope of the services offered by family planning programs and re-examine the priorities for contraceptive technology development. These issues must be part of the agenda as we debate the next decade of population and development policy.

R U T H   M A C K L I N

Ethical Issues in Population and Reproductive Health

....

Programs designed to slow population growth raise a number of ethical questions — both about the objectives of such programs and the methods employed to achieve them. Even among those who endorse the goals of population programs there is concern about methodology. Indeed, the history and ongoing practices of many population programs illustrate how efforts to reach an ethically acceptable goal can go astray through the use of ethically impermissible means. However, it is possible to implement ethically sound population programs, by focusing on the reproductive health of women and couples rather than on narrow demographic goals.

In essence, ethics has to do with the way human beings treat one another. Although different nations, cultures, and religious or ethnic groups may adhere to different norms of behavior, individual behavior and social practices can be measured against general ethical principles. Whether those principles are universal, applying to all societies at all times in history, is a matter of ongoing debate. But without ethical principles to serve as an ideal to strive for, there could neither be a basis for moral judgment nor a concept of moral progress.

Our ethical analysis of population programs begins with the premise that slowing the rate of world population growth is a desirable end. Whether it is for the sake of citizens of developing countries, where rapid population growth is outstripping economic development and

Ruth Macklin is a professor of bioethics in the Department of Epidemiology and Social Medicine at the Albert Einstein College of Medicine.

natural resources, or to forestall global environmental degradation, few deny that a slower rate of population growth would be beneficial. Indeed, some contend that slowing population growth is not only permissible but ethically obligatory. That perspective reflects the common-sense ethical notion that right actions and practices are those that result in a balance of good consequences over bad. If reducing population growth will have the likely outcome of lessening pollution of air and water, diminishing the rapid consumption of nonrenewable resources, ensuring adequate food supplies, and improving the quality of life for those who live in poverty, then surely those good consequences would justify efforts to slow population growth — particularly when weighed against the grim alternatives that may result if population growth continues at its present rate.

THE PRINCIPLE OF BENEFICENCE

The ethical principle that supports this conclusion is known as *beneficence*. Right actions, practices, and policies are those that tend to bring about a balance of good consequences over bad, or of human welfare over "illfare." To apply the principle, it is first necessary to come to agreement on which consequences of actions are good and which are bad. Fortunately, there is little room for disagreement over whether poverty, famine, air and water pollution, overcrowding, and ill health are good or bad consequences. However, there is disagreement about whether reducing population growth would bring about these desirable ends, and about whether family planning programs are the most effective means to reduce population growth. For the purposes of this discussion, however, we will accept the widely held belief that slower population growth rates would produce more good than harm, and that family planning programs are an important component of efforts to reduce fertility and slow population growth.

Like any ethical principle, *beneficence* requires interpretation and careful thought in order to apply it to actual situations. To focus population programs solely on reducing fertility rates is to ignore some consequences in favor of others. In practice, population policies and programs have tended to ignore the broader value of health — in particular, women's health — in their use of targets, quotas, disincentives, incentives, and other means selected to achieve the desired end. A proper application of the principle of *beneficence* must take into account not only the consequences of reducing population growth but also the additional beneficial consequences that could result from an alternative approach. A program or policy that achieves an improvement in health while lowering fertility rates is ethically superior to a policy that focuses only on reaching demographic targets.

China's family planning program is an infamous example of the narrow, target-driven approach. For example, the rule in China's family planning program specifies that after having two children all women are to be sterilized. This practice is carried out even among women using the IUD — and all women are told to have an IUD inserted after having one child. One Chinese physician charged that this policy is wrong on two counts: first, it is medically contraindicated as "unnecessary surgery," which places women at unnecessary risk; second, the practice causes psychological harm to women. Clearly, in this respect the Chinese program does not meet the demands of the ethical principle of *beneficence*.

RESPECT FOR PERSONS

Another objection to population programs stems from violations of the ethical principle known as *respect for persons*. This principle may apply to "persons" as individuals or as members of a group, such as "women who have borne more than two children." At both levels, the *respect for persons* principle has been violated by population control programs in many countries.

What does *respect for persons* require? Stated in the most general terms, it requires that human beings not be treated as mere means to serve the ends of others. It requires that people are not coerced by those having greater power or authority unless there is an imminent danger of threat of harm to others. Recognizing the dignity of persons means not treating them as instruments in the service of a goal, such as a state-imposed quota of IUD insertions. In the medical or health-care setting generally, *respect for persons* requires that before physicians invade the bodies of patients — even for the patients' own good — they must first inform the patient of what is to be done and obtain the patient's freely granted permission to proceed. That is the ethical and legal doctrine of informed consent.

In the context of family planning, *respect for persons* clearly requires informed consent before sterilization, insertion of IUDs or Norplant, or administration of other long acting contraceptives. *Respect for persons* recognizes the right of individuals to control their own reproductive lives, without interference by the state or by other powerful agencies or individuals. When an ethical concern is framed in the language of rights, it signals a human value of overarching importance.

Unfortunately, identifying the reproductive rights of individuals and couples is only the first step toward ensuring the exercise of those rights. For example, Article 4 of the Mexican Constitution asserts the right of each individual or couple to decide freely and responsibly on the number and spacing of their children. But, in reality, the rights of Mexican women and their doctors are trampled in the effort to meet demographic goals set by the Mexican government.

At a recent meeting of Mexican health workers, two physicians described the ethical dilemmas they face in their work.¹ During their medical training, the doctors were told to insert an IUD in every woman who has delivered three children. Since many women might refuse the IUD if asked, physicians typically insert the IUDs without informing their patients. The patients are not the only ones deprived of choice; one group of doctors was told that if they did not insert the IUDs, they would lose their jobs. Government-imposed quotas thus deny the personal autonomy of women and compromise the professional autonomy of physicians.

Perhaps the clearest example of violations of *respect for persons* is the program implemented in India in the years 1969-74 and continuing on into the late 1970s. The Indian government adopted a strenuous policy of reducing the birth rate and with the added boost of external aid from the U.S. Agency for International Development (USAID) and other sources, family planning received the highest priority. To achieve the predetermined goals, mass "sterilization camps" were set up and hundreds of sterilizations per "camp" were carried out each day.² According to one report:

> ...service providers were threatened with punitive measures for non-fulfillment of targets....Not surprisingly, program implementation included measures that blatantly violated individual rights and well-being. Not only women, but a large number of men were the victims of compulsory sterilization during the period of political emergency in 1975-77.³

The *respect for persons* principle prohibits the state from compelling a pregnant woman to have an abortion. Similarly, it prohibits the state from preventing a woman from having an abortion. Although abortion remains a controversial topic throughout the world, this ethical principle supports granting women the right to control their reproductive lives by allowing them access to safe, legal abortion.

Respect for persons refers to all and only those human beings whom no one could deny are persons, that is, children, women, and men who are already born. Although there is considerable debate about the point in prenatal life at which personhood can be said to begin, no controversy exists over the status as persons of living adults and children. (A different sort of debate surrounds the status of comatose individuals and anencephalic infants. However, that debate is related to the permissibility of terminating the life of individuals who are already born but who never had or have lost all mental capacity.) Although some believe that a fetus is a person from the moment of conception, a far greater number of people and religious groups do not adhere to that view. Other points along a continuum at which personhood has been held to begin are: implantation of the fertilized ovum; the onset of genetic individuality; the time of "ensoulment" according to different religious traditions; the appearance of electroencephalographic activity (brain waves); quickening (fetal movement felt by the pregnant woman; fetal viability; birth; and some point after birth perhaps as late as the end of the first year of life. Given this array of different views on the beginning of personhood and the arguments in support of them, it is arbitrary to select fertilization as the only possible correct interpretation of what *respect for persons* entails.

It must be emphasized, however, that ethical objections to laws that prohibit abortion do not derive solely from the *respect for persons* principle. An array of bad consequences results from the continued prohibition of legal, safe abortion in many countries. It is estimated that between 100,000 to 200,000 of the annual deaths worldwide from pregnancy-related causes are due to improperly performed and usually illegal abortions.[4] [See "Abortion and the Global Crisis in Women's Health" by Jodi Jacobson, page 177.] These figures demonstrate without question that prohibition of legal abortion, which denies women the opportunity to obtain medically safe termination of pregnancy, results in negative health consequences of such severity and magnitude as to outweigh any good consequences that could flow from keeping such laws in force.

While the principle of *respect for persons* is clear enough in theory, it can be difficult to apply in practice. If it is wrong to force women to undergo sterilization or to have abortions, is it also wrong to use disincentives or incentives to encourage or discourage those practices? Is it wrong to punish couples for having more children than the number mandated by the state? When family planning programs give money or other material goods to impoverished women who agree to undergo sterilization, is that not coercion, and therefore wrong? How can we determine when an individual's response to incentives or disincentives is truly voluntary and when it is coerced? No formula will suffice to cover all cases, but a general rule of thumb can serve as a guideline. If outright coercion is a violation of *respect for persons*, so too is any penalty or incentive that is the equivalent of coercion. If an amount of money offered to a poor peasant woman is an offer she cannot refuse, it is tantamount to coercion. This conclusion requires drawing the line, always a difficult task. But the fact that it is difficult to specify the precise point at which an incentive becomes coercive does not mean that it is hopeless or misguided to attempt to locate and justify that point.

According to this view, incentives are more ethically permissible if they do not foreclose the right of individuals or couples to control their reproductive lives. Therefore, an offer of a reversible long-acting contraceptive might be acceptable, because it is less restrictive than, say, sterilization. Of course, on a practical level, a contraceptive is only reversible if women have access to safe and timely removal. Unfortunately, the history of both the IUD and Norplant bear testimony to the fact that it is easier to have a long-acting contraceptive inserted than to have it removed. The difficulty may stem from poor women's lack of funds, or from the unwillingness or inability of physicians to perform the removal. In either case, programs that offer incentives for reversible methods are ethically acceptable only if women have real access to removal. Thus the correct application of the *respect for persons* principle depends on a set of background circumstances as well as on features of the individual case.

Incentives which offer some benefit are likely to be less harmful than disincentives that take the form of penalties, and are therefore more ethically acceptable. Nevertheless, some types of incentives are ethically problematic for other reasons. For example, in China a system of incentives supporting the "one family, one child" policy denies certain benefits (access to the best educational opportunities) to subsequent children following the first child in a family. This system penalizes the child for the decisions and actions of the parents. In a hierarchy of ethical alternatives, it would be better to deny privileges to the parents themselves, although that, too, might be deemed an ethically unacceptable penalty.

From an ethical perspective, it is not sufficient for population programs to adhere to the *respect for persons* principle. It is also necessary to ensure *equal respect for persons*, which requires applying policies and practices equitably to men and to women, to poor as well as to wealthier citizens. Almost all programs that have sought to impose quotas of one sort or another have been implemented in a manner that viewed women as "targets" of population control (with the exception of the now-abandoned sterilization programs in some Indian states, which compensated men who elected to undergo sterilization).[5] A striking example is the demographic goal set in Nigeria to reduce the fertility rate from six to four births per woman by the year 2000. Since the society is polygynous, a man might have four wives and thus father 16 children. A population policy that imposes different quotas on men and women clearly violates the principle of *equal respect for persons*. A more even-handed approach would be more ethically acceptable and a lot more effective in achieving the goal of slower population growth.

JUSTICE

The concept of *equal respect for persons* is closely related to a third important ethical principle: *justice*. In the reproductive health context, *justice* relates to the distribution of family planning methods, including access to safe abortion in cases of contraceptive failure. The principle of *justice* mandates that all individuals — regardless of income, education level, or geographic location — should have equitable access to family planning and health services.

However, legal guarantees of access are not sufficient to ensure access to reproductive health services in societies that do not recognize a right to health care. And access is meaningless without widely available information about the existence and nature of the services. Governments have a moral obligation to ensure that couples have the information as well as the means to obtain family planning services.

These conclusions apply with particular force in the developing countries. Poor women everywhere bear a disproportionate burden from restrictive abortion laws and inadequate or nonexistent family planning services. *Justice* dictates not only that equal respect be shown to women, but also that the needs of the least advantaged members of society be addressed.

CONCLUSION

By focusing attention on reproductive health rather than on demographic targets, future efforts to reduce population growth could remedy the ethical problems associated with ongoing and past programs. Not only do reproductive health programs comply with the principles of *beneficence, justice,* and *respect for persons,* they are also likely to pay off in improved fertility reduction. Couples are more likely to continue using freely chosen contraceptive methods than those imposed on them by government policies inattentive to the needs of individual users. Promoting reproductive health is compatible with the goal of reducing population growth and is an ethically sound means of achieving that goal.

It might be argued that an appeal to ethical principles in a discussion of world population policies poses a problem of cultural and ethical relativism. Perhaps even more than in other areas of medicine and health care, the introduction of new practices related to human reproduction gives rise to ethical controversy stemming from social, cultural, and religious differences. [This issue is discussed in depth in "Population Policy and the Clash of Cultures," by Judith Lichtenberg, page 273.]

To resist ethically mandated change because of long held beliefs or practices is a philosophical error. The error lies in concluding that because a state of affairs has existed in the past, it ought to continue into the present and future. The flaw in that reasoning can easily be seen by reflecting on the fact that manifestly unjust social institutions, such as slavery and colonialism, would still be with us if history and tradition served as an infallible moral guide. Moral progress requires a critical evaluation of past practices and institutions. Of course, many social practices and institutions will withstand such critical evaluation, but others will not. Subjugation of women, denial of the right to self-determination in choosing an acceptable method of family planning, and lack of access to safe, legal abortion are practices that cannot withstand critical ethical evaluation.

LYNN P. FREEDMAN

Law and
Reproductive Health

·····················

You can't legislate good reproductive health. Nor has anyone ever conducted a controlled experiment to prove that changes in law will yield improvements in health. Yet virtually everyone who works in the reproductive health field would ultimately agree that law and health are closely connected, because law helps structure the social, political, economic, and cultural context in which reproduction takes place. Indeed, a good deal of social science and epidemiological research has been devoted to understanding these connections. For example, we know that women's status — as measured by such indicators as literacy, education, decisionmaking power within the family, control over income and property, and power over sexual relationships and bodily integrity — correlates positively with their ability to control fertility. Both independently and through the intermediate variable of fertility control, women's status also correlates positively with improvements in their own health and that of their families. Of course the converse is true as well: Women's ability to control fertility and to protect health contributes to advances in their status and overall well-being.[1]

But it is a mistake to see health as solely an outcome of individual behavior or status. Indeed, we know that the ability of individuals to effectuate improvements in health is inextricably tied to the ability and willingness of societies — of states, communities, and families — to meet basic needs on an equitable and non-discriminatory basis. This has been

Lynn P. Freedman is an assistant professor of clinical public health at the Columbia University Center for Population and Family Health.

studied most carefully in relation to infant and child health, but it is no doubt true of women's reproductive health as well.[2]

Taking these correlations between social factors and health as a starting point, the connections between law and health can then be described and analyzed from two distinctly different perspectives. The first approach is the one conventionally used: We start with the categories in which laws are made and by which legal systems are generally organized. By surveying those categories of law, we can identify myriad rules and regulations that can be shown to influence women's reproductive health. The second approach begins not with law, but with health or, more accurately, with the experiences of the women whose lives are at stake. As we try to show briefly below, this perspective yields a rather different picture of the relationship between law and health.

Each perspective has strengths and weaknesses. Each contributes something different to our understanding of the dynamics of changing health. Both perspectives, taken together, provide the essential background from which to develop strategies for improving the well-being of women and their families.

CATEGORIES OF LAW

Population policies: Formulated at the national or state level, population policies typically set the overall direction for more specific laws related to the three variables affecting a country's demographic profile: fertility, mortality, and migration.[3] As noted elsewhere in this volume, China's population policy is perhaps the world's most controversial. It has long included a one-child-per-couple fertility target, implemented through an elaborate system of incentives and disincentives that has reportedly led to practices as patently coercive as forced abortions and forced pre-term induced deliveries. Moreover, it soon may also include a eugenics component, with government-ordered sterilizations of people with disabilities, as well as forced abortions of fetuses predicted to be "abnormal."[4] But many other countries also use less explicitly coercive measures to change birth rates by influencing individual fertility decisions. Such measures can be as direct as payments in cash or kind for undergoing sterilization or as indirect as tax deductions for each additional child.

It is important to stress that scrutiny of population policies should not be limited to the population policies of southern countries with high growth rates. A law such as one recently enacted in the state of Georgia that would deny benefit increases to certain women who have children while receiving public assistance is every bit a "population policy" as it targets women in a particular segment of the U.S. population in an effort to change their birth rates.

Laws governing provision of and access to goods and services necessary to meet basic needs: Despite the growing tendency to medicalize and individualize health issues, thereby lodging responsibility and blame for poor health on individual behavior, it is clear that poor health in the North and South alike is also deeply rooted in social, economic, and political systems. Thus laws and policies that affect the provision of and access to goods and services necessary to meet basic needs can have both direct and indirect effects on reproductive health. Such laws can include those adopted pursuant to structural adjustment programs; those that encourage, prohibit, or fail to address discrimination on the basis of gender, race, and class in access to society's resources; as well as the multiplicity of laws and policies that create and perpetuate broad systems of inequality.

Laws governing the provision of health services and information about them: While not minimizing the importance of overall living conditions to health status, good reproductive health cannot be achieved in the absence of specific kinds of health services and information about them.[5] This includes services and information related to reproductive tract infections, sexually transmitted diseases (STDs, including AIDS), as well as emergency obstetric care, and fertility regulation. With respect to the last of these, it would include not only general laws determining the legality of particular methods of contraception and abortion, but also specific ones affecting their availability and accessibility: e.g., regulations governing pharmaceutical manufacturing, marketing, advertising, and research and testing; laws governing sex education, censorship, rights of minors, provider-patient confidentiality, informed consent, and personal injury; as well as laws related to funding of health services, licensing and training of medical professionals, and coverage by medical insurance.

Marriage and family law: A woman's ability to control sexual encounters and their potential health consequences (including pregnancy, STDs, and violence) is often directly linked to laws governing her rights, obligations, and behavior both inside and outside the marital relationship, as well as to laws governing other aspects of marriage, divorce, inheritance, and child custody. Importantly, this includes not only formal, written laws enacted by legislatures but also unwritten, customary laws, often grounded in or justified by religion. Beliefs about women's sexuality and its relationship to their roles and obligations inside and outside marriage also underlie particular cultural practices, such as female genital mutilation, that can have a profound influence on a woman's reproductive health.

Other laws affecting women's status: This category would include, for example, laws mandating primary education, encouraging or discouraging political participation, or otherwise determining women's legal capacity. It might also include laws regulating the conditions under which women work, including workplace exposure to toxic substances that can damage their health and compromise their ability to conceive and deliver healthy children.

Criminal laws: Reproductive health is further influenced by laws that criminalize and punish — or that fail to criminalize or to prevent — acts that invade women's bodily integrity, such as rape, domestic violence, and trafficking in women. It is also potentially influenced by laws criminalizing particular forms of sexual expression or particular sexual practices.

Although these categories may provide a fairly accurate description of the substantive law in many parts of the world, there are at least three different kinds of problems with attempting to understand the relationship between law and health from this perspective. First, despite the similarity of substantive categories (or, indeed, even of the content of the rules themselves), the way the law actually works in practice can vary dramatically from country to country, and even within countries, in part due to differences in the structure of the legal system and in the "legal culture" surrounding its operation and use.[6] Second, women do not experience the effects of the law in the kinds of discrete categories set out above; nor do all women experience the effects of law in similar ways. Third, describing the substantive areas in which law influences health tells us nothing about what principles should guide the content of the laws. To say simply that governments should enact those laws most apt to improve health begs the question because, as we demonstrate below, biological "good health" is inseparable from the social dimensions of an individual's life.

The second approach begins to address some of these deficiencies. Women's articulations of their own experiences provide the critical lens through which to understand the way that law actually works in people's lives and through which to assess and mediate its relationship to health. Indeed, if the true goal of reproductive health policy and programs is to improve the quality of women's lives, then both law and health itself need to be approached from women's points of view.

APPROACHING LAW AND HEALTH THROUGH THE EXPERIENCES OF WOMEN

Health is not simply an objective, biological state of being — *reproductive* health even less so, because sex and childbearing are embedded in the wider context of a woman's life. This point becomes clear when we consider the concept of "lived risk."[7] Nearly every aspect of a woman's reproductive life — sex, pregnancy, childbirth, and the means used to facilitate, prevent, or control them— entails risk. There is the risk of disease or bodily injury including STDs, contraceptive side effects, and injury, even death, caused by pregnancy or childbirth. In fact, much health research is devoted to identifying and quantifying the chance that a particular "risk factor" (either interventions, such as use of an IUD, or individual characteristics, such as maternal age) will lead to a particular health outcome (such as pelvic inflammatory disease or death in child-birth). But such statements of risk provide only a limited view of health consequences because they do not describe the way a woman experiences the event, assesses her own risks and benefits, understands her own best interests, or determines the place she wishes to accord her personal inter-est in the whole calculus of social factors that affect her life.[8] For each woman, the risk of an adverse "biological" outcome will be incorporated into the many other personal risks and benefits she experiences from her sexual and reproductive life. Together this will constitute her "lived risk."

For example, in deciding whether to use a particular contraceptive, a woman might not only weigh its potential side effects against the physical risks of abortion or pregnancy and childbirth, she might also consider such factors as social or religious attitudes toward contraception, what effect its use will have on her relationship with her sexual partner, what implications the birth of a child will have for her status and power in the family and community, the economics of her household, or her future security. For evidence that women make choices based on "lived risk," whether or not social and legal systems formally give them the right to do so, one need only consider the fact that millions of women every year knowingly risk death or serious injury from illegal and unsafe abortions rather than carry their pregnancies to term. Or consider the failure of people to follow "safe sex" practices even when they are fully aware of the mechanics of HIV transmission and its consequences. Indeed, each of us need only think about our own risk-taking behavior and how we ourselves process and act on information about physical risk in deciding how to conduct our lives.

Just as health interventions can not be evaluated solely on the basis of seemingly objective scientific data while ignoring women's own percep-tions of their health and well-being, so law and policy positions can not be determined strictly on this basis either. Consider the case of *UAW v. Johnson Controls*, decided by the United States Supreme Court in 1991. Johnson Controls, a company that manufactures batteries for automo-biles, instituted a "fetal protection" policy that banned all women of reproductive age— regardless of whether or not they were actually preg-nant or intended to be pregnant— from jobs in which they could be exposed to lead. The purported rationale for the policy was the fact that some epidemiological and animal studies had found a correlation between a pregnant woman's prenatal exposure to lead and possible harm to the fetus. On this basis, women were told they must either undergo sterilization or lose their jobs.

Even if we assume, for the sake of argument, that the studies were well conducted and provided a reasonable basis for hypothesizing a causal connection between a woman's exposure to lead and the health of her fetus, there are serious problems with the way such data were translated into policy. The problem was *not* that the policy professed concern for the fetus; the birth and growth of healthy children is obviously an important goal. The problem is that no regard was given to the life circumstances of the women who were the policy's targets. No regard was given to economic necessities that these women faced or the importance of their earnings and employment benefits to their own health or that of other children in their families, and no consideration was given to their own calculus of the risks and benefits of their work, taken in the context of their entire life circumstances. In short, the policy failed to consider the women employees' "lived risk" of potential effects on a fetus or what actions they might take based on it.

Indeed, the policy also failed to consider the effect of lead on men's sperm or the consequences that a father's exposure can, and sometimes does, have for the health of the fetus. Rather than addressing the real public health problem exposed by the scientific studies— the need to clean up reproductive health hazards in the workplace — *Johnson Controls* used one hypothesized correlation (between maternal exposure and fetal changes) to justify a policy meant to "protect," and thereby to control, individual women whose lives had taken an unconventional route — i.e., entering the workforce by taking scarce, traditionally male, and relatively high-paying jobs. Ultimately, the policy was premised not *just* on scientific data, but also on a worldview in which all men have as their primary responsibility the financial support of their families and all women have as their foremost duty and responsibility the birth of healthy babies. All women were viewed as potentially pregnant and were deemed unable and unwilling to control that condition. Under this policy, unless a woman (but not a man) could demonstrate that she had been sterilized, her employer would simply make the choice for her— and ban her from the workplace.[9]

The concept of "lived risk" and the case of *Johnson Controls* illustrate the fact that, while law and health can be closely connected, there is rarely such a thing as a reproductive health law or policy that is based exclusively on objective scientific evidence without incorporating, explicitly or implicitly, some value-laden choices about women's lives. These examples also demonstrate the difference it can make to approach law and health — and the connections between them — not just on the basis of theory or even statistics but also through the experiences of women. Indeed, we could go systematically through each of the categories of law outlined in the first section and demonstrate how an approach that valued women's experiences or points of view might yield laws or policies different from those that have actually been adopted — including laws that were justified by their potential effect on women's reproductive health.

Obviously we do not expect reproductive health policymakers and programmers to go out and ask every woman in the population how she experiences each aspect of law and health. But what we *can* do is to develop legal and ethical principles that value women's experiences and then use such principles in formulating and implementing laws, policies, and programs.

REPRODUCTIVE RIGHTS AND HUMAN RIGHTS

The categories of law listed in the first section tell us where to look for laws that influence reproductive health. But, as we pointed out above, listing the categories does not tell us what values or principles should guide their content. Moreover, resorting strictly and exclusively to a technical health rationale only disguises further the values implicit in policymaking because it fails to acknowledge that reproductive health itself is never simply an objective, biological state of being. This is very definitely *not* to say that considerations of biological or physical health are unimportant or that technical data are irrelevant; much to the contrary, they are absolutely critical elements of laws and policies related to reproduction. But our understanding and use of them must be informed by other legal principles that create a space for and place value upon the experiences and voices of women.

International human rights instruments, especially the Convention on the Elimination of All Forms of Discrimination Against Women, begin to provide a basis for some of the reproductive rights principles that can guide work in the reproductive health field.[10] This includes rights of non-discrimination, such as were used to invalidate the Johnson Controls policy, as well as principles of bodily control and integrity built on the fundamental values of human dignity articulated in human rights instruments, ratified by dozens of countries from every continent and echoed in many cultural and religious traditions around the world.

Human rights law can be especially valuable in assessing policymaking and program development at the international level, particularly in the population and reproductive health field where international private and public actors — including U.N. agencies, the World Bank, bilateral aid organizations, international service providers, and multinational corporations that develop and market contraceptives — are so influential. But human rights can also be useful, both legally and philosophically, for those working locally to define and articulate reproductive rights in ways that can be effectively incorporated into the particular legal systems within which they live. The precise content of such rights and the way they intersect with other parts of the law may vary from place to place, but the perspective that they bring to the reproductive health field can ultimately improve the well being of women, their families, and their communities everywhere.

Population, Gender, and Culture

. . . .

NAFIS SADIK

Investing in Women: The Focus of the '90s

·····················

This chapter demonstrates some of the costs of ignoring the needs of women: uncontrolled population growth, high infant and child mortality, a weakened economy, ineffective agriculture, a deteriorating environment, a generally divided society and a poorer quality of life for all. For girls and women it means unequal opportunities, a higher level of risk and a life determined by fate and the decisions of others rather than choice.

Many women, especially in developing countries, have few choices in life outside marriage and children. They tend to have large families because that is expected of them. Investing in women means widening their choice of strategies and reducing their dependence on children for status and support. Family planning is one of the most important investments, because it represents the freedom from which other freedoms flow.

Investments in women include, besides family planning, "social investments"—services such as health and education. The chapter demonstrates that such services help women to do much better what they are already doing and open the door to new possibilities. But investing in women must go beyond such services, and remove the barriers preventing them from exploring their full potential. That means granting them equal access to land, to credit, to rewarding employment—as well as establishing their effective personal and political rights.

Nafis Sadik is the executive director of the United Nations Population Fund. This chapter is adapted from Investing in Women: The Focus of the '90s, *published by the United Nations Population Fund in 1992.*

Making investment in women a development priority will require a major change in attitudes to development not only by developing countries but by financial and lending institutions. Under increasing economic pressure, in the last four years 37 of the poorest countries have cut health spending by 50 percent, and education by two percent. This burden falls hardest on the poor, and hardest of all on poor women.

Taking the long view, investing in women has a finite if unquantifiable economic value: the return will be an approach to development which will make the most effective use of the world's limited resources; slower, more balanced growth in the labor force; security for the family; and—most important—the possibility of better health, education, nutrition, and personal development not only for women but for all people.

Investing in women is not a panacea. It will not put an end to poverty, remedy the gross inequalities between people and countries, slow the rate of population growth, rescue the environment, or guarantee peace. But it will make a critical contribution towards all those ends. It will have an immediate effect on some of the most vulnerable of the world's population. And it will help create the basis for future generations to make better use of both resources and opportunities.

WOMEN AND THE SEARCH FOR SELF-DETERMINATION

In many societies, a young woman is still trapped within a web of traditional values which assign a very high value to childbearing and almost none to anything else she can do. Her status depends on her success as a mother and on little else. Increasing a woman's capacity to decide her own future—her access to education, to land, to agricultural extension services, to credit, to employment—as an individual in her own right has a powerful effect, not least on her fertility.

There is a sharp contrast between persistently high fertility rates in some countries—most of the Arab world, for instance, and many parts of south Asia—and decreasing fertility rates in countries with equivalent levels of socioeconomic development—China, for example, Sri Lanka and the state of Kerala in South India. One of the key differences between the two groups is the status of women, whether among themselves or relative to men.

Women who have managed to acquire a measure of self-determination tend to have smaller families than women with no such assets. Broadly speaking, in the second group of countries women have a wider range of choice. Their status and security do not depend only on bearing children.

OLD SOURCES OF INSECURITY

Many sources of women's dependence used to be seen as positive—they offered a woman some measure of protection, in return for her main contribution—producing children, especially sons. Most of these practices have existed for centuries and are woven into the fabric of society. Under the stress for change, the fabric is falling apart: what seemed to offer security in the old times is revealed as a deception in the new. These practices are often found together with the greatly increased risks to women's lives and health imposed by inadequate health care and family planning services.

Marrying out, marrying young. In many traditional communities a newly married woman is expected to move into the household of her in-laws. She is totally dependent on her new family—her husband's family—for her survival.

The age at marriage in many developing countries is very young for women—around 50 percent of African women, 40 percent of Asian women and 30 percent of Latin American women are married by the age of 18, according to the World Fertility Survey. Men tend to marry at older ages. In Sudan, for instance, husbands are on average more than eight years older than their wives. In Pakistan the average age gap is around six years; in Colombia and Paraguay it averages between four and five years. Such wide age gaps have several consequences. For a start they mean that a new wife is likely to be less experienced and less confident than her husband. But, perhaps more important, it means that the likelihood of a woman being widowed is very high and—in the absence of alternative economic means of securing her future—this increases her potential dependence on her children.

Infertility. But many women may never bear children at all. Infertility is a curse in any society—but it is worst in societies where families are traditionally large and children are perceived as the source of security. The World Health Organization (WHO) estimates that, worldwide, one in 10 couples is involuntarily infertile. What is known as "primary infertility"—where a woman is never able to conceive—can be as low as 1.5 percent in some areas. But there are countries with pockets of primary infertility running as high as 40 percent.

"Secondary infertility"—where disease or birth damage means that a woman with one or more children is unable to have any more—afflicts much higher numbers of women. Among the Mongo people of Zaire, for instance, over 40 percent of women are sterile in some communities and between 35 and 50 percent of pregnancies end in stillbirth or spontaneous abortion.

Old age. Three fifths of the world's people produce their livelihood with their own hands. Up to 80 percent of people in developing countries have neither a wage nor a pension. In fact the International Labor Organization (ILO) predicts that, worldwide, only 25 percent of retired men and six percent of women will be receiving a pension by the year 2000.

In the industrialized world, economic security is provided by income from jobs, savings and pensions. Children are not expected to support their parents: Only four percent of United States parents interviewed in one study mentioned "economic support" as a reason for having children; it was the major reason given by 73 percent of Mexican parents.

One United Nations survey of 50 countries found that government spending on pensions was associated with lower fertility rates. China's birth rate was halved in less than 10 years when people knew they would be cared for in the communes when they grew old. The less parents' future depends on their children, the fewer children they need to ensure their security.

Discrimination against girls. Eight out of nine cultures who express a preference want more sons than daughters. Parents expect little from a girl once she marries and leaves home. Before she marries, though she starts to work at an earlier age than her brothers, and works harder and longer, her economic contribution is seen as less valuable because it contributes less to the family's income. In Mali and Afghanistan, for instance, 89 and 97 percent respectively of all formal employment goes to men. Women's wages are usually lower than men's, even when they do similar work.

Greater expectations lead parents with limited resources to invest more in their sons than their daughters. This discrimination begins very early in life. Research in Bangladesh found that under-five-year-old boys were given 16 percent more food than girls and that girls were more likely to be malnourished in times of famine. A study in India found that boys were given far more fatty and milky foods than girls. Not surprisingly girls were over four times as likely as boys to be suffering from acute malnutrition, but more than 40 times less likely to be taken to hospital. Another Indian study found that sick boys were more likely than girls to be taken to the city hospital when they failed to recover from illness.

Deaths between the ages of one and five are higher for girls than boys in Bangladesh, Nepal, Pakistan, Sri Lanka, Egypt, Jordan, Mauritania, Morocco, Syria, Sudan, Turkey, the Yemen Arab Republic and in half of the Latin America countries studied by the World Fertility Survey.

Discrimination against girls continues as they grow older. In the developing world as a whole, 65 percent of girls were in primary school in 1985 compared with 78 percent of boys. In secondary school, 37 percent of girls and 48 percent of boys attended. In Pakistan and the Yemen Arab Republic, there are over three times as many boys as girls in secondary school. Parents with little income are apparently less willing to invest it in girls' education than boys'; but poorly educated girls are much less likely to find well-paid employment or marry well-educated men, so the attitude that girls are a poor investment perpetuates itself.

A woman brought up to feel weaker, less useful to the family and less valuable to society in general suffers a crucial loss of self-esteem, which in a vicious circle further reduces her potential to contribute. Her only possible route to status and respect is marriage and childbirth; but the disadvantages with which she starts reduce her ability to do this successfully—and pose and added threat to her life and health.

Cycles of malnutrition. There is no maternity leave for a tea-picker on a plantation in Malawi or for a woman working on a building site in India. But if a pregnant woman is overworked and underfed she is more likely to have a small and weak baby. Low-birth-weight babies are more vulnerable to infection: They are 13 times more likely to die of infectious disease than normal-weight babies.

Between 20 and 45 percent of women of childbearing age in the developing world do not eat the WHO-recommended amount of 2,250 calories a day under normal circumstances, let alone the extra 285 a day they need when they are pregnant. In Thailand, for example, the average woman gets only 1,900 calories a day; in the Philippines the average is 1,745—despite the back-breaking work women in those countries do at peak times of the year in the paddy-fields. Anemia associated with poor nutrition affects as many as two thirds of Asian women, half of all women in Africa, one sixth of Latin American women, and 60 percent of women worldwide.

Heavy manual work combined with poor nutrition in mothers results in low-birth-weight babies and an inferior supply of breastmilk. Latest figures reveal that 16.7 million babies—that is one in six worldwide—are born weighing less than 2,500 grams. The totals are worst in Asia, where one in five babies are of low birth weight. In Africa the proportion is one in six; in Latin America it is one in nine. In Europe only one in 17 babies are this tiny, and because they are premature, not because of their mothers' poor nutrition.

Low birth weight and poor nutrition makes for sickly infants and stunted children. The low-birth-weight girl is particularly vulnerable. In societies where girls are valued less than boys, she is fed less, and never has a chance to catch up.

This is the point where the story starts to come full circle—because small women tend to have narrow pelves and many have considerable trouble delivering their babies. In India, for example, there are two mortality peaks for women during their lives: the first is in early childhood; the second during their childbearing years.

One eighth of deaths in childbirth in Bangladesh are from obstructed labor or ruptured uterus—usually because the pelvic opening is simply too narrow for the baby's head to pass through it. Equivalent figures were one woman in 13 in a Tanzanian study and one in 19 women in Addis

Ababa, Ethiopia. In Zaire 18 percent of hospital deliveries are obstructed labors; worldwide the figure is between five and 10 percent.

Poor nutrition also contributes to three other major causes of maternal mortality—pregnancy-induced hypertension, hemorrhage and septicemia. Women and their daughters brought up under these conditions are trapped in a cycle of low birth weight and small adult stature in which they run higher risks at all stages than women who are better fed.

Teenage pregnancy. "Society will condemn us if our daughters are not married off by the age of 15," say women from two Indian villages where the average age of marriage for girls is just 14.3 years. It has been estimated that 40 percent of all 14-year-old girls alive today will have been pregnant at least once by the time they are 20. In Bangladesh four out of five teenage girls are mothers; three out of four teenagers in Africa as a whole. Africa has the highest rate of births to very young mothers: 40 percent of teenage births are to women aged 17 or under, compared with 39 percent in Latin America, 31 percent in Asia and 22 percent in Europe.

In many developing countries the majority of births to teenage mothers take place within marriage. The average age of marriage in countries with high teenage fertility rates is often very low precisely in order to ensure the resulting children's legitimacy. In Pakistan and Sierra Leone, for instance, the average age of marriage is 15.3 and 15.7 respectively; while in Jordan 58.5 percent of teenage girls are married.

In industrialized countries, however, large numbers of teenage births are the result of premarital sexual activity—three out of four in Denmark and Sweden, for example. In the United States and in England and Wales over 50 percent of all illegitimate births in 1982 were to teenage mothers.

An illegitimate pregnancy can be catastrophic in rich countries and poor alike. One study in Zaire found that the typical hospital patient being treated for a septic or botched abortion was a 15-to-16-year-old unmarried schoolgirl who had never used contraception and who had tried to abort her pregnancy herself. In the United States 30 percent of all teenage pregnancies end in abortion; in Norway a staggering 87.5 percent of pregnancies in the under-18s are aborted. Teenagers account for half of all late—and therefore more dangerous—abortions in the United States. One reason for these high rates is that family planning programs tend to be aimed at married women; worldwide, three quarters of girls under 15, and half of those 16 or over, have no access to family planning information.

Dangerous though a late abortion may be for the young mother, the evidence from the United States is that a teenage birth is five times more dangerous. Around the world, the teenage mother and her baby face a worse combination of risks to their health than any other age group. The mother herself is likely to be anemic and her body not yet fully developed; she is less likely than older women to seek prenatal treatment, and she faces possibly complicated and prolonged labor. Her baby is more likely to be premature and underweight.

In countries as different as Malaysia, Japan, the Dominican Republic and Bangladesh, the United States, Tanzania, Nigeria, Jamaica and El Salvador, 15-to-19-year-old mothers are twice as likely to die in childbirth as mothers aged between 20 and 24. Risks to the younger mother are even greater: in Bangladesh the under-15-year-old teenager is five times more likely to die in childbirth than a mother aged between 20 and 24; in the United States she is three times more likely to die.

Babies born to a teenage mother are more than twice as likely to die in their first year of life, and they run double the extra risks associated with births spaced less than two years apart. In Sri Lanka and the Republic of Korea, babies born to under-16-year-old mothers are three times more likely to die than those born to mothers aged between 20 and 24. If the first child dies, subsequent children to a teenage mother run three times the normal risk.

Teenage motherhood damages more than health. A teenage mother has much less chance of continuing her education and of ever becoming anything other than a mother. The pattern is likely to be passed on: Studies in Denmark and the United States show that teenage mothers beget teenage mothers. They also tend to be poor and to remain poor.

With an early start, teenage mothers go on to have large families. If their daughters also marry early, then the gap between subsequent generations is shortened.

Mothers in danger. Childbirth anywhere in the world has its risks, particularly without proper prenatal care and attention during and after delivery. For most women in developing countries the risks are multiplied. One in five women die before menopause in Afghanistan, Benin, Cameroon, Malawi, Mali, Mozambique, Nepal, Nigeria, and North Yemen.

Yet every one of those deaths—over 500,000 a year, or one every minute of every day—could be prevented. A woman in Africa is 200 times as likely as a European woman to die as a result of bearing her children. More women die of maternal causes in India in one month than in North America, Europe, Japan and Australia put together in one year. The last recorded maternal death in Iceland was in 1976.

A single prenatal visit could identify three quarters of the pregnancies at risk. But many women die because they are unable to get to hospital in time, or obtain proper treatment once there. The WHO estimates that only 55 percent of births are attended by a trained person; 34 percent in Africa, 64 percent in Latin America and 49 percent in Asia (excluding China and Japan). Once there, the hospital or clinic may not have the right supplies or equipment. In Tanzania, for instance, shortage of drugs or blood for transfusions is a factor in half of all maternal deaths.

Too young, too old, too many and too close. The youngest and oldest mothers are most at risk. According to evidence presented to the Safe Motherhood Conference in 1987, women over 35 are two to five times more likely to die as a result of childbirth. Having many previous pregnancies also greatly increases the risks of the next one. Jamaican women having their fifth through ninth child are 43 percent more likely to die than women having their second: in Portugal the risk is over 300 percent greater.

The leading cause of maternal death—postpartum hemorrhage—is most common in women who have already had several closely spaced pregnancies. One country study found that the typical case was a woman over 30 years of age, unemployed, with three or more existing children, who had never used contraception and had no prenatal care. She simply bled to death after delivering her baby.

Babies born in quick succession, to a mother whose body has not yet recovered from a previous birth, are the least likely to survive. Latest research shows clearly that of the three most important "risk" factors for infant death, closely spaced pregnancies are more dangerous for the resulting babies than either the age of the mother (under 18 or over 35) or the number of other children in the family (more than four). It has been found that a baby born less than two years after its sibling is 50 percent more likely to die between the ages of one and five. Yet around 50 percent of all births in Jordan, Colombia, Costa Rica and Jamaica—and in many other countries—are less than two years apart.

Every death brings the next birth a little nearer. United Nations figures from 25 developing countries reveal that couples experiencing the death of one child are likely to have larger families than those whose children all survived. The more recent the death, the greater the likelihood of an additional child. The mothers' experience of loss means that they will be the least likely to use family planning—which could save not only their children's lives but their own. Nearly half of maternal deaths could be avoided simply by avoiding unwanted pregnancy.

The Safe Motherhood Conference pointed out that most developing countries allocate less than 20 percent of their health budgets to maternal and child health programs, and the majority of that goes to child health. With no action on maternal mortality, there could be 600,000 deaths in the year 2000. With action to halve the rate of maternal mortality, and fertility 25 percent lower than United Nations projections, the number could be reduced to 225,000.

NEW SOURCES OF INSECURITY

Environmental instability. "When we were young we used to go to the forest early in the morning without eating anything. There we would eat plenty of berries and wild fruit and drink the cold sweet water from hollows among the banj roots. In a short while we could gather all the fodder and firewood we needed. Now with the going of the trees everything else has gone too." These women from the Uttarrkhand Hills in India tell a familiar story. Once—not so long ago, well within living memory—they could rely on the land around them to provide for their needs. Today in many developing countries that sense of security is threatened.

Today tropical forests are being felled at a rate of 11 million hectares a year; topsoil is being washed away by wind and rain at the rate of 26 billion tons a year; new stretches of desert are appearing at a rate of six million hectares a year; and build-up of salt and stagnant water threatens half of the world's irrigated cropland.

The effects are felt most acutely at the lowest level: by individual people, often the poorest and least powerful of people, most often by the women. It is they who have to walk further and further each year to fetch firewood from the dwindling woodlands; they who must search for hours for a stretch of unpolluted water; they who must cope with the effects of environmental degradation and pollution on their own and their family's health.

Land hunger, scarce fuel, pollution and migration deepen women's sense of uncertainty about their future. Already many are being forced into actions that they know are likely further to jeopardize their security. In countries with a shortage of firewood, for example, women use manure as a fuel instead of fertilizer—mortgaging tomorrow's food to cook today's. Women are also beginning to conserve fuel by cooking less often. Reports are already coming in from places as far apart as Korea, Bangladesh and the Sahel that only one meal—as opposed to two—is being cooked each day. In Rwanda 62 percent of families cook just once; 33 percent even less frequently. In Mexico and elsewhere there has also been a move away from nutritious food like beans because they use so much fuel. It can only be a matter of time before other reports—of increased malnutrition, diarrhea, infant and child death rates—start to follow.

Children may be needed even more in countries where the environment is the most threatened. In Bangladesh girls spend over six hours a day collecting fuel; in Nepal and Java the task takes them over three hours; in Tanzania girls spend an average of 1.9 hours a day collecting firewood.

Migration. Migration is a growing phenomenon in the developing world. The population of the world's cities doubled between 1950 and 1980 and will have doubled again by the year 2000, when city-dwellers in the developing world will outnumber those in the developed world by two to one. In poor countries this expansion of the urban population is partly the result of men—husbands or sons—leaving their homes in the countryside to find employment in one of the burgeoning cities.

One study of 74 developing countries discovered that 22 percent of households in Africa, 20 percent in the Caribbean, 18 percent in Asia, 16 percent in the Near East and 15 percent in Latin America were headed by women. Worldwide it has been estimated that as many as one in three households are headed by women. While a proportion of those women will be widows—48 percent in Indonesia, for instance—in most countries the majority are without men because of migration.

Head of the household. It hardly needs saying that households headed by women are the poorest in the world. Women are not permitted to own land in Colombia, Nepal, Kenya, Ethiopia, Panama, Chile, Iraq and Egypt. New land-reform laws exclude divorced women from land ownership in Zambia, Tanzania, Ethiopia and Nepal. In India 35 percent of the rural landless population are comprised of woman-headed households, compared with 20 percent in the country as a whole. In Guyana equivalent-sized man-headed households are twice as wealthy as their woman-headed counterparts. And studies in Botswana, Lesotho, Chile, Brazil and in parts of the Caribbean have all found that households headed by women are poorer than the male-headed equivalents.

Less than 50 years ago this situation would have been unthinkable in most societies—somehow such households would have been absorbed into the wider kin network. But evidence from Zambia, Pakistan, Syria, Democratic Yemen and the Yemen Arab Republic indicated that changes in the rural economies of many developing countries have made it more and more difficult for extended families to cope with the growing number of poor women.

The catch in cash crops. Changes in agriculture have also tended to undermine women's security. The widespread shift from the growing of subsistence food (mostly grown by women) to the growing of cash crops has often meant that control of the crop passes to men, because agricultural training, credit and technology are routinely given to men rather than women. The women farmers lose an important source of income (and possibly of nutrition too) as well as their chance of benefiting from development. "Development" ends up as a net loss to the community.

PLANNING THE FUTURE

Fate and fertility. Any planning process starts with the assumption that there is some possibility of choice: according to the World Fertility Survey, many women in developing countries would like to delay or stop childbearing, but have no effective choice in the matter. They need some assurance about what they can expect for themselves and their children; they also need the services which would allow them to exercise choice.

This assurance springs from many aspects of life. Among them are social stability and reasonably steady income. For a woman it is apparently additionally derived from her status within the family and before the law, the quality of health care available, and particularly from her education. Most importantly, it is derived from some sense of control over her own fertility.

Power to choose. In just two decades the percentage of people using modern family planning has more than quadrupled in the developing world—from around nine percent of the childbearing population in 1960-1965 to 45 percent of a much larger population in 1983. One wide-ranging survey of 39 developing countries found that half of women who want no more children—temporarily or permanently—are able to have their wish. That same survey discovered that young married women today want, on average, around two children fewer than their mothers. In the space of a single generation, ideal family size has dropped from 5.7 children to 3.8.

The simple presence of effective and accessible family planning can change people's attitudes towards contraception and cause parents to reconsider their attitudes towards family size. It also has a snowball effect—people talk among themselves about its possible dangers and advantages; next someone actually goes to the clinic or pharmacy; if all goes well, many of their neighbors will follow. The availability of some services outside the usual health channels can also encourage their use, particularly if accompanied by useful and positive information.

HEALTH CARE: THE GLOBAL GOALS ARE SET

Three quarters of the health problems in the developing world could be solved by a simple combination of prevention and cure: enough of the right food, clean water to drink, safe sanitation, access to family planning, immunization, and around 200 basic drugs.

This is primary health care in a nutshell: an initiative that tries to shift the focus of attention away from doctors and nurses and towards parents and village health workers; away from hospitals and operating theaters and towards communities and homes. By its nature it focuses attention on women. Women are responsible for children from embryo to adulthood. Increasingly they are the sole breadwinners of the household. Because they are the providers as well as the recipients of health care, women's own health needs safeguarding above all.

Paying for prevention and cure. The recent international focus on safe motherhood has concentrated on practical, affordable solutions to the problem of maternal and infant mortality. The World Bank has proposed a three-pronged approach: stepping up health care in the community, improving the referral system, and providing transport between the different levels in the system. It calculates that if better-off countries allocated an extra two dollars per person per year to upgrading their health services along the lines they suggest, there would be a 50 percent drop in maternal mortality rates. For poorer countries, and outlay of just one dollar per person per year would reduce maternal mortality by 25 percent. In a separate series of calculations, World Bank economists have also estimated that an extra two billion dollars spent annually on providing family planning to women who want it would avert the deaths of 5.6 million infants and 250,000 mothers each year.

EDUCATION: LESSONS IN LIFE

Education is perhaps the single strongest influence on women's control of their own future. Every large-scale survey in developing countries has discovered that the education women receive is one of the most universal and reliable predictors both of their own fertility and of their children's health. This effect holds regardless of school curricula and different cultures, and even though other factors, such as income and employment opportunities, come into play.

Today 65 percent of girls and 78 percent of boys in the developing world are in primary school—20 and 11 percent more than a decade ago. Secondary school enrollment rates are lower—with only 37 percent of girls and 48 percent of boys in school—but this is still 29 and 16 percent more than 10 years ago. If current trends continue, literacy will continue to win the race against population growth.

Fewer children, healthier children. In almost every country studied in recent years, educated women have been found to have fewer children than their less educated or uneducated sisters. A study of four Lain American countries discovered that education was responsible for between 40 and 60 percent of the decline in fertility in the last decade. The authors suggested that this effect might soon be seen around the world.

The families of educated mothers are likely to be healthier as well as smaller. Mothers' education may be even more important to her children's health than flush toilets or piped water, or even food intake. Studies put the difference in child mortality (deaths of children between one and five) as high as nine percent for every year the mother was at school. In Peru, for instance, educated women had healthier children regardless of whether there was a clinic or hospital nearby.

The impact on deaths of children in the first year of life is also significant, though less dramatic. Research from 46 developing countries has discovered that a one percent rise in women's literacy has three times the effect of a one percent rise if the number of doctors. Two other major reviews of information—from 15 and 33 developing countries respectively—estimate that four to six years of education is associated with a 20 percent drop in infant deaths. The effect is less pronounced probably because around half of infant deaths occur in the first few weeks of life and have to do with the birth itself.

A foot in the door. Educated women tend to be older when they get married are more likely to be employed and live in the city than uneducated women. The World Fertility Survey, for instance, discovered that women with seven or more years of education married an average of four years later than less educated women. Women living in the city in Latin America and parts of the Arab world have between two and three fewer children than their rural counterparts.

Education has a powerful effect in its own right. It may be not so much what girls learn when they are at school that makes the difference, but simply that they have been to school at all. Education, for many girls, is their foot in the door of the modern world—the world of clinics, post offices and banks; of books, medicines and buses. With increased confidence and self-esteem, they can begin using the resources of the modern world to make their own choices and improve their lives.

Educated women are more likely to stand up for themselves. In Kerala, for instance, educated women—however poor—seem to believe they have a right to good health care. Conversely, lower-class Nepalese women expected and received worse treatment from health staff; illiterate women in Ibadan fared similarly. And detailed interviews with 29 literate and illiterate wives in Uttar Pradesh found that the latter would not take their sick children to the doctor before obtaining permission from their husbands or in-laws.

If women's confidence is higher in clinic and hospital waiting rooms, it is likely to be higher at home too. Research from Nigeria, Bangladesh and Mexico confirms that educated women tend to communicate more with their husbands, to be more involved in family decisions and to be more respected: more able in other words to plan what happens in their lives.

WOMEN AND WORK

The distinction the modern world likes to make between "productive" and "reproductive" work—that is to say between economic activity and household work—makes no sense when applied to women's lives in much of the developing world. Their productive work has to be fitted in around their reproductive work and vice versa. This is one reason why women fare so badly in the formal economy—the other demands on their time make it difficult for them to operate within the rigid confines of a job. A more useful distinction might be made between "visible" and "invisible" work.

Visible work. There are now more women in the labor force than ever before. According to figures from the International Labor Organization (ILO), 676 million women had jobs in 1985 and that number will rise to an estimated 877 million by the year 2000. Their share of the total labor force will remain relatively unchanged over that period, at around 35 percent. Men will continue to dominate the labor market well into the next century.

Many things were expected of women's entry into the labor force. Employment was supposed to build on the foundations laid by education to allow a woman to become an independent, respected and self-sufficient member of society. The additional income would be spent on food and clothes for her children, with a little left over each week as savings: The demands of the job would make for later marriage and smaller families. An independent income would make women less reliant on husbands and children and reduce the need for sons as props for old age. Women would value daughters as much as sons and share the benefits of extra income equally. The result would be a new generation of healthy, confident, secure and forward-looking young women ready to take their place in society.

The expectations were exaggerated. There is evidence from many countries that each of the links in this chain of increasing security can be forged and will hold. But in most countries the chain has been broken at many points.

The evidence linking women's employment with lower fertility and mortality is contradictory: One study of 60 developing countries found that women working outside tended to have fewer children than those working at home or in the fields and plantations, even when other factors—such as education, urban living and industrialization of the economy—were taken into account. But surveys of Turkey, Thailand and other countries have given opposite results.

The World Fertility Survey established that women working in the modern sector—that is those formally employed as teachers, factory workers, nurses, shop assistants and so on—marry an average of 2.4 years later than women doing domestic and agricultural work. Later marriage has been found—in surveys from 77 developing countries—to be associated both with enhanced access to resources and—in a round-up of information from 21 developing countries—with more economic independence. A smaller age difference between spouses—which tends to go along with education for women, employment in the modern sector, and later marriage—also tends to lower fertility rates.

On the other hand, a major review of the evidence, which put together results of surveys from all over the developing world, concluded that employment for women reduces fertility rates by an average of only around 0.5 of a child. And it has no predictable effect on child survival rates. Most women's jobs—particularly poor women's jobs—are neither secure nor well-paid. They go along with rather than substitute for child-bearing—yet another call on a woman's time.

Women are not able to put in as many hours as men because their work in and around the home makes competing demands on their time. They may be unable to work the hours many employers require, such as shift-work or overtime, and may be forced to accept badly paid part-time work. In Sri Lanka, for instance, women in manufacturing industry work an average of 10.5 hours less than men. This also makes it much less likely that a woman will be promoted. Only 1.2 percent of women with jobs in Singapore were employed in managerial positions in 1985, as opposed to 8.2 percent of a much larger number of men.

The issue of child care is crucial. Its absence may be serious enough to undermine almost all of the potential benefits of extra income. It may even help to keep fertility rates high. In the countryside children are needed to fetch water and firewood and care for younger siblings so that their mothers are free to work in the fields. In the town a child's labor may be equally important in allowing her mother—and father—to go out to work. In Colombia, for example, girls as young as six are kept away from school and left in charge of younger siblings during the day.

Women are also limited by the jobs they do. Women's work is over-whelmingly concentrated in "pink-collar" occupations—cleaners, secretaries, clerks, nurses, teachers, waitresses, textile workers—employing the skills women pick up as part of their domestic role. In Latin America and the Caribbean, for instance, 82 percent of health service workers and 74 percent of teachers are women.

These occupations can be secure and well-paid, but most are paid at a much lower rate than the equivalently skilled male "blue-collar" job. The discrepancy can be quite shocking. In Brazil in 1980, the average woman with a job had one third more education than the average man, but earned one third his wage. One study in Santiago, in Chile, found that the discrepancy was even greater further up the occupation ladder. Primary-school-educated men and those who had never been to school were paid 71 percent more than women with the same education; those with secondary education 84 percent more; while male graduates took home nearly three times as much as women with the same qualifications.

Reliance on women for cheap, unskilled or partly skilled labor represents a massive waste of human and economic resources. Better education and employment at a higher level would reduce the social and economic cost of large families and enable women to make their full contribution to development.

Invisible work. The ILO's statistic for women's participation in the work-force—28 percent—is a cruel misrepresentation of the actual amount of work that women do. A look at regional participation rates shows why. In Africa, for instance, only 22.9 percent of women were considered to be in the labor force in 1985, defined as either "performing some work for wage or salary, in cash or in kind," or "self-employed" performing "some work for profit or family gain, in cash or in kind." But women are responsible for between 60 and 80 percent of the food grown in Africa. Somehow their hours of planting, weeding, picking, threshing and winnowing have become invisible when national statistics are being compiled.

The same is true of their care of livestock, their vegetable growing, their trading, as well as the myriad activities normally classified as "housework": fetching water and firewood, pounding grain, preserving fish and meat, cooking, cleaning, sewing, weaving, washing and mending clothes, teaching children and nursing them when they are ill, caring for old or disabled family members. In Pakistan, where one survey of village women found them putting in 63 hours of work a week, only 13.9 percent of women aged between 25 and 54 are considered to be "economically active." In Rwanda, where women were found to work three times as much as men, only 55 percent of women are considered to be "economically" active—exactly the same percentage as men.

Women predominate in the "informal sector" in many countries—the exchange of goods and services that oils the economic wheels of the poorer parts of the developing world and provides many a subsistence family with a small income for essentials. In West Africa, the Caribbean and South Asia, between 70 and 90 percent of all farm and marine produce sold is traded by women. In Central Peru 61 percent of women are involved in trade; in the Philippines the figure is 16 percent. It is estimated that the informal sector generates over 30 percent of urban wealth.

Though the volume of their trade is considerable, and wealthy traders not unknown, individual traders are generally operating near the poverty level. Usually women working in the informal sector earn perhaps half as much as a male laborer.

Work in the informal sector may allow women to combine their income-earning with childbearing, child care and domestic work. But it does nothing for their lack of education, lack of access to credit, and other more traditional restrictions on their activities. So women have little opportunity to expand into more profitable enterprises.

Women's earnings from these activities are rarely taken into account when national statistics on economic activity are compiled. Like housework and subsistence agriculture, women's business activity simply disappears. Its contribution to a nation's wealth and well-being is overlooked, and never appears in development plans. In Asia generally, women are only 10 percent of those being trained in small-scale co-operative organization or business management. As of 1985 only 12 countries had set up organizations to help women with trade.

Recognizing women's work. Trying to develop without acknowledging or involving the people who do two thirds of the work is inviting failure. The first step is to quantify women's work—in the home, in the fields, in the marketplace—and acknowledge its value. Even ostensibly "non-productive" work like child care and cooking has an economic value if the health and creativity of generations of future workers can be brought into the equation. It certainly has far-reaching effects on infant and child mortality and on family size.

Development plans for women, when they exist, tend to assume mistakenly that women have free time to devote to them. Women in Uganda work 50 hours a week compared men's total of 23 hours. In the Philippines they work and average of 66 hours a week to men's 41 hours; in Indonesia women average 78 hours a week compared to men's 61 hours. Plans for women which involve extra work should include relief from current burdens or a compensating income.

It may also be a mistake to assume that women wish to continue their current or traditional roles. Though women grow food crops in much of Africa, many would like to diversify into cash crops, hitherto the province of men.

Sometimes even quite minor interventions can have a dramatic effect. In Peru, for instance, the introduction of grinding mills saved women so much time they were able to expand their trading activities. Providing a grinding mill and a standpipe might create so many extra hours in a day that women might be able to undertake a second weeding of the maize or sorghum fields, which has been shown to improve yields by 20 percent. It may well be that two such simple improvements can have as powerful and effect on agricultural productivity as a full-scale irrigation scheme—not to mention the beneficial effects on health and the quality of life generally. Women's work is seldom assessed in these terms.

Reducing women's work-load and making their labor more profitable might also help reduce family size, which would reduce the load still further. Benefits for mothers would spill over into benefits for daughters. In most developing countries little girls work harder than boys—and that means they have less time to concentrate on their schoolwork. In Java, for instance, girls work an average of 8.1 hours a day compared with boys' 5.2; in Nepal the figures are 9.8 and 7.8 hours respectively.

If small interventions can make a difference, bigger ones can make even more, if women's contributions are known and acknowledged. It cannot be assumed that the benefits of a development scheme designed with men in mind will be passed on to women. It may even have the opposite effect, if it means extra work for women or removes a source of income.

Taking women seriously means consulting their wishes and allowing them to set the priorities—whether it is for family planning or farm credits. A study in Kenya found that women forced to manage their farms alone because their husbands had migrated or divorced them harvested the same amount per acre of land as men who had been involved in a local agricultural project. When such women were included in the project they did even better. Yet Kenya is one country where women's traditional rights to land have been undermined by projects which require the land to be registered in a man's name.

Security of tenure has long been known to increase a male farmer's efficiency and productivity. So it should not be surprising to find the same effect among woman farmers. In Zimbabwe, for instance, maize production quadrupled after women's access to land, agricultural training and credit was improved. And this is one reason why Zimbabwe has one of the highest rates of family planning usage and the lowest infant mortality rates in Africa. There are many other examples demonstrating the multiplier effects of taking women seriously as breadwinners and improving their access of productive resources.

CONCLUSION

This chapter has demonstrated that a change in any one aspect of women's lives—for good or ill—affects every other aspect. Increasing the availability of family planning will have its full effect on fertility when both women and men are prepared to use it; improving girls' education will make its full impact when it is accompanied by better employment opportunities; better employment opportunities are irrelevant if women are too burdened by childbearing and domestic work to take advantage of them.

Making the necessary changes means recognizing women not only as wives and mothers, but as vital and valuable members of society. It means that women themselves must take power into their own hands to shape the direction of their lives and the development of their communities. It means rethinking development plans from the start so that women's abilities, rights and needs are taken into account at every stage—so that women's status and security are derived from their entire contribution to society, rather than only from childbearing.

RUTH DIXON-MUELLER

Women's Rights and Reproductive Choice: Rethinking the Connections

....

"How is one to understand a woman's attitudes — any woman's — on such a personal issue as the planning of births in her family without first delving into the reality of her life?" asks Perdita Huston in her book, *Message from the Village*.[1] "What is the situation in which she lives? What is the range of her freedom to make decisions? What are the burdens she bears, in numbers of children already born to her or in the tasks she must perform?" Reproductive policies and programs must be tailored to the diverse realities of women's experience if they are to engage women's trust.

If we accept the premise that a woman's ability to exercise her human rights is enmeshed with other aspects of her life, then how can the linkages between reproductive rights and other social and economic rights be most usefully investigated? Where is the locus of control over women's sexual, reproductive, and productive capacities? Who makes the decisions, and how are women's opportunities structured?

HOW THE RIGHT TO FAMILY PLANNING AFFECTS OTHER RIGHTS

The use of safe and effective methods of fertility regulation, especially those that can be used by a woman without the knowledge of her sexual partner, breaks the link between sexuality and reproduction in powerful ways. The knowledge of how to delay a first birth, space additional births,

Ruth Dixon-Mueller is a former professor of sociology at the University of California, Davis, and now works as a consultant to the International Women's Health Coalition. This chapter is excerpted from Population Policy and Women's Rights: Transforming Reproductive Choice *(Westport, CT: Praeger, 1993).*

stop childbearing, or avoid pregnancy altogether gives women the means to shape their lives in ways undreamed of by those who have never questioned the inevitability of frequent childbearing or who have resorted in desperation to cumbersome, ineffective, and often dangerous methods to stop unwanted births. Knowledge of pregnancy prevention can weaken patriarchal controls over female sexuality and reproduction by enabling women to circumvent them. Indeed, this potential underlies the oft-cited fears of men in many societies that contraception will encourage women to be sexually "promiscuous" or too independent in other ways, thus threatening the very foundations of social control. Alternatively, as feminists have pointed out, contraception, sterilization, and even abortion can be used to reinforce patriarchal controls over women's reproductive capacity on the part of the state, the community, the family, or the male partner if they are imposed against the woman's will.

The potential of effective birth control for personal liberation — or, more modestly, for facilitating the exercise of other rights — depends on the environment of risk and uncertainty in which a woman lives. From what wellsprings does a woman's survival, security, or mobility flow? How does she perceive the costs and benefits of different sexual, marital, and reproductive outcomes? Are the opportunity structures in place to provide alternatives to early marriage and motherhood, for example, or to broaden her options if she has three children instead of six? Will contraceptive use and controlled fertility improve her physical health and alleviate some of the physical and emotional stress of competing role obligations? Are various options even permitted to her, or are other facets of her life — and perhaps the use of birth control itself — determined by persons or forces beyond her control? Because sexuality, contraception, and childbearing have such personal and social significance in all societies, it is inevitable that they will be regulated by institutional arrangements linking them with other social relations.

At the individual level, fertility regulation has the capacity to facilitate the exercise of a woman's other rights in at least six ways. The extent to which this abstract capacity is actualized in a specific context depends in large part on whether the woman has some control over the decisions that affect her and on whether the structure of opportunities in the family and community offers some choice.

First, where premarital sexual relations are common, the ability to avoid pregnancy could delay marriage and place a woman in a better position to choose a spouse or choose not to marry at all, that is, to marry only with free and full consent. As for premarital childbearing, at least one-tenth of recently married women had experienced a first birth before their first marital or consensual union in nearly one-third of all countries surveyed by the WFS.[2] The impact of effective birth control would be greatest where a high proportion of early first marriages is triggered by an unplanned pregnancy or where out-of-wedlock births are met with social criticism or economic hardship. If a premarital pregnancy is intended to ensure marriage, however, as in societies where a woman's fecundity must be proven before she is acceptable as a bride, then there would be little advantage in preventing pregnancy with an appropriate partner at this stage.

Second, effective birth planning following marriage can make it easier for couples to exercise their right to marry without incurring the costs of having a child right away. Under these conditions, the average duration between marriage and first birth tends to lengthen. Contraceptive practice could also improve a woman's chances of marrying eventually in societies where out-of-wedlock births are common, as in the Caribbean, yet where a woman without children is in a better bargaining position for marriage than is a woman who has had children by another man.[3]

Third, fertility limitation within a marriage or consensual union could improve a woman's ability to terminate an unsatisfactory relationship with less personal cost if she has the social, economic, and legal option. Insofar as having a large family intensifies women's economic dependence, it limits her capacity to exercise equal rights with men during marriage and at its dissolution. Of course, having many children can be a valuable security strategy for women without alternative sources of support. And in societies where the husband has unilateral power to divorce his wife and take another, or where a man simply abandons a woman to whom he is not married, the fear of repudiation can motivate a woman to have many children to try to bind her partner to her. Under these conditions of risk and uncertainty, a woman with no children or with only one or two may expose herself to a higher risk of unwanted dissolution than a woman with many children. Mead Cain's research in rural areas of India and Bangladesh identifies a sequence of tragic outcomes for couples experiencing "reproductive failure."[4] Itself a partial consequence of poverty which produces smaller family sizes and higher infant and child mortality, the absence of a surviving son increases the vulnerability of both parents to economic crisis and premature mortality. Women are particularly vulnerable because of their economic dependence: for them, reproductive failure raises the risk of divorce and widowhood as well as destitution and early death.

Fourth, fertility control should support the right to education. The most important factor here is the delay of a first birth, whether by postponing marriage, postponing the first birth within marriage, or avoiding a nonmarital pregnancy or birth. The effects on school enrollment of postponing a first birth should be strongest where two conditions are met: (1) there is a high probability that a girl will continue her education beyond the typical age of first intercourse (whether marital or otherwise)

— that is, the opportunity structure for schooling is otherwise favorable — and (2) marriage or the birth of a child would effectively preclude her chances of staying in school. For girls who have never been to school or attend only through the primary level — a situation typical of countries such as Papua New Guinea, Tanzania and Uganda, Pakistan and Bangladesh — postponing a first birth would make little difference. Where continuation through secondary and perhaps into post-secondary education is the norm, however, as in Sri Lanka and Hong Kong, Kuwait and South Korea, or Argentina and Chile, then a woman's ability to avoid or terminate an untimely pregnancy can be crucial. At the individual level, birth control is crucial for any woman who wants to continue: studies in several West African countries have identified a high frequency of abortions among female secondary school students who are desperate to stay in school following an unwanted pregnancy.[5] The policy of expelling pregnant adolescent girls from school (but not boys who have fathered children) intensifies the pressure.

Parental control over the decision as to which of their children will attend school, and when and whom their children will marry, may preclude a young woman from making an individual decision that attaches a higher priority to education than to early motherhood. Indeed, a girl may be withdrawn from school in order to prepare her for marriage or to keep her from becoming too independent. Yet, as literacy or higher education for women become more generally accepted and valued, parental priorities can change. Parents may keep a daughter in school longer to raise her value on the marriage market as an educated bride, which has been cited as one reason for postponing marriage in southern India,[6] or to raise her market wage and contribution to family earnings before she marries, which is a common pattern among the Chinese in Southeast Asia.[7, 8]

Fifth, where opportunity structures favor female employment and where the roles of employee and mother are incompatible, women practicing effective birth planning can have significant economic advantages over those for whom early or frequent childbearing is inevitable. Delaying the first birth may enable women to complete their education and vocational training to qualify for more highly skilled jobs or to establish themselves in a profession. Controlling the timing of births permits women to combine employment and childbearing in the least disruptive way. Contraception can enable a woman in a sexual union to plan for an uninterrupted investment in an occupation (again, where such opportunities exist), perhaps even breaking out of stereotyped female jobs.[9] Keeping family size small makes it easier to work outside the home, especially in those countries where child care assistance is scarce or expensive.

Viewed in this light, a woman's ability to determine the number and spacing of her children can have a direct impact on the exercise of her economic rights. Yet, there are circumstances in which a woman may not be able to improve her chances in the labor market by delaying, spacing, or limiting her births. Employment opportunities for women are often scant in any case because of truncated or segmented labor markets. Women may face discrimination in formal sector employment on the *assumption* that they are (or will be) married and have children — regardless of their actual marital or childbearing status — and thus have less right to a job or an income than a man.[10] In informal sector employment (e.g., crafts production in the home, small-scale trading) or subsistence agricultural production, the number or spacing of children may have little effect. In some cases, higher fertility may actually facilitate her work if children assist in farming, crafts, or trading operations, as in West Africa.[11,12] In other cases, as suggested previously, high fertility may force her to enter the labor force or extend her work time in order to support her growing family.

Sixth, birth control should also enable women to exercise their political rights more fully, insofar as these involve active community work or public sector employment. Again, the usual qualifications apply with respect to opportunity structures and the locus of behavioral controls.

Several general points are also worth making here. One is that the collective impact of contraceptive use and controlled fertility can be greater than the sum of its parts. As women increasingly delay, space, and limit their births and spend shorter portions of their lives rearing children, their claims to equality in education, employment, and political life are likely to become more persistent. Reproductive rights and social, economic, and political rights are synergistic in this respect: pressure for more responsive public policies intensifies as women overcome the "biological imperative" of unavoidable motherhood. A related point is that individual actions are magnified through their intergenerational effects. A girl with few siblings, for example, is more likely to be enrolled in school (all other things being equal) than a girl with many. Fertility limitation *by her parents* improves her educational prospects, in part because resources are distributed among fewer children and in part because she is less likely to be kept home to care for younger siblings.[13] In turn, her own fertility limitation improves the prospects of her daughters.

In sum, given the right conditions, a woman's ability to regulate her fertility can have a profound impact on other aspects of her life and on the lives of others. Recognizing this, one can also turn the question around. In what ways are a woman's contraceptive knowledge and practice affected by her schooling, employment, position in the family, and the exercise of her civil and political rights? How might public policy interventions in these spheres transform the conditions of reproductive choice?

EDUCATIONAL RIGHTS: SCHOOLING AS A PERSONAL, SOCIAL, AND ECONOMIC RESOURCE

A woman's schooling is among the strongest determinants of her contraceptive knowledge and use and of family size, especially in high fertility countries. The relationship is not a simple one, however, nor is it inverse in all cases.[14] The question is how, and under what conditions, does a woman's education make a difference to her childbearing? And what is it about schooling that has an effect?

It may be that the number of years of formal schooling is simply the most visible and quantifiable element in a cluster of interdependent forces affecting fertility. Education beyond the primary level is often associated with factors such as openness to new ideas, higher standards of living, exposure to an urban environment, higher occupational achievement, and a greater range of other options and interests outside the home. Any or all of these could be responsible for the apparent influence of education on fertility. Interestingly, most studies show that the educational level of the wife is more strongly and inversely correlated with a couple's fertility than is the educational level of the husband after controlling for other influences.[15,16] Whatever way the causal mechanism works, investment in female education appears to have a greater impact in reducing family size than the same investment in schooling for males. Mother's schooling also has a greater impact than any other variable typically considered in improving child health and reducing infant deaths.[17,18] Conclusions such as these offer persuasive arguments for affirmative action policies in countries where girls and women are currently educationally disadvantaged relative to boys and men, quite apart from the equal rights rationale per se.

World Fertility Survey (WFS) results reveal generally strong associations between female education and marriage age, desired family size, and contraceptive use in developing countries even when other variables influencing these behaviors are controlled.[19] For example, women in 28 southern countries with seven or more years of schooling married on average four years later than those with no schooling, that is, at 23 rather than 19 years. Secondary schooling had the largest impact on marriage age in sub-Saharan Africa where only a small minority of women were enrolled (a five-year difference) and less impact in Asia and Latin America (3-4 years). Of course, decisions about marriage age and schooling tend to be jointly determined — whether by the woman herself, her parents, or family elders — and so one must be cautious about drawing causal connections. As noted earlier, parents may withdraw a girl from school as her marriageable age approaches in order to protect her reputation and improve her marriage prospects; alternatively, if literacy or higher education are valued on the marriage market, she may be required to stay.

Women with seven or more years of schooling are also from two to four times more likely than women with no schooling to be currently using a contraceptive method. The differentials are widest in sub-Saharan Africa where overall levels of contraceptive use are low. Female education in most southern countries is closely related to family size. Among 30 countries with WFS data, women with no schooling had an average total fertility rate (TFR) of 6.9 children per woman, three children more than women with seven or more years of schooling.[20] Contrasts between the two educational groups were least marked in sub-Saharan Africa (a two-child difference, which persists in Demographic and Health Surveys [DHS]) and most marked in Latin America and the Caribbean (a difference of 3.6 children, which is also replicated in the DHS). Moreover, the educational differential appears to be widening as highly educated women experience more rapid fertility declines than those with little or no education.

Where higher education is confined to a small elite, its impact on overall birth rates is bound to be slight. In Burundi, for example, only one percent of rural women and 30 percent of urban women had some secondary education; in Guatemala, four and 32 percent, respectively. But even the transition from zero to three years of schooling or from three to six years has a substantial (although not necessarily linear) effect on family size in most southern countries. The exceptions are those cases where pronatalist pressures can obliterate the effects of even six or eight years of schooling. For example, Kenyan women with at least some secondary schooling interviewed in the WFS had a TFR of 7.3 children, just one child less than women with no education. In Bangladesh, the comparable figures were 5.0 and 6.1.[21] Discovering similar situations in Latin America, Joseph Stycos[22] suggested that a certain amount of urbanization may be necessary "to activate the effect of education on fertility." More recent data confirm that although the general relationship between female education and family size is similar in rural and urban areas, the impact of female education is usually stronger in cities.

The urban catalyst may be due partly to a connection between female education and employment. Higher education may not motivate a woman to want fewer children if her training does not lead to salaried employment outside the home. Higher education does increase the likelihood of women's employment in most countries, although women with little or no schooling are often pushed into the labor force in even larger proportions because of economic pressures.[23] These class-specific push-pull forces typically result in either a positive or U-shaped relationship between female education and labor force participation that has implications for fertility outcomes. Moreover, not only the number of years but also the type or quality of schooling can make a difference.

Women trained in traditionally "female fields" have been shown in some studies to have more children than women in less "feminine" fields with equal or fewer years of schooling.[24]

RIGHTS RELATING TO EMPLOYMENT AND EARNINGS

To what extent might the exercise of women's equal rights in employment contribute to greater sexual and reproductive choice? If a consistent causal effect were to be found, the implications for development strategies would be clear: Ensuring women's rights to equal work and equal pay would reduce birth rates while simultaneously facilitating economic development and raising household incomes. The relationship depends not on the simple fact of gainful employment, however, but on such factors as the sector of the economy in which the woman is employed; her occupation, income, and work commitment; whether she is a wage earner, self-employed, or unpaid worker in a family enterprise; the duration and continuity of employment; the location of work and whether it is full or part time; and the availability of child care, among other influences. Different motivations for working, and the vast distinctions among women throughout the world in the nature and conditions of their work, carry inevitable implications for the relationship between female employment and fertility.

Three major arguments can be found in the literature about the influence of women's employment on fertility. The *role incompatibility* hypothesis contends that female employment would reduce fertility where contraception is widely available and where the roles of worker and mother are most incompatible, that is, where the job is away from home, which may pose practical problems relating to child care, and where people believe that mothers should care for their own children rather than relying on substitute caregivers of "lower quality."[25, 26] The *opportunity cost* hypothesis also assumes an incompatibility of maternal and occupational roles but is more narrowly economic. A woman should be most motivated to restrict her childbearing where the income she foregoes (if any) by having a child and staying home to rear it is highest.[27] Expected wage rates are generally imputed from her years of schooling modified by the probability of finding employment. The *female autonomy* hypothesis contends that by providing alternative sources of social identity and economic support, employment outside the home could reduce women's dependence on men and children (especially sons); broaden girls' and women's social horizons, thus helping to counter family-based pronatalist pressures; increase women's desire to delay marriage (or to avoid or terminate an unsatisfactory union) and to space and limit births; and contribute to greater sexual and reproductive autonomy. [28, 29]

Note that although at least some of these conditions are expected to apply to women's work in industrialized countries and in urban sectors of developing countries, they do not apply to the work that most girls and women do in agrarian settings.[30,31] Regarding the autonomy hypothesis, for example, demographer Helen Ware points out that the idea of work as liberating "would probably seem very strange to an Indian woman working sixteen hours a day breaking stones in a quarry... or an Indonesian wife harvesting rice stem by stem."[32] Yet, with the right conditions, even the work of illiterate rural women may produce such effects. The most substantial social, economic, and demographic benefits are likely to accrue when the employment provides income over which women have direct control, offers basic economic security, and is considered status-enhancing or status-neutral rather than status-degrading.[33] Jobs located outside the home in a central workplace, labor-intensive light industries, cooperative work organizations with decision-making roles for women, direct access to production credit and marketing, and additional job-related services and incentives such as health education, child care, and family planning all offer possibilities for transforming women's options and enhancing their autonomy.

What kinds of employment are likely to have the greatest impact? Again, one must consider the question of who controls female sexuality, reproduction, and labor power and what the opportunity structures for employment are. In general, the least likely candidates for transforming gender relations and fertility decision making are those that do not confront patriarchal family relations of production and reproduction. These include unpaid work in the family fields or livestock herds (no matter how productive or valuable to the household's consumption), unpaid work in other family enterprises such as cottage industries, and self-employment in enterprises that have low returns to labor or where women depend on men for their capital, materials, or marketing.[34] In countries such as Burundi, Rwanda, Mali, Iraq, the Philippines, Thailand, and Turkey, for example, the majority of all economically active women are unpaid family workers, typically in agriculture. (If unpaid and informal-sector activities such as these were more adequately enumerated, such findings would apply to many other countries as well.) In these situations, wealth and decision-making power continue to flow from the young to the old and from women to men.

In contrast, work for wages (what Caldwell calls the labor market mode of production) is far more likely to provide girls and women (as well as young men) with moral leverage to challenge patriarchal controls over their sexual and reproductive lives *if* the earnings contribute a substantial share of the household income or could provide an independent livelihood. (Patriarchal ideologies may demand the appropriation of the earnings of daughters or wives by elders or husbands as part of a "family wage," however, at least in transitional stages from familial to labor-market modes of production.)[35,36] This is where women's access to modern-sector jobs at equal wages is critical. Yet, in North Africa and the Middle East, women hold fewer than 15 percent of all waged and salaried jobs, on average, and in sub-Saharan Africa, from 15 to 25 percent.[37] Corresponding figures for Latin America are 20 to 40 percent, for Asia (excluding Muslim South Asia) 30 to 40 percent, and for the Caribbean, 40 to 50 percent. These differentials are clearly related to fertility rates at the aggregate level, although both rising wage employment and declining birth rates are functions of overall development and per capita incomes.[38]

RIGHTS ON ENTERING MARRIAGE

The principle of equality in marriage has elicited considerable controversy in international debates over women's rights. The idea that daughters and sons should have the right to choose whom and when to marry, and that wives and husbands should have the same rights and responsibilities in marriage, collides headon with the customary rights and privileges of men and elders. In many countries equality within the family has not yet been recognized in civil law and upon marriage a woman may be deprived of many rights such as the independent ownership of property or the right to work or travel without her husband's consent. Even in countries where legislation favors equal rights, traditional cultural patterns of male dominance in marriage and the family are slow to change.

The decades since 1960 have seen a dramatic rise in female age at first marriage and a slight rise in proportions never marrying in many southern countries under the influence of urbanization, education, and economic change.[39,40] Although many young girls continue to marry as children and bear children as children, especially in sub-Saharan Africa and in South Asia, contrasts are marked in every region. In the 1980s from half to two-thirds of women ages 15 to 19 in Afghanistan, Bangladesh, and Nepal were already married, for example, but only one in ten in Sri Lanka married this young.

How does the timing of marriage affect the ability of girls and women to exercise their sexual and reproductive rights, apart from its association with higher education and premarital employment? First, early entry into a marriage or consensual union often contributes to (or results from) limited contraceptive knowledge and practice, at least in the early years. It is also associated with higher completed fertility: a gross difference of two to four children, on average, between those who marry before age 18 and at 25 or older.[41] In this sense, delayed marriage appears to facilitate control over the timing and number of children.

Second, early marriage, especially before age 16, escalates the health risks of pregnancy and contributes disproportionately to maternal death and disability where medical resources are scarce.[42] In Bangladesh at least one woman in five gives birth before she is 16, for example, and reproductive complications can be severe. In Ethiopia, half of a sample of women interviewed at health and family planning clinics said they first had intercourse with their husbands before having menstruated, and one in five had done so before age 13.[43] Early intercourse was associated in the sample with a higher prevalence of sexually transmitted diseases and of cervical cancer. Thus, delayed marriage can facilitate the exercise of a woman's right to reproductive health.

Third, in most southern countries the earlier a woman marries, the more likely it is that her husband is considerably older. This is especially true of arranged marriage systems where a daughter may be married off to a widower or to a polygynous husband who already has another wife, where patrilocal residence is the rule, and where contractual transfers of wealth between families in the form of dowry or brideprice serve to legitimize the claims of the male lineage to the couple's offspring.

Surveys in several sub-Saharan African countries found that recently married women were six to eight years younger than their husbands, on average, compared with about 2.5 to five years in Latin American and Asian countries but up to nine years in Bangladesh.[44] Early marriage for girls combined with a wide age gap can perpetuate a situation of powerlessness between the daughter whose marriage is arranged and the parents or elders who arrange it, and between the young bride and her husband and his kin. Indeed, this is precisely the intention.

Cain's analysis of the consequences of the age gap between spouses in a number of southern countries demonstrates that it acts as a "filter" through which social and economic changes in society affect the couple's fertility. Regions of strong patriarchal family structures where husbands are at least five years older than their wives on average (sub-Saharan Africa, South Asia, and the Middle East) have higher fertility than do weak patriarchal regimes with smaller age differences (Southeast and East Asia, Latin America) even when other factors affecting fertility such as infant mortality, per capita income, female secondary school enrollments, and female age at first marriage are statistically controlled. Fertility rates in weak patriarchal regimes respond in the expected way to variations in the other factors whereas in strong patriarchal regimes they are resistant. In this sense, then, both the timing of marriage *and* the age difference between spouses appear to be important determinants of fertility behavior as well as of other rights.

The loss of parental control over the arrangement of their children's nuptials, insofar as such control is associated with early and universal marriage in an extended family system, should serve to delay marriage on the average and to increase the probability of nonmarriage for some women, either voluntarily or involuntarily. Courtship, after all, takes time, and an independently contracted marriage requires a degree of maturity not needed or even desired of a young girl whose primary obligation is to obey the wishes of her husband and her elders. The free choice of a spouse — even with a considerable amount of parental guidance — also implies a degree of equality between husband and wife at the time of their marriage that may be important for effective communication about family size desires and the practice of family planning.

GENDER EQUALITY WITHIN MARRIAGE

Under what conditions are women likely to decide equally with their husbands on contraceptive use and the timing and number of children? What impact does (or might) such equality have on reproductive outcomes? This area is a rich and fascinating one to explore, for it is in the everyday interactions between women and men that extraneous factors such as education and employment are translated into actual reproductive patterns through birth planning behavior.[45, 46, 47] Unfortunately, we can touch on it only briefly here.

Resource theory argues that the greater are the material and social resources that a woman brings into her marriage relative to those of her husband (e.g., education, paid employment, personal assets, a strong network of social support from her relatives), the greater is her power or influence over family decisions.[48, 49] Legal rights are also resources. Gender inequalities in the public sphere are thus mirrored in the private sphere, with implications for the nature of family decision making. Yet, there are at least two possibilities of overriding their effects. First, gender role ideologies may support or negate the importance of relative resources. A "modern" educated couple may believe strongly in the ideal of equality in decision making despite a large difference in their incomes, for example, whereas a more traditional couple may believe that the husband should make the major decisions even though his wife earns as much or more than he does. Second, the possibility of using contraception, sterilization, or abortion surreptitiously permits a woman with fewer resources to prevent an unwanted pregnancy or birth unilaterally without having to confront her partner or other decision makers directly. (The fear of discovery may inhibit this option, however; in addition, evidence suggests that contraceptive acceptance and continuation are related to the husband's approval.)[50, 51]

A related approach looks to the degree of gender role segregation as a key to understanding the decision-making process. Some studies suggest that married couples who share a more equitable division of labor within the home — that is, who have joint rather than segregated role relationships — are more likely to communicate with one another about sex, family size desires, and birth planning, which in turn leads to more empathy with the partner's feelings, greater sexual satisfaction, a desire for fewer children, and more effective contraceptive use.[52, 53] Empirical support for these findings in developing countries is sketchy, however, and the causal paths are unclear.[54] A fuller understanding requires knowledge of the cultural and structural conditions of each society.

RIGHTS TO PROPERTY AND SOCIAL SECURITY

Whereas a great deal of attention has been paid to the effects on women's sexual and reproductive choices of education and employment, the effects of rights to property and other forms of support have not been systematically addressed. How does the denial of the right to own, inherit, or bequeath property affect a woman's perceptions of the need to stay married, for example, or her desired family size? Is a woman who bears no children, or only one or two, or only daughters, disadvantaged under some legal systems more than others?

The question of property rights and public support is crucial for both married and unmarried women. According to DHS findings from 21 countries, fewer than two-thirds of all women of reproductive age (15 to 49) in the late 1980s were currently married or cohabiting. Over one-fourth had never (or had not yet) married or entered a consensual union, and an average of one-tenth were widowed, separated, or divorced. At ages 60 and over, the proportions of women not in unions typically rise to 50 to 70 percent.[55] Because women's life expectancy exceeds men's in almost every country and because women typically marry at younger ages than men, women whose marriages are terminated by the death of a spouse can generally expect to spend from five to 15 years or more as a widow.

It is often assumed that widows in most societies are protected by strong extended family systems: by their sons, perhaps, or by their husband's kin. This may be true among the affluent, but research suggests that households facing economic crisis are increasingly unable or unwilling to support additional dependent members.[56] In short, rather than veneration a widow may face ill treatment, exploitation, or neglect at the hands of her relatives. Alternatively, she may be forced into exploitative labor markets, often with limited skills and experience. With highly imbalanced sex ratios at older ages, most older women's chances for remarriage are slim. Considering the probability of divorce as well as widowhood, and the higher propensity for divorced and widowed men in most cultures to remarry, the number of divorced, separated, or widowed women can greatly exceed the number of men. In Bangladesh, for example, formerly married women of all ages outnumber formerly married men by nine to one.[57]

Many of these women — as well as those who have never married or cohabited — are heads of households living alone or with their children. Sometimes termed the "ignored factor in development planning,"[58] the growing number of households headed by women in both northern and southern countries derives from a mixture of rising rates of delayed marriage, non-marriage, divorce, desertion, and out-of-wedlock births in many countries; the breakdown of kinship obligations that formerly absorbed single or previously married women into ongoing households; male outmigration from rural areas that leaves many wives as de facto heads of households; and the inability or unwillingness of husbands to support their families in times of economic crisis. The proportions of all households headed by women run typically from 10 to 20 percent in Asia, 15 to 30 percent in sub-Saharan Africa (but 45 percent in Botswana), and 20 to 40 percent in Latin America and the Caribbean.[59] In view of these trends, the elimination of discrimination against women in all aspects of property rights and social security as well as in education and employment becomes especially compelling.

Nondiscrimination in property rights is also compelling from another point of view: its impact on women's marital and reproductive goals. The limited research that has been done on this subject — primarily in rural Bangladesh and India — suggests that the production of children, especially sons, is a form of insurance for women who are denied independent access to income and land.[60, 61] Facing a high probability of divorce, desertion, or widowhood, and with no public assistance for the destitute and no pensions in old age, rural women in resource-poor households have little option but to turn to their sons for support. Discrimination against women in labor markets and property rights exacerbates their vulnerability.

SUMMING UP: THE NEED FOR AN INTEGRATED REPRODUCTIVE POLICY

The relationship between the status of women and fertility has attracted considerable interest at international population conferences and among some multilateral and bilateral donors and non-governmental organizations involved in population assistance. Much attention has been paid to the possibility that improving women's positions in the family and society — especially through the expansion of female educational and employment opportunities — could significantly reduce birth rates in developing countries with rapid population growth. The World Population Plan of Action adopted at the World Population Conference in Bucharest in 1974, for example, urged "the full integration of women into the development process, particularly by means of their greater participation in educational, social, economic and political opportunities… "as a development goal that would also create conditions favorable to smaller family size."[62] Similar recommendations were made at the 1984

International Population Conference at Mexico City.[63] In this context, raising the status of women represents one means among many of achieving the demographic goal of lower birth rates.

Less attention has been paid to the impact of fertility regulation on women's rights, that is, on family planning as one means among many for reducing gender inequalities as a social goal. The 1974 Bucharest population conference touched on this briefly with its statement that "...the opportunity for women to plan births also improves their individual status,"[64] while the 1984 Mexico City conference noted more strongly that "The ability of women to control their own fertility forms an important basis for the enjoyment of other rights...."[65] The heavier emphasis placed on the demographic rather than the human rights side of the equation is understandable in view of the concern with population dynamics expressed at international meetings and in population program assistance. Yet, as we have noted in this chapter, a woman's ability to determine the number and spacing of her children can influence in fundamental ways her ability to complete her schooling, to take full advantage of employment opportunities, and to acquire greater control over the terms of marital or nonmarital sexual relationships.

The crucial factor shaping the relationship between reproductive choice and other rights is the structure of opportunities that girls and women face. A rural woman who has never been to school gains no educational advantage from delaying her first birth, for example, nor does an illiterate woman forced by economic hardship to labor in the fields or factories for minimal wages improve her chances for advancement by having fewer children. By the same token, a woman kept secluded in her husband's household is denied the right to work regardless of her reproductive behavior. Thus, it makes little sense to promote policies encouraging contraceptive use and smaller families without simultaneously addressing other legal, social, and economic constraints on women's rights. Similarly, it makes little sense to promote policies advancing women's integration in development without addressing their fundamental need for sexual and reproductive choice.

The role of public policy in this context is twofold. First, women's rights to participate equally with men in policy-making and implementation need to be fully recognized. And second, substantive policies need to be designed to eliminate gender discrimination in health care, schooling, employment and income generation, property rights, family law, and public life. Government ministries can set specific targets in all sectors and invest selectively in promoting women's participation until equal access has been achieved. Within this framework, a reproductive health program offering women the means to choose how many children to have and when to have them takes on real meaning.

CYNTHIA B. LLOYD

Family and Gender Issues for Population Policy

....

Implicit in the debate over population policy are certain assumptions about the family and about the roles women and men play within it. The central assumption for population policy is the long-term stability of the conjugal family as a closed physical, economic, and emotional unit within which children are planned, borne, and reared. When the policy debate centers on family planning and the supply of contraceptive methods, it is often further assumed that meeting a couple's needs for fertility regulation is synonymous with meeting the individual needs of men and women. This assumption is implicit in the term *"family* planning," which frames decisions about childbearing exclusively within a family context. Indeed, the 1974 World Population Plan of Action and the 1984 Mexico City recommendations for its further implementation assert the centrality of the family as "the basic unit within society," while at the same time declaring that both "couples" and "individuals" should have "the basic right to decide freely and responsibly the number and spacing of their children."[1]

When the debate turns to development policies designed to affect the demand for children, in particular policies targeted to improving the "status of women," several additional assumptions about the roles

Cynthia B. Lloyd is senior associate and deputy director of the Research Division of The Population Council. This is a revised version of a paper presented at the United Nations Expert Group Meeting on Population and Women, Gaborone, Botswana, 22-26 June 1992. It is forthcoming in Proceedings of the UN Expert Group Meeting on Population and Women, *to be published by the United Nations.*

of women and men within the family and the intra-family distribution of resources are implicit in the linkage typically drawn between rising costs of children and declining demand for children: (1) that improvements in women's livelihoods outside the family provide them with greater economic mobility and thus lessen reliance on children and other family members for future economic support; (2) that fathers share with mothers joint responsibility for their children's maintenance and upbringing; and (3) that parents support each of their children to an equal extent.

These assumptions structure the collection and analysis of demographic data and the design of population policy. The goal of this chapter is to examine the empirical evidence surrounding these assumptions and to draw out the implications of that evidence for future research and policy. In so doing, the chapter builds a case for a much broader framework within which to view population issues, one in which family organization and gender relations are central.

Women acting on behalf of the family are seen as agents of change in all aspects of population and development policy, whether it be the adoption of family planning, the provision of health care for children, or the acquisition of independent economic livelihoods. This chapter argues that women cannot bring about the demographic transition alone, particularly within the context of existing family structures and gender relations in many of today's high-fertility countries. Men have much to contribute as well. Indeed, the extent of women's autonomy and men's family responsibility will likely dictate the pace at which economic and social change as well as population policy are able to affect demographic behavior. The effectiveness of population policy would be much enhanced if we knew more about men's reproductive and familial roles and about how the costs and benefits of children are distributed.

THE FAMILY: WHO IS A MEMBER AND FOR HOW LONG?

The family has different meanings in different cultures, but at its core in every society are parents and their biological children. Simple models of the family rely on the assumption that these core family members reside together in the same household and function within a unified household economy. Parents are assumed to plan, bear, and rear children jointly with a long-term view of their costs and benefits. The head of the family is assumed to be an altruist acting on behalf of this core family unit, organizing production among family members so as to maximize efficiency and distributing resources fairly. Reality differs from this model in ways that have important implications for population and development policy.

Family relationships operate within a framework of culturally accepted notions about the division of rights and responsibilities by sex and generational position. As a result, families, according to the degree of their connectedness as well as the distribution of power within them, mediate the effects of policy on intended individual beneficiaries through the redistribution of resources and responsibilities among family members. Three factors — coresidence, multiple membership, and longevity — affect the strength of family bonds and therefore the extent to which family members function as a unit and act altruistically toward one another.

Connections: residential or relational?

For a variety of reasons, a child's biological parents do not always live together. These include job migration, polygamy, divorce, and remarriage, as well as childbearing outside wedlock. While this is widely known, it is more surprising to realize the extent of a mother's reproductive years in certain settings that are not spent living in the same household with their children's biological father(s). Among those women who are mothers, the estimated proportion of time during the reproductive years spent living with first husbands (in most cases the first child's father) is as low as 38 and 39 percent in Ghana and Botswana, 49 percent in Zimbabwe, and 59 and 60 percent in Senegal and Kenya respectively. The non-coresidence of couples due to custom or migration as well as to high rates of divorce (and remarriage) are important factors in explaining these low proportions. In contrast, mothers in Sri Lanka (the only non-African country for which data are available on spousal coresidence) spend most (89 percent) of their reproductive lives living with their first husband. Thus, the assumption that family members live together in the same household may be appropriate in some settings but very far from reality in others.

While physical distance between spouses does not preclude the exchange of financial support — indeed family separation is often motivated by economic reasons — distance can make economic links less secure, particularly with the passage of time. A recent review of the literature on migration points to the fact that, while the chief reason for men to migrate away from their families is to generate support for the family, the subsequent flow of remittances is typically uncertain and highly variable, often leaving women to support themselves and their children inefficiently or to rely on other family members.[2] For example, in Lesotho, poor agricultural practices and yields have been identified as consequences of the unpredictability of remittances of male migrants to women and children.[3] Research based on U.S. data shows that a father's presence in the home is closely linked to the extent of his financial and emotional commitment to his children.[4]

Financial exchange between parents is even more precarious when parents are not linked to each other through marriage, either because of childbearing outside of marriage or because of separation and divorce. The extent of a father's support for his children appears to be affected by, among other factors, his sexual access to the mother of those children. Research on child-support arrangements in the U.S. demonstrates that, when parents are divorced or separated, relatively few children receive financial support from their father.[5] Much less information is available for developing countries on fathers' contributions to children's support in cases of divorce or nonmarital childbearing. This is mainly because so much of the data we rely on are derived from women-based surveys, such as the Demographic and Health Surveys, which link children with their biological mothers but not explicitly with their biological fathers.[6]

The proportion of a mother's reproductive years after age 20 spent unmarried varies from a low of four percent in Tunisia to a high of 46 percent in Botswana. In certain countries, women spend substantial amounts of time married to husbands who are not their first. There is substantial intercountry and interregional variation in both of these indicators; but, on average, mothers in sub-Saharan Africa spend roughly one-third of their reproductive years outside of marriage or in second or higher order marriages. The average for Asia and North Africa is less than half the African average, with Latin America falling between the two extremes. The Dominican Republic, the only Caribbean country for which data are available, shows its own distinctive pattern — with only slightly more than 50 percent of a mother's reproductive years spent in a first marriage.

Obviously, when parents do not live together, children cannot live with both biological parents. Even when parents do live together, however, children sometimes live apart from them, particularly in sub-Saharan Africa where child fostering is common. Unfortunately, because most data collected on fostering are based on interviews with women, there has been no attempt to establish the relationship between a woman's children and her current marital partner; therefore, it is not possible to assess the extent to which children live with both biological parents.

To get a fuller picture of children's living arrangements, we need household-based data. Such data are available for Ghana, where a recent study found that the proportion of school-age children living in a household without their biological father is 43 percent, with as many as 22 percent of children living in a household with neither biological parent.[7] While Ghana presents an extreme contrast to the traditional assumptions about the coresidential core family unit, it highlights the importance of not taking these assumptions for granted and of exploring the particular social arrangements in each setting that may lead to violations of commonly held assumptions.

Boundaries: exclusive or overlapping?

One reason core family members cannot all live together in the same household is that some of them are members of more than one core family. Men and women do not always form exclusive bonds that last throughout their respective reproductive years, but sometimes have children with more than one partner. From a child's perspective, this means that the set of siblings they share with their mother may not be entirely the same as the set they share with their father. Thus, rather than functioning within a unified family economy, many children compete with one set of siblings for their mother's resources and with another set of siblings for their father's resources. Currently available women-based data do not tell us the prevalence of such divergent sibling sets among children in different settings. Again taking Ghana as an example, however, one study found that the average number of siblings school-age children share with their father is substantially greater (5.9) than the number they share with their mother (4.1), suggesting that divergent sibling sets among children are not uncommon in a polygamous society or one in which divorce and remarriage are prevalent.[8]

The fact that men's reproductive years extend over a much longer period than women's is an additional factor contributing to the phenomenon of overlapping sibling sets. Male and female fertility rates often diverge at older ages, with men experiencing higher fertility than women. While little data have been collected on overall male fertility as compared with female fertility, a few recent surveys in West Africa (Ghana and Mali) allow a direct comparison of the cumulative fertility of currently married women and their husbands by age. In Mali, women end their reproductive careers with 4.3-4.5 living children, while those husbands in the sample between ages 50 and 55 have on average 8.0 living children. In Ghana the story is similar, with women ending their reproductive careers with 5.7 living children on average while those husbands in the sample over age 50 had on average 8.5 living children.[9]

Stability: long or short term?

A life cycle perspective provides a more wide-angled view of the stability or instability of the core family unit. What percentage of men and women will remain with the same spouse throughout their reproductive lives? What percentage of children will live with both biological parents throughout their childhood years? What percentage of children will acquire a stepparent before they become adults? These are questions that need answers if we are to understand the family context within which reproductive decisions occur and childrearing takes place.

FERTILITY REGULATION: AN INDIVIDUAL OR FAMILY STRATEGY?

Couples form family units within which to bear and rear children. At the same time, however, individual men and women (and boys and girls) engage in sexual relations outside of marital unions — sometimes leading to pregnancy and childbirth — without planning for a family. Furthermore, some of these men and women are simultaneously spouses in marital relationships. Thus, individual needs for fertility regulation relate simultaneously to the achievement of fertility goals within families and the control of fertility outside of families. To complicate things further, husbands and wives may not share the same fertility goals for their family unit or for their reproductive careers. In such cases, differences can be resolved within the marriage if both members compromise, if one member subsumes her/his wishes to those of the other, or if the husband or wife realizes his/her excess fertility goals with another partner outside the marriage. Differences in fertility preferences, however, can also be a cause of marital dissolution. The threat of marital dissolution is a factor determining how differences are resolved. The partner who is more dependent on the marriage is the one more likely to defer to the preferences of the other.

OUTSIDE OF MARRIAGE: WHO IS RESPONSIBLE?

Research on fertility regulation behavior has been typically based on currently married women of reproductive age. These are the women assumed to be exposed to the risk of pregnancy and childbirth. While there is growing recognition of the wide variation across societies in sex outside of marriage, studies estimating current levels and projecting future trends in contraceptive prevalence traditionally ignored the behavior of unmarried women, for the practical reason that many surveys have not asked unmarried women about their contraceptive practice.[10]

Estimates of "unmet need" for contraception, "couple years of protection,"[11] future contraceptive demand, and the costs of contraceptive commodities[12] are largely based on the assumption that unmarried women are not sexually active.* Recent data on contraceptive use among unmarried women from the United Nations, however, reveal that they (and their male partners) are a potentially important group of users in many countries. Indeed, in several African countries — notably Ghana, Liberia, Mali, Togo, and Uganda — use among formerly married and/or single

---

* *In those countries where all women were asked about their contraceptive practice, Mauldin and Ross (1992) reclassified unmarried women reporting contraceptive use as married and their use was thus accounted for in the estimates of future contraceptive use and commodity costs.*

women exceeds use among the married group. In some sense, this should not be surprising given the strong motivation of unmarried people to avoid pregnancy despite the obstacles they may face in obtaining information and supplies. Among those countries for which data are available, contraceptive prevalence among the formerly married averages roughly 17 percent in sub-Saharan Africa, 20 percent in Latin America, and 18 percent in Thailand and Sri Lanka — the only Asian countries for which data are available.† Use among single women in Africa averages 11 percent but is notably lower in Latin America and no data are available for Asia. When men and women engage in sexual activity outside a marital union, their needs for contraception are individually based rather than couple based. An assessment of contraceptive needs requires information on the behavior of both men and women.

Rising ages of marriage have been the most important factor leading to increases in women's exposure to the risk of pregnancy outside of marriage. One sample of 14 countries from different parts of the developing world indicates that the proportion of a woman's reproductive years spent unmarried has risen from roughly 29 percent to 32 percent over the last two decades. With the exception of Senegal, where women spend less than one-fifth of their reproductive years unmarried, proportions range from 26 percent to 40 percent, with the sharpest increases having occurred in Africa.

These data highlight the substantial and growing proportion of women's reproductive years that escape our attention when we focus exclusively on couples. The contraceptive service needs of adolescent men and women are particularly noteworthy. Furthermore, the growing risks of unwanted pregnancy and extra-union childbearing point to the importance of giving greater attention to male sexual behavior and contraceptive needs in the interest of helping men become more responsible partners.

Within marriage: who decides?

The assumption underlying estimates of unmet need is that, if a married woman expresses a desire to limit fertility but is not currently using contraception, she would adopt family planning if it became available. The potential influence of her husband (and/or other family members) on her contraceptive behavior is ignored. The question here is whether women and men (or other family members) share the same fertility preferences and the same attitudes toward family planning and, if not, what are the implications for contraceptive use and fertility.

---

† *Use among the formerly married in some cases includes terminal methods previously adopted within a marital union.*

A recent review of the literature found no evidence that women consistently want fewer children than men on average.[14] However, it is not always clear whether questions about desired family size are interpreted by respondents to relate exclusively to their current conjugal relationship or more broadly to their lifetime, which might include the possibility of multiple partners. Furthermore, agreement at the societal level may mask disagreement at the level of the couple. In Ghana, for example, only 23 percent of couples report the same desired family size.[15] Data on husbands' and wives' fertility preferences in Ondo State, Nigeria show little agreement between husbands and wives on future fertility.[16] In other settings, disagreement is less extensive but nonetheless problematic for a potentially pivotal group of couples. For example, 20 percent of Thai couples gave contradictory answers to the question on whether more children were desired.[17] In Egypt, 17 percent of urban couples and 22 percent of rural couples gave different responses on the question on desire for more children.[18]

Attitudes toward family planning form an important link between fertility preferences and actual contraceptive behavior. Even when husbands and wives agree on their fertility preferences, they may not share the same attitudes toward the actual practice of family planning. For example, in Mali the opinions of wives and husbands were very different, with 16 percent of husbands and 62 percent of wives approving of the use of family planning for limiting or spacing births.[19]

What happens when partners disagree? Data on male attitudes from Zimbabwe[20] and Sudan[21] document that men perceive that they should have the major role in the decision to use family planning. In Ghana, both men and women who participated in recent focus group discussions expressed the view that, when differences in fertility preferences occur between a husband and wife, the man's preferences usually dominate.[22] As a young mother from the Volta region said, "When I wanted to do family planning, my husband did not allow me so I didn't do it."

Only a few studies of husbands' and wives' attitudes toward family planning attempt to link these attitudes with actual behavior. Results from the Egyptian World Fertility Survey showed that husbands' fertility preferences are substantially different from those of wives and more closely linked to women's use of contraception.[23] Women-based data from metropolitan Indonesia show that a wife's perception of her husband's approval of contraceptive use is the most important determinant of her actual use.[24] An analysis of the recent Demographic and Health Survey in Ghana shows that the consistency of husbands' and wives' attitudes toward family planning is an important factor in the level of contraceptive use. Whereas 38 percent of couples in which both partners approve of family planning are currently using some form of contraception, less than 11 percent of those where only one spouse approves are doing so.[25]

Clearly, women have an important say, but not always the final say, in contraceptive use. These data reveal, however, that the formation of preferences, the reconciliation of differences, and the specific actions taken by individual men or women are the outcome of a complex negotiation process. The relative power of women in sexual relationships will depend on their access to and control over resources as well as on the strength of the ties that bind their male partners to them. The types of family planning methods most suited to individual needs in terms of privacy, secrecy, and the need for cooperation between partners will depend on the nature of these relationships.

THE DEMAND FOR CHILDREN: DO RISING COSTS LEAD TO DECLINES IN FERTILITY?

No one denies that the costs of rearing children are rising rapidly in most parts of the developing world. Growing labor market opportunities for women provide them with alternatives to children as sources of support and fulfillment, thus raising the opportunity cost of their time and the indirect costs of children. As economies modernize and diversify, child maintenance requires increasing access to the cash economy for the purchase of nutritious foods, essential medicines, and school books and uniforms. While primary health clinics and schools are now accessible to most parents, the direct costs of these services have been rising rapidly as many governments seek to privatize them.

With various elements of the costs of children rising, in part due to the withdrawal of government subsidies, parents may be induced to consider changes in their fertility behavior. However, other responses are also possible. Parents may seek ways to shift some of the increased costs of childrearing onto other relatives and older children. Fathers may seek to shift more of the burden onto mothers. Alternatively, when family resources are inadequate or credit is unavailable and when returns on child investment appear low or uncertain, parents may select only certain children to be the beneficiaries of their investments or distribute their investments unequally. Less favored children can stay at home, fend for themselves, and help support their parents and siblings without the benefit of education or proper health care. Few developing country governments make altruistic parenting institutionally compulsory,[26] and in many cultures the idea of equity among household members — in particular, siblings — is completely foreign.

Improved livelihoods for women: who benefits?

Improvements in the status of women and the equalization of their rights with those of men were identified as essential prerequisites for the effective implementation of the World Population Plan of Action in the recommendations of the 1984 International Population Conference in Mexico City. Included among the recommended improvements were women's equal access to and control over resources, in particular education and employment in the paid labor force.[27]

However, unless women can capture directly the gains from improvements in education and livelihoods and translate them into greater personal autonomy and economic mobility, their economic gains are likely to be dissipated within the family. The result is apt to be a reduction in men's family responsibilities, leaving women dependent on men for essential complementary resources and dependent on children for long-term security. Such an outcome would lead to a reduction in men's economic contribution, an increase in women's labor market work, and relatively little change in the demand for children, despite the increased personal costs of children shouldered by women.

A growing body of evidence from developing country settings indicates that women's economic activity is not necessarily incompatible with the bearing and rearing of children.[28] However, certain types of work are more difficult to reconcile with the bearing and rearing of large families — in particular, work for cash in the modern sector.[29] As a result, there has been much effort in the donor community to promote income-earning opportunities for women through training and access to credit. Indeed, some of these interventions (e.g. in Bangladesh, Indonesia, Thailand) have been directly linked to the design and implementation of family planning programs.[30]

There is evidence, however, that even women's gains from certain types of self-employment in the modern sector, specifically in commerce, do not necessarily lead to increased mobility in the form of business expansion, given the constraints imposed on women by family institutions and prevailing gender ideologies.[31]

In three case studies drawn from disparate cultural settings — Ghana, India, and Thailand — anthropologist Susan Greenhalgh identifies the factors limiting women's business opportunities.[32] In the case of vegetable vendors in Madras, women's limited physical mobility required them to depend on men to perform crucial business transactions. Women turned over their income to the men. Business success led to greater male inputs, with the inevitable consequence that growing businesses gradually became male-controlled. In the case of market women in Ghana and Thailand who had full independence in the conduct of business, their business mobility appeared to be limited by claims made on their income by other family members. The financing of children's education was an important element of women's financial "obligations" in both instances. Husbands made additional financial claims in the case of Thailand, as did extended kin in the case of Ghana. In the Indian case, women's success led to increasing male control; in Thailand and Ghana, women's success led them to take over a larger share of family obligations. In each case, men reaped the surplus from women's entrepreneurship, either directly for themselves or indirectly in the form of reduced obligations to other family members.

These examples indicate that, in talking about women's livelihoods, it is important to identify those elements that are likely to enhance their access to and control over the necessary inputs for their work as well as the profits derived from such work. Men's dominant position within the family and situational advantage in the world of work must be directly recognized in the design of interventions to enhance women's income-earning opportunities if these are to become genuine economic alternatives to the family as a source of women's economic support. More research will be needed to understand the factors that encourage men's economic support of families — in particular, their children — in a world in which women are playing an increasingly important role in the cash economy.

Locating the costs of children: who pays?

In an idealized traditional division of labor between mothers and fathers in the maintenance and support of their biological children, mothers contribute their time directly to care and nurturing while fathers provide necessary material inputs. In societies where the support of children is not confined to the nuclear family unit, female relatives share the mother's roles and male relatives the father's roles. In the process of development, the costs of children to their parents (and other family members) inevitably rise, primarily because of increased aspirations for children's education. Children's enrollment in school involves not only direct monetary outlays but also a reduction in the family labor force. How do changes in the cost of children affect the traditional division of parental and more broadly shared family responsibilities? Few studies have addressed the question but the answers are likely to depend on, among other factors, the structure of the family and the role of family members other than parents in the support of their children.

In polygamous societies, men view wives as a source of wealth. Wives are expected to provide largely for themselves and their children through their own labor and through their access to the resources of other family members. Evidence suggests that in sub-Saharan Africa, where polygamy is prevalent, mothers provide food for themselves and their children.[33] Concrete evidence of mothers' and children's independence from fathers for economic support emerges from a comparison of children's nutritional status in monogamous and polygamous unions in Ghana, Mali, and Senegal. One study found that child malnutrition was no more apparent among children in polygamous unions than in monogamous unions, despite the fact that women in polygamous unions have claim to a smaller proportional share of their husband's resources than women in monogamous unions.[34]

Because co-wives operate with a certain amount of financial autonomy and without full knowledge of their husband's resources, it is relatively easy for a man with several wives to shift a substantial share of the material costs of his children onto their mothers. This process is facilitated by the breakdown of the traditional marriage contract with the introduction of Western monogamous marriage forms in much of urban West Africa. This has resulted in the spread of informal polygamy (*deuxieme bureau*), which is outside the reach of customary law and practice.[35] The result is that men can choose the amount of support they provide to their children and discriminate between them according to the status of their relationship with the child's mother.[36]

In Latin America and the Caribbean, the role of the extended family in child support is not as extensive and, hence, the consequences of weaker conjugal bonds for children's welfare are more apparent. There is evidence of a substantially higher prevalence of stunting among children whose mothers are in consensual unions compared with mothers in legal unions, controlling for family size and parental resources.[37] This suggests lesser child investment by parents in consensual unions, probably the result of a smaller paternal contribution in less stable unions. Demographer Mead Cain referred to this as the "free rider" problem, in which men can father children with few, if any, economic consequences for themselves.

There is evidence of rising numbers of female-headed and -maintained households in developing countries,[38] causing concern that a growing proportion of children may not be receiving their full share of support from their biological fathers. However, households become female-headed for a variety of reasons. While female-headed households with children are, on average, poorer than male-headed households, well-off women and children are found in a variety of household types. Children's welfare is conditioned less by headship or even residence than by whether the costs of children are primarily borne by women alone or are shared.[39] The growing evidence that the internal distribution of resources in female-headed households is more child-oriented than in male-headed households suggests that mothers and fathers have very different expenditure priorities. This may be due to the very different benefits mothers and fathers expect from investments in their children. In a polygamous society, for example, older men can expect to be supported by younger wives, while older women may have to rely largely on the support of their own children. This material support may come in exchange for further childcare responsibilities, as grandmothers become the caretakers of the next generation.

It is important to note here that suggestions of possible cost shifting between fathers and mothers are not based on direct observation but rather on observed differences in resource allocation that occur between family and household types to which men (in particular, fathers) have varying degrees of commitment. We have not yet begun to explore the factors influencing the bargain that parents strike with each other and other family members over the division of childrearing responsibilities or how it is renegotiated over time in response to the arrival of additional children and changing family relationships. These should be future research priorities.

Resources for children: who is the lucky child?

An implicit assumption in much research on intra-family resource allocation is that parents are altruistic toward their children — in other words, that they distribute resources "fairly" among them. The negative effects of family size on child outcomes have been interpreted as the result of parents having to distribute their given resources "fairly" among more children.[40] However, if parents are able to accommodate the rising costs of health care, food, and education by choosing to invest in some children but not in others, then the link between rising costs and declining demand would not be as direct or as strong. Empirical evidence of systematic differences between boys and girls within the same sibling sets in such outcomes as mortality,[41] nutrition,[42] child care,[43] and education[44] suggests one of the ways parents may be at least partially accommodating the rising costs of children — that is, by increasing their investments in sons but not in daughters. A few studies that have explored differences in educational outcomes for boys and girls in high-fertility settings have found that the negative consequences of having many siblings — or more specifically younger siblings — is much greater for girls than for boys.[45]

Not all parents can afford the luxury of being fair, and one easy way to differentiate between children is according to sex. Both mothers and fathers may see very different payoffs to investments in their sons vis-a-vis investments in their daughters, and these will vary across societies according to marriage customs and family organization. Differential investments in sons and daughters are likely to make economic sense from a parent's perspective.

 Birth order is also a factor in differential care. In some settings older children are advantaged relative their younger siblings, whereas in others they may be expected to carry an extra burden of caring for younger siblings. A child's value in Nepal, for instance, is calculated in terms of future household needs and, once those needs have been met, children become redundant.[46] In Taiwan the oldest children in large families, especially females, do particularly poorly with respect to education.[47]

Foreshortened education and early marriage are often the fate of older girls from large families, with the result that patterns of early and high fertility are perpetuated and intergenerational inequality is accentuated.[48]

Clearly, the organization of families and gender relations within the family can affect the distribution of material and opportunity costs of children among family members. In settings where decisionmaking authority rests largely with men who do not carry the bulk of the childrearing costs, there is the likelihood of a lag in the fertility response to rising costs. The possibility that some parents will choose not to pay the costs, to shift them onto others (including their own children), or to pay them for some children only (with a preference for boys) further weakens the expected relationship between rising costs and fertility decline.

Future research will need to probe the determinants of altruistic behavior within families, particularly between parents and children. Why are women more child-centered in their expenditures than men? How could government policy induce mothers and fathers to treat their sons and daughters equally? The answers to these questions would provide the framework for more equitable and more effective population policy.

SUMMARY AND CONCLUSIONS

This chapter has illustrated the ways in which assumptions about family organization and gender relations inform our thinking about population policy. Our empirical view of the family is needlessly limited, largely because of the biases introduced by our women-centered approach to the collection of demographic data for policy analysis. The cross-country diversity in family forms is only hinted at in the data presented here. The full extent of the diversity will only be fully known once proper account is taken of father-child links and of membership in multiple families. The expectations of men and women about the stability and economic cohesiveness of their own families are surely important factors influencing their family size goals and their choice of individual versus couple-based strategies for achieving those goals. Yet we currently have only fragmentary data on the stability of the family in different cultural settings.

Population experts seem to have implicit faith in women's capacity to act as demographic innovators. From either a supply-side or demand-side perspective, they are seen as having the most immediate stake in fertility regulation. However, within a variety of family systems, women face constraints that limit their options outside the family and circumscribe their roles within it. Whether the limitations are on their physical mobility, as in parts of South Asia and the Middle East, or on their economic mobility, as in much of the developing world, the consequence is that, even when the costs of bearing and rearing children are high, the benefits may also be *relatively* high where women have limited access to alternative sources of support. Given men's dominant position within the family, increases in the costs of children will have their most direct and immediate impacts on fertility behavior in families where fathers carry a significant portion of the financial costs.‡ Given men's dominant position within society, improvements in women's livelihoods will have the most direct and immediate impact on fertility behavior where women can retain a fair share of their economic gains for their personal benefit. Finally, given the fact that parents do not necessarily treat all children "equally" but invest in them according to expectations of return, increases in the costs of children will have their most complete impact on fertility behavior in settings where the economic prospects for girls are as promising as for boys.

Bringing men into the picture, adding an individual perspective to the more traditional couple perspective, developing a more realistic view of "the family" in all its manifestations, gaining a sense of fairness about the process of change — these are all essential steps that will lead to more effective population policies. Nothing about families should be taken for granted if these new inquiries are to yield the understanding necessary to achieve the international and national goals originally adopted in 1974 by the World Population Plan of Action.

‡ *Indeed, in low-fertility societies, it could lead to an increase in fertility, which may be seen as desirable in some countries.*

M . TERESITA DE BARBIERI
Translated by Marge Berer

Gender and Population Policy

··················

ost population policymaking has gone on at a far remove from
the people these policies aim to influence. Beyond the labora-
tories where contraceptives are developed and apart from the
data on fertility trends are women and men, sex and procreation and social customs. Without an understanding of these issues, population debates and dynamics may never make sense. This chapter looks
through the lens of gender to try to improve our understanding of population policies and trends and the forces behind them.

GENDER: A DIMENSION OF SOCIAL INEQUALITY[1]

Gender, a frame of reference used more and more in the social sciences, was first employed by feminist activists in the early 1970s in an
effort to address theoretical issues raised by social inequalities based on
sex, issues that had not been explained by theories of class and social
stratification.

Yet the term is used in different ways. In this article, I understand gender as the social construction that defines and gives meaning to sexuality
and human reproduction. A first definition might be:

---

*M. Teresita De Barbieri specializes in research on gender and women's issues at the
Institute of Social Research at the Independent National University of Mexico. This
chapter was published (in slightly different form) in* Reproductive Health
Matters, *May 1993, and in* Conscience, *Autumn 1993. It was written in memo-
ry of Alaida Foppa, a friend and feminist who "disappeared" in Guatemala City in
December 1980.*

the totality of arrangements whereby a society transforms biological sexuality into a human activity and in which human needs are both satisfied and transformed.[2]

From this perspective, gender is a system of power over certain capacities of the human body: sexuality and reproduction. I work within this particular framework because I believe it best allows us to understand the inequality between men and women in Latin America, as well as social inequality more broadly. Theoretically, there are three possible systems of gender power — male dominance, female dominance, equality between men and women — but research has focused on understanding the actual domination that men exert over women in today's societies.

Central to the system is the control that men exercise over women's sexuality and reproductive lives and, subsidiary to that control, the power that men exert over women's capacity to work. To assure the closest control over women's bodies, the social division of labor assigns women certain essential reproductive and nurturing duties and yet, at the same time, devalues these tasks. In the same way, women are afforded little or no access to certain occupations nor to political power.

Gender power operates most forcefully at the reproductive stage of the life cycle, where mechanisms for controlling sexuality, reproduction, and access to work function most sharply and clearly. But the rules and customs at work during the reproductive years, when they are most apparent, also illuminate what happens earlier and later in women's lives and reveal, too, how age and gender intersect as axes of social power.

For example, in societies like ours in Latin America, a child — boy or girl — assumes its father's name and keeps that name until death. The nurture and preservation of that child's life, however, is the mother's responsibility. Being a mother with no father at home (wrongly called "single motherhood") means degradation for both child and mother. The child has no lineage; the woman obviously has exercised her sexuality outside of socially accepted boundaries.

Gender systems are intertwined with specific social structures, beginning with norms of kinship, descent, inheritance, and suitability (or unsuitability) for marriage — all the relationships that determine basic loyalties among individuals of different generations and genders. In addition, gender works together with age to shape the social division of labor. Gender systems even affect citizenship, the rights and responsibilities of individuals, the organization of the state, the exercise of political power, and political culture. Further, gender shapes the psyche, including the formation of subjects and objects of desire.

In the way of all power relations, gender isolates and separates women from men, and women from each other, through a vast array of written and unwritten rules, customs, and values.

These systems of control over women's physical capacities are not meant to eliminate those capacities, however. Therefore the controls can always be transgressed. Women can claim the power of their bodies, disobey the norms, feign obedience, act up, and resist domination. In fact, because women's bodies are the first source of our earliest identity and of the satisfaction of our emotional needs, they are extremely potent; they are intimately linked with life, death, and self-esteem.

Gender differences have been, and continue to be, interpreted and rationalized as belonging to the natural, immutable biological order.[3] They are not seen as socially and historically determined — in which case they would be subject to change through social forces and political action.

Rarely do gender differences appear in isolation. They are interwoven with systems of power and social hierarchy based on the dividing lines of age, class, ethnicity, and race. It is useful to unravel these threads to analyze them, but we must remember that in reality they are interconnected and that social changes directed toward one area can have unforeseen and unintended effects on the others.

GENDER AND THE ROOTS OF POPULATION POLICY

Can the study of population disregard issues of gender? While demography can identify the rates, trends, and determinants of fertility, death, marriage, and migration, a gender perspective is required to understand their meaning. In fact, the main objects of demographic and sociodemographic study are laden with some of the most profound gender meanings. It is perhaps due to the influence of gender dynamics that fertility, marriage, and, to a lesser extent, mortality change so slowly.

We must ask, too, whether one can disregard gender as a system of power in analyzing policies that aim to slow population growth by means of birth control, or in analyzing the reasons (beyond demographic and economic determinants) for declining fertility.

Recent population policies in Latin America have arisen in the context of a tension, a conflict. On one side, there is a recognition of individuals' right to decide on the number and timing of their children — that is, the recognition that the private sphere should remain private. On the other side is the interest of the state, and other social forces, in reducing rates of birth and fertility in order to slow population growth.

We must remember that population policies were introduced to slow population growth without altering the fundamental dimensions of social inequality: the unequal distribution of income and differential access to goods and services that have such a major impact on the quality of peoples' lives. The unequal distribution of wealth has appeared to be taken for granted in the introduction of population policy.

In this way, the state can avoid increasing its expenditures on health, education, housing, and urban infrastructure; the state can avoid raising taxes on the rich and diverting resources from profitable investments. In fact, twenty years of declining birth rates in Latin America have not ameliorated poverty or mitigated the viciously inequitable concentration of wealth.[4]

Population policies also have not attempted to alter other axes of social inequality, such as age or ethnicity or gender. The inequality between men and women globally, at the family level, and in interpersonal relationships has gone unquestioned. Quite the contrary: Population policies seem to aim to preserve the existing social order, with its hierarchies and divisions.

True, some population policies, and the family planning programs through which they are implemented, to their credit have aimed at reducing maternal mortality, preventing the need for abortion — and the harm caused by clandestine abortions — and improving the health and lives of poor women, which are undermined by frequent pregnancy. Some proponents of expanded family planning services have spoken and acted from a recognition of women's inalienable rights to determine their own sexuality and reproduction — and by "proponents" I refer mainly to men, because the leading voices and actors in policymaking have been, for the most part, male.[5]

Yet certain forces and actors have carried more weight in forming population policy than have the women and men whom the policies seek to affect, and certain issues have been given more attention than others. Attempts to alter reproductive patterns have drawn a huge variety of actors into the scene: the churches, political parties, doctors and their professional organizations, social service organizations, and newer social movements, especially, in the last two decades, the feminist and gay liberation movements. Among these, the Catholic hierarchy, corporate and union officials, politicians and experts in industrialized countries, and technicians from international organizations have exerted more influence than ordinary women and men.

The immense majority of women have been consulted only through surveys asking a few questions about sexuality and reproduction: the number of children they want, the frequency of sexual relations, their contraceptive use, and access to family planning services. What sexual relationships and motherhood mean to women has been poorly researched, and current and potential users of family planning services have been given few opportunities to talk about their own experiences and needs.

Even less is known about men's sexual and reproductive practices, views, and values. The first survey of contraceptive knowledge, attitudes, and practices among Mexican men (in particular, working class men in Mexico City) was not conducted until 1990.[6]

In short, the design of population policies and family planning programs has been dominated by a male perspective and cut from a techno-
cratic cloth, which underestimates the complex processes at work in the
phenomena these policies target for control.

LIMITATIONS OF POPULATION PROGRAMS

Many women, especially those with little access to private health care services, clearly have welcomed the extension of family planning services and
contraceptive availability. At the same time, however, a gender perspective
allows us to see and critique the limitations of population control policies
in three central areas: family planning, sexuality, and reproduction.

Family planning. Some of the critiques in this area are well known.[7]
First, family planning and fertility control rest almost exclusively on
women. This has consequences for women's health, because so-called
"modern" methods are not tolerated equally well by all women's bodies.

Men, meanwhile, are not called upon to control their fertility. In
Mexico, "responsible fatherhood" — the slogan with which population
policies were launched in the early 1970s — was supplanted by demographic goals. Since then, the father and men in general have disappeared from family planning campaigns as people who would receive and
be responsible for contraception. The male figure reappears hazily in the
appeal to couples to plan the number of their children. It is only since
the appearance of AIDS in the mid-1980s that condoms have been promoted — and then only for health reasons and not as a method of birth
control. The data from a 1987 survey in Mexico are quite revealing: 1.9
percent of the sexually active couples polled use condoms while 10.2 percent rely on IUDs; 35 percent rely on female sterilization versus 1.5 percent resorting to vasectomy.[8]

A second criticism of family planning programs is that, while access to
contraceptives is straightforward for married women, single people are
ignored and excluded from information and services. Policies have been
oriented toward helping married couples space their children and cease
childbearing once they have their "desired number of children."

Third, contraceptive options are limited. In particular, barrier methods are rarely distributed, and little or no research is conducted on them.

Fourth, there has been pressure recently to limit childbearing to the
years from age 20 to 35, shortening a woman's reproductive life to only
fifteen years.[9] The rationale is prevention of "adolescent motherhood." But the tendency overlooks the fact that, at 18, one acquires citizenship and full civil and political rights and that people in the region
tend to marry at an early age, which cannot appropriately be called
adolescence. Thus health agencies alter, or try to, how people see their
life cycle.

On the other hand, it is said that preventing early and late pregnancies will reduce the incidence of genetic disorders such as Downs syndrome. Yet at the same time, a review of the literature published in the wealthier industrialized nations has shown that the risks of pregnancy after age 35 are not very different from those before; pregnancy outcomes depend more on women's health and the quality of prenatal and childbirth services than on age.[10]

Fifth, the criminalization and clandestine nature of abortion prevents women from exercising in full their right to determine freely the number and spacing of their pregnancies. Women are forced to assume the burden of unplanned, even unwanted, pregnancies and children — not only after using contraception incorrectly or not at all, but also in circumstances beyond their control, such as when methods have failed or exceeded their expiration date. Clandestine abortions of every sort are practiced, killing an appalling number of those women with the least resources.

Sixth, a number of the responses on fertility surveys, as well as statements to the press and unpublished anecdotes, leads one to suspect abuse: female sterilizations and insertion of IUDs without consent, and hormonal methods prescribed without the necessary medical examinations. In the 1987 Mexican survey, 12.4 percent of sterilized women said someone else had decided they should undergo the operation — including 8.7 percent who said they were consulted and agreed with the decision and 3.7 percent who did not.[11] Lawyer Alicia Pérez Duarte reports that she has represented women sterilized without their consent on a number of occasions, appearing before an agency of the Ministry of Public Justice in Mexico's Federal District, and in every case the petition was refused. It has been impossible to bring suit for abuses of this kind.[12]

In addition, in some Latin American countries, there are abuses in clinical contraceptive trials conducted mainly among poor urban and rural women, including indigenous women, without their fully informed consent.[13]

Finally, women are subject to mistreatment by health care providers, especially poor women who are uneducated, unaware of their rights, and low in self-esteem. Male chauvinism, racism, and classicism permeate the relationships between service providers and clients.[14]

Sexuality. The early debates over the suitability of formulating and implementing family planning and birth control policies occurred just as the "sexual revolution" was beginning in Latin America. The ideas and values spread by this movement worked to the advantage of the governments and institutions that supported a reduction in fertility. Young people liberalized old customs, prejudices eroded, and taboos against talking about sexuality eased, enabling changes to extend beyond the urban, educated middle class.

The feminist and lesbian and gay liberation movements that followed facilitated broader understanding of the value of self-determination and privacy in sexuality and reproduction. People began to demand free and unrestricted abortion and family planning services, comprehensive and unbiased sex education, reform of the laws on sex crimes, and so on.[15]

Still, for many, sexuality remains a dark and forbidden topic — a hard habit of mind to change because it is instilled so young. In elementary and secondary schools, students learn, at best, only the basic anatomy and physiology of reproduction not the facts they would find truly useful.[16] Nor do the media, when they dare broach the subject at all, speak about sex in common, concrete language. For vast sectors of the population, sex is a realm of myth and fantasy that is discussed privately, generally in an exchange of fantasies and braggadocio.

Only a few private organizations, schools, and teachers impart appropriate, useful information to adolescents and adults. Interestingly, some of the main points of entry into the feminist movement for working class urban women are workshops and classes on sexuality and the body. This shows a real unmet need among many adult women.[17]

For many people, there are few alternatives to an objectified sexuality that is practiced and symbolized as phallic power, with a reductive focus on the genitalia and penetration rather than the exchange of bodily pleasure. Women's bodies continue to be "used" by men, and by women themselves, as erotic objects rather than as subjects. These patterns extend into the roles that people are expected to play — wife/husband, lover, fiancé/fiancée, boyfriend/girlfriend — in place of achieving real mutual knowledge or profound communication.

For those who do not adjust to the norms — gays and lesbians, people who have no partner or change partners frequently, those who sell sexual services — suspicion, marginalization, and repression are routine.

Sexual misery continues to thrive and to be expressed, at its extreme, in violence and abuse: harassment, blackmail, and rape in all its forms, frequently including familial violence and acquaintance rape.

Reproduction. The emphasis on self-interest in our culture carries over into gender roles and reproductive behavior. Population policies have not changed this. They have reduced birth rates, but neither the stereotypical roles of mother and father nor the meaning of motherhood and fatherhood have been questioned.

Women and men continue to try to affirm themselves by having children. Life choices are not valued unless they include transcending death by having children. This idea reinforces women's need to seek completion and men's need to demonstrate virility through children.

In addition, in a world that offers women few routes to recognition apart from procreation, population control policies pose a new dilemma. Women may want to limit childbearing, but they cannot afford to lose that avenue of recognition and of manipulative power.[18] For example, a woman may feel the need to get pregnant to get a man to marry her, to avert separation or divorce, to confirm a second marriage or new relationship, or to replace older children who have grown up and no longer depend on her.

Early sterilization can have serious repercussions for the stability of any relationship, but particularly for women who have married more than once, who have been separated, abandoned, divorced, or widowed, or who have had children that have died. Studies show a high rate of regret among the youngest women sterilized; in one study that asked women whether they would have the operation again if they had the choice, 10.5 percent said no, and 3.3 percent said that they were unsure or that it would depend on the situation.[19] The options that have opened up for women in the work place, in politics, and in society are as yet inadequate to offer the necessary social and personal recognition now provided by motherhood. If in these years of crisis and economic deterioration, the region has seen more and more women join the paid work force, this has occurred in the context of long-standing devaluation of traditionally female work and occupations, and often it has been no more than a way to meet household expenses.[20] Joining the labor market has not exempted women from domestic work, and jobs have opened up to women only in those fields where the salaries are so low and the conditions so deplorable that only women who are very needy would accept them.

In politics, women are far from achieving an equal or significant level of participation, notwithstanding the existence of some female ministers, governors, representatives, and senators. In the urban social movements, women make up the majority of the rank and file, but among the leadership, the proportion is inverted.[17, 21]

Reducing the number of children has been said to turn abundant motherhood into better motherhood: "fewer children, to give them more." But motherhood and fatherhood in themselves — that is, the wonder and challenge of generating life and creating a person — are no more highly valued. Children continue to be born as instruments to fulfill the needs of their fathers and mothers.

The failure to reassess motherhood and fatherhood has, for certain women and their children, catastrophic consequences. Once children are born, responsibility for raising them, caring for them, socializing them — not just keeping them alive but making them human — falls on the mother. Hers is the extremely demanding job of attending to them constantly. She must ensure that she has the resources — financial, physical, and emotional — that the child needs. If she fails to meet these needs, the child may develop lifelong emotional problems that will be laid at her door. Despite the demands that society places on mothers, women receive little or no reliable, trustworthy support in child rearing.[21]

POPULATION POLICIES FOR THE 21ST CENTURY

We should think about population policies, then, in quite a broad sense, not merely as measures to alter the behavior of one or more demographic variables, but in terms of meeting peoples' needs around reproduction, which are also society's needs. My hope is to provoke a broad-based, deep, and unprejudiced debate on sexuality, reproduction, and society's division of labor along gender lines.

Quite likely, some structures of social power relating to age and gender already are changing, such as how male-female relationships operate in the different stages of life. In Mexico, for example, the penal code and procedures in the Federal District have been reformed to better address rape and sexual harassment and to make the system more sensitive to rape victims. Similarly, more and more women of every social background are demanding political and interpersonal respect for their freedom and bodily integrity. Such changes, which address sexual violence — the worst manifestation of the power imbalance between men and women — may reduce that imbalance, even as the demographic structure is being transformed.

A reexamination of the meaning of motherhood and fatherhood must be at the heart of the debate that I would provoke, because it is these concepts that currently frame sexuality, reproduction, and the social division of labor between the genders. Therefore they are the doorway to assuring a better and fuller quality of life, from conception, for existing and future generations.

Parenthood is an inalienable and essential right, one that entails obligations and responsibilities for both mothers and fathers. But this right goes hand in hand with the inalienable and essential rights of any new life that is born: to be sustained with affection, fed, educated, and nurtured, to know its genetic and social history. Although having children is a private decision, it must be supported and protected by society.

The debate I seek requires a greater knowledge than we now have of the underpinnings of sexual and reproductive behavior, especially at the level of symbolic meaning and imagination. This requires knowledge of the many subtle aspects of the culture and subcultures of each segment of society. It requires an evaluation of national and personal experiences in our countries and in other sociopolitical contexts. It requires discussion that speaks the truth and recognizes the many people — above all, women — who are taken into little account even as policies target them for manipulation. If all these factors were considered, I am sure that the recent fertility decline that is attributed to existing policies would be greater, maybe much greater, than it has been so far.

With better knowledge and understanding, we might avoid Manichean confrontations such as "Abortion, yes: a woman's right" *versus* "Abortion, no: we must protect life" or "The state must direct population policy" *versus* "The state must not intervene." Through this debate we could lay a more humane foundation for reproduction and human development. From this foundation, we could alter radically existing relationships of inequality based on gender, generation, class, and race.

Women's Voices '94

....

WOMEN'S DECLARATION ON POPULATION POLICIES
IN PREPARATION FOR THE 1994 INTERNATIONAL CONFERENCE ON
POPULATION AND DEVELOPMENT

Just, humane, and effective development policies based on principles of social justice promote the well-being of all people. Population policies, designed and implemented under this objective, need to address a wide range of conditions that affect the reproductive health and rights of women and men. These include unequal distribution of material and social resources among individuals and groups, based on gender, age, race, religion, social class, rural-urban residence, nationality, and other social criteria; changing patterns of sexual and family relationships; political and economic policies that restrict girls' and women's access to health services and methods of fertility regulation; and ideologies, laws, and practices that deny women's basic human rights.

While there is considerable regional and national diversity, each of these conditions reflects not only biological differences between males and females, but also discrimination against girls and women, and power imbalances between women and men. Each of these conditions affects,

---

In September 1992, women's health advocates representing women's networks in Asia, Africa, Latin America, the Caribbean, the United States, and Western Europe met to discuss how women's voices might best be heard during preparations for the United Nations' 1994 International Conference on Population and Development and in the conference itself. The group agreed that it was important for women around the world to issue a strong statement that could help to reshape the population agenda so as to better ensure reproductive health and rights. With the New York-based International Women's Health Coalition acting as secretariat, the group drafted the following declaration, which has been signed by more than one hundred women's organizations around the world.

and is affected by, the ability and willingness of governments to ensure health and education, to generate employment, and to protect basic human rights for all. Governments' ability and willingness are currently jeopardized by the global economic crisis, structural adjustment programs, and trends toward privatization, among other factors.

To assure the well-being of all people, and especially of women, population policies and programs must be framed within and implemented as a part of broader development strategies that will redress the unequal distribution of resources and power between and within countries, between racial and ethnic groups, and between women and men.

Population policies and programs of most countries and international agencies have been driven more by demographic goals than by quality of life goals. Population size and growth have often been blamed inappropriately as the exclusive or primary causes of problems such as global environmental degradation and poverty. Fertility control programs have prevailed as solutions when poverty and inequity are root causes that need to be addressed. Population policies and programs have typically targeted low income countries and groups, often reflecting racial and class biases.

Women's fertility has been the primary object of both pronatalist and antinatalist population policies. Women's behavior rather than men's has been the focus of attention. Women have been expected to carry most of the responsibility and risks of birth control but have been largely excluded from decision-making in personal relationships as well as in public policy. Sexuality- and gender-based power inequities have been largely ignored, and sometimes even strengthened, by population and family planning programs.

As women involved directly in the organization of services, research, and advocacy, we focus this declaration on women's reproductive health and rights. We call for a fundamental revision in the design, structure, and implementation of population policies, to foster the empowerment and well-being of all women. Women's empowerment is legitimate and critically important in its own right, not merely as a means to address population issues. Population policies that are responsive to women's needs and rights must be grounded in the following internationally accepted, but too often ignored, ethical principles.

FUNDAMENTAL ETHICAL PRINCIPLES

1. Women can and do make responsible decisions for themselves, their families, their communities, and, increasingly, for the state of the world. Women must be subjects, not objects, of any development policy, and especially of population policies.

2. Women have the right to determine when, whether, why, with whom, and how to express their sexuality. Population policies must be

based on the principle of respect for the sexual and bodily integrity of girls and women.

3. Women have the individual right and the social responsibility to decide whether, how, and when to have children and how many to have; no woman can be compelled to bear a child or be prevented from doing so against her will. All women, regardless of age, marital status, or other social conditions have a right to information and services necessary to exercise their reproductive rights and responsibilities.

4. Men also have a personal and social responsibility for their own sexual behavior and fertility and for the effects of that behavior on their partners' and their children's health and well-being.

5. Sexual and social relationships between women and men must be governed by principles of equity, noncoercion, and mutual respect and responsibility. Violence against girls and women, their subjugation or exploitation, and other harmful practices such as genital mutilation or unnecessary medical procedures, violate basic human rights. Such practices also impede effective, health- and rights-oriented population programs.

6. The fundamental sexual and reproductive rights of women cannot be subordinated, against a woman's will, to the interests of partners, family members, ethnic groups, religious institutions, health providers, researchers, policymakers, the state, or any other actors.

7. Women committed to promoting women's reproductive health and rights, and linked to the women to be served, must be included as policymakers and program implementors in all aspects of decision-making including definition of ethical standards, technology development and distribution, services, and information dissemination.

To assure the centrality of women's well-being, population policies and programs need to honor these principles at national and international levels.

MINIMUM PROGRAM REQUIREMENTS
In the design and implementation of population policies and programs, policymakers in international and national agencies should:
1. Seek to reduce and eliminate pervasive inequalities in all aspects of sexual, social, and economic life by:
   • providing universal access to information, education, and discussion on sexuality, gender roles, reproduction, and birth control, in school and outside;
   • changing sex-role and gender stereotypes in mass media and other public communications to support more egalitarian and respectful relationships;

- enacting and enforcing laws that protect women from sexual and gender-based violence, abuse, or coercion;
- implementing policies that encourage and support parenting and household maintenance by men;
- prioritizing women's education, job training, paid employment, access to credit, and the right to own land and other property in social and economic policies, and through equal rights legislation;
- prioritizing investment in basic health services, sanitation, and clean water.

2. Support women's organizations that are committed to women's reproductive health and rights and linked to the women to be served, especially women disadvantaged by class, race, ethnicity, or other factors, to:
   - participate in designing, implementing, and monitoring policies and programs for comprehensive reproductive health and rights;
   - work with communities on service delivery, education, and advocacy.

3. Assure personally and locally appropriate, affordable, good-quality, comprehensive reproductive and sexual health services for women of all ages, provided on a voluntary basis without incentives or disincentives, including but not limited to:
   - legislation to allow safe access to all appropriate means of birth control;
   - balanced attention to all aspects of sexual and reproductive health, including pregnancy, delivery, and postpartum care; safe and legal abortion services; safe choices among contraceptive methods including barrier methods; information, prevention, and treatment of sexually transmitted diseases, AIDS, infertility, and other gynecological problems; child-care services; and policies to support men's parenting and household responsibilities;
   - nondirective counseling to enable women to make free, fully informed choices among birth control methods as well as other health services;
   - discussion and information on sexuality, gender roles and power relationships, and reproductive health and rights;
   - management information systems that follow the woman or man, not simply the contraceptive method or service;
   - training to enable all staff to be gender sensitive, respectful service providers, along with procedures to evaluate and reward performance on the basis of the quality of care provided, not simply the quantity of services;
   - program evaluation and funding criteria that utilize the standards defined here to eliminate unsafe or coercive practices, as well as sexist, classist, or racist bias;

- inclusion of reproductive health as a central component of all public health programs, including population programs, recognizing that women require information and services not just in the reproductive ages but before and after;
- research into what services women want, how to maintain women's integrity, and how to promote their overall health and well-being.

4. Design and provide the widest possible range of appropriate contraceptives to meet women's multiple needs throughout their lives:
   - give priority to the development of women-controlled methods that protect against sexually transmitted infections, as well as pregnancy, in order to redress the current imbalances in contraceptive technology research, development, and delivery;
   - ensure availability and promote universal use of good quality condoms;
   - ensure that technology research is respectful of women's right to full information and free choice, and is not concentrated among low income or otherwise disadvantaged women, or particular racial groups.

5. Ensure sufficient financial resources to meet the goals outlined above. Expand public funding for health, clean water and sanitation, and maternity care, as well as birth control. Establish better collaboration and coordination among the United Nations, donors, governments, and other agencies in order to use resources most effectively for women's health.

6. Design and promote policies for wider social, political, and economic transformation that will allow women to negotiate and manage their own sexuality and health, make their own life choices, and participate fully in all levels of government and society.

NECESSARY CONDITIONS

In order for women to control their sexuality and reproductive health, and to exercise their reproductive rights, the following actions are priorities:

1. *Women decision makers.* Using participatory processes, fill at least 50 percent of decision-making positions in all relevant agencies with women who agree with the principles described there, who have a demonstrated commitment to advancing women's rights, and who are linked to the women to be served, taking into account income, ethnicity, and race.

2. *Financial resources.* As present expenditure levels are totally inadequate, multiply at least four-fold the money available to implement the program requirements listed in this Declaration.

3. *Women's health movement.* Allocate a minimum of 20 percent of available resources for women's health and reproductive rights organizations to strengthen their activities and work toward the goals specified in this declaration.

4. *Accountability mechanisms.* Support women's rights and health advocacy groups, and other nongovernmental mechanisms, mandated by and accountable to women, at national and international levels, to:

   • investigate and seek redress for abuses or infringements of women's and men's reproductive rights;

   • analyze the allocation of resources to reproductive health and rights, and pursue revisions where necessary;

   • identify inadequacies or gaps in policies, programs, information, and services, and recommend improvements;

   • document and publicize progress.

Meeting these priority conditions will ensure women's reproductive health and their fundamental right to decide whether, when, and how many children to have. Such commitment will also ensure just, humane, and effective development and population policies that will attract a broad base of political support.

JUDITH LICHTENBERG

Population Policy and the Clash of Cultures

....

When people talk about population as something for which "policies" are needed, they are likely to be talking about population policy for developing countries. According to one recent report, 123 developing countries have adopted such policies over the past two decades. These have often involved the participation of developed countries, which, through public agencies as well as private donors, have provided resources and expertise to help establish family planning programs.

Yet much wariness on the part of people in developing countries surrounds these international family planning efforts. Critics suspect that they represent a continuation of the colonialist legacy, and that Western nations are once again attempting to exercise power over the fate of peoples who have been subject to them for centuries. Concerns of this kind might be valid even if it could be shown that the motives of the donor nations were largely admirable; for if we acknowledge the worth of national self-determination, then high-minded paternalism is one of the forces it must be defended against.

But of course the opponents of population programs do not often agree that Western motives are benign. They are likely to argue instead that developed nations are pursuing their own economic or political interests when they support efforts to limit population growth in developing countries. Other critics, even more suspicious, may claim that pop-

Judith Lichtenberg is a research scholar at the Institute for Philosophy and Public Policy at the University of Maryland. This chapter is reprinted from the Report from the Institute for Philosophy and Public Policy, *Fall 1993.*

ulation programs are racist or even genocidal in intent. And some may insist that indigenous cultures must remain inviolable, particularly in an area as intimate as people's reproductive choices.

How telling are these criticisms? Let us examine them more carefully.

IDENTIFYING THE SELF IN "SELF-DETERMINATION"

Concerns about Western involvement in population programs in developing countries are sometimes expressed in terms of national self-determination: the idea that nations have the right to determine their own destiny without interference or domination by foreign elements. Those who frame the issue in these terms may worry that developing countries cannot throw off the yoke of colonialism without repudiating Western involvement in their population programs and other internal affairs.

In light of their historical experience, developing countries' suspicions are understandable. And for this reason alone, quite apart from any others, it is clear that external efforts to influence fertility, if they are to have any hope of succeeding, must be undertaken with care and sensitivity.

Yet there is still a substantive question lurking beneath these claims of self-determination. Is there an ineradicable conflict of interest between developed countries who would provide family planning assistance and developing countries who would receive such aid? If not — if developing countries want what developed countries want to give them — the claim of self-determination might pose no bar to international efforts.

But this way of putting the issue is obviously inadequate. It oversimplifies in at least two respects. First, it suggests that the relationship in international population programs consists of two and only two parties, "foreign initiator" and "domestic target." And second, it implies that each of these parties possesses a distinct and unified set of interests.

In fact, much of the assistance for population programs is channeled through multilateral organizations, including U.N. agencies that are by no means dominated by developed nations. On the other side, many developing countries have taken the initiative in creating their own family planning programs, and it is widely agreed that the success of such programs often hinges on the strength of internal initiatives and on local participation in program design and management. Accusations that population policies violate a nation's right to self-determination sometimes overlook these facts.

Just as the agents initiating population programs constitute a diverse group with different interests, so too do developing countries. National boundaries in the world today do not — as strife in Central and Eastern Europe reminds us — necessarily coincide with cultural or ethnic boundaries. Even within distinct cultures, we find conflicts of interest between different political and ethnic groups and between rich and poor, men and women, adults and children. The question, one might say, is "Who is the *self* in self-determination?"

When we recognize that there is no single self but rather a variety of selves with different and often conflicting interests, two difficult kinds of tasks remain. One is to sort out these interests and, sometimes, to decide which among them are most pressing. How we do that will depend on the particulars of each case. For example, in many traditional societies women tend to be more interested than men in limiting family size. We might argue that in such instances women's health and freedom outweigh men's interests in preserving their authority to make such decisions and in conserving the existing social order. In addition, we will be influenced by how this view fits with other arguments for limiting population size.

At the same time, while recognizing the illusion of the single self in self-determination, we should not forget the real issues of perception and politics surrounding relations between developed and developing countries. Developing countries have every reason to be on their guard when developed countries "take an interest" in them. We must take the political leaders of those countries seriously, and, unless they engage in gross violations of human rights or otherwise overstep the bounds of tolerable behavior, we must allow them to speak for the nation — even though we know that, like political leaders everywhere, often they do no such thing.

CHARGES OF RACISM AND GENOCIDE

In some African countries, such as Kenya, the accusation of genocide has been made against population programs. The charge may seem misplaced, because birth control measures do not destroy life, but simply prevent its coming into being. Thus, it seems, these measures can hardly be said to be morally equivalent to murder. (For present purposes, we may exclude abortion from consideration; the claim that population programs promote genocide is not grounded in views about the moral status of the fetus.) Yet this point will not mollify those who believe that even if the means are different, the *aim* of population policies is indistinguishable from the aim of genocide: the disappearance of a targeted people. Quite apart from the intricacies of defining "genocide," the specter of racism is difficult to dispel.

Nor is it much help to point out that Westerners who promote family planning in, say, Mexico or Kenya bear no particular animus against the Mexicans or the Kenyans. For what arouses the fears of some in the developing world is a sense that the West regards its own populousness differently than it does that of its poorer, and generally darker or racially "other," neighbors. Perhaps Westerners have nothing against Mexicans or

Kenyans *in particular;* but is it false to say they have something against these groups as instances or parts of the larger class of nonwhite peoples? How, it may be asked, can developed countries such as France and Germany support pronatalist policies at home while arguing that developing countries are producing too many people?

Of course, one might argue that economic, geographical, and environmental factors determine the optimal population size in different regions, and that although France can support more people than it now has, Mexico cannot. Such arguments shift the ground of population policy in an important way; they suggest that the problem is not too many people in the world, but rather too many people in some parts of the world.

On the other hand, the growing animosity towards immigrants in many developed countries with low fertility can only reanimate suspicions of racism. Is it more people that these pronatalist countries want, or just a particular kind of people — their own?

CULTURAL INTEGRITY AND CULTURAL CHANGE

To alter existing population trends in a country often means abandoning traditional practices and so, it seems, changing irreversibly the way of life of its people. Concerns of this kind — about threats to a set of traditions that characterize or define a given community, society, or culture — are often framed in terms of cultural integrity. If, for example, large families are essential to an indigenous culture, there appears to be an inescapable conflict between cultural integrity and policies aimed at reducing fertility: One cannot enact such policies without at the same time violating the culture. This does not mean that the policies are therefore unacceptable, but it does suggest that they come with an unavoidable cost.

One trouble with this view is that it rests on a static conception of culture. A culture is not an eternal, unchanging entity, impervious to influences from without or within. Rather, it is a complex set of practices in which we find constant tension between the old, as expressed in ideological or social norms, and the new, represented by people's attempts to create new patterns of thought and action. Some of these new patterns arise from cross-cultural contact which, in the contemporary world of telecommunications and mass transportation, is pervasive and inescapable. To criticize family planning programs on the grounds that they violate cultural integrity ignores the heterogeneity within cultures, and the creation of social relationships across them.

Given the capacity of cultures both to resist intrusion and to adapt to new circumstances, issues such as the transfer of reproductive technology turn out to be more complex than is commonly supposed. On the one hand, people in developing countries often actively resist technological innovations that they believe are inimical to their interests or that do not meet their needs. Though these people may be stereotyped as "backward" or "conservative," closer study indicates that in fact they are behaving in accord with "Western" notions of rationality: they are acting to hold on to resources or improve their chances of survival. Their behavior helps to explain the failure to win acceptance for innovations ranging from modern agricultural implements to large-scale projects such as dams. It also accounts for some of the documented failures in the dissemination of contraceptive technologies.

Moreover, even when people do accept technological innovations from foreign sources, it does not follow that they accept the values attached to the culture originating the technology. Sometimes people incorporate new technologies into the meaning systems and social organization of their own cultures. In Northern India's Punjab region, for example, women who employ amniocentesis will sometimes decide to have an abortion if they learn the fetus is female, reinforcing the indigenous value attached to male children in a highly patriarchal society. Similar use has been made in China of sonograms (despite their lesser reliability in predicting the sex of the child). Far from changing values, then, this technology has reinforced local belief patterns and practices — no doubt in ways contrary to the values of those who introduced the technology.

It seems clear, then, that an undifferentiated appeal to cultural integrity ignores the plasticity of culture, and distracts us from more complex and important questions concerning the degree of change a given policy will create, and how far a culture may adapt or innovate without ceasing to be the *same* culture.

DEFENDING TRADITION

How seriously should we take inroads into cultural integrity, anyway? To answer this question we need to know why traditions ought to be respected in the first place. That something has been done in a certain way for a long time may give it value in the eyes of a traditionalist, but this alone hardly seems an adequate justification. The mere longevity of discrimination or poverty carries no moral weight at all against efforts to overcome them. Traditions must have some intrinsic value, or at least no intrinsic disvalue, if they are to be worth preserving.

A more plausible account of the appeal of tradition rests on what we can broadly characterize as aesthetic grounds. Traditional practices often seem to possess a richness and depth, a meaning and spiritual quality lacking in industrialized mass society. We regret the loss of these traditions both because of their intrinsic appeal, and because we value diversity over the encroaching homogenization that modernization seems to bring.

Such concerns are of the utmost importance, but they can never be decisive in themselves. In gauging the impact of cultural change, we must always ask what is at stake besides the aesthetic value of the tradition — including, most centrally, the interests of those within the culture. Otherwise, we will find ourselves in the position of a certain French anthropologist who, in Martha Nussbaum's description, regretted that "the introduction of smallpox vaccination to India by the British eradicated the cult of Sittala Devi, the goddess to whom one used to pray in order to avert smallpox."

In general, our assessment of a tradition requires that we examine the divergent interests of the individuals and groups who are involved in or affected by it. Thus it is crucial to ask whether and how men and women fare differently in cultures where large families are the norm. After all, it is women who carry and bear children, and women who are primarily responsible for rearing them. In societies with inadequate nutrition and health care, the burdens associated with raising a large family are especially acute, even after the countervailing benefits are taken into account. Yet the needs of women, and their greater interest in family planning, are likely to be overlooked in arguments for cultural integrity. Such arguments are inherently conservative, and often the cultures they would conserve are patriarchal ones that subordinate the interests of women to those of men.

Of course, how women view their interests will itself be culturally determined, at least in part. In Hindu and Muslim societies, for example, where menstruation has a complex cultural and religious significance, women may feel it is in their best interest not to use modern contraceptives that interfere with the menstrual cycle. Because some of these contraceptives are associated with breakthrough bleeding, they will be unacceptable in a culture where bleeding renders a woman ritually unclean and interferes with normal marital relations. At the same time, because a woman's value to her husband in these societies depends upon her capacity to reproduce, cessation of the menstrual cycle will seem to deprive women of an essential element of their identity.

One way of addressing the needs of these women is to continue research into alternative contraceptive technologies. Another is to try to recoup and revitalize indigenous strategies of fertility control — long periods of breastfeeding, which postpones ovulation; postpartum abstinence from sexual relations; herbal birth control techniques — that have been lost under Western influence. (Indeed some of these approaches might be adopted in industrialized countries, where "alternative medicine," with its connections to traditional cultures, is growing in popularity.) Health workers may also begin encouraging people to think about reproduction differently — not by preaching to them from a Western point of view, but by invoking concepts and values from their own systems of belief. It is this option we must now consider more closely, as we look at the strategies for implementing population policies in the developing world.

THE *HOW* OF POPULATION POLICY

Thus far we have been considering the view that the very *existence* of a population policy is problematic. Assuming that such policies unavoidably alter traditions, we have asked how we might weigh that fact against other considerations. But the threat posed to cultural integrity by the mere existence of a population policy may be easily overstated. The more plausible criticism might be not that population policies as such violate a culture's integrity, but rather that such policies *as they have been promoted and implemented* have often been insensitive to the specific practices and traditions they encounter.

Critics have argued, for instance, that population programs in some countries have employed a top-down bureaucratic style that does not adequately involve indigenous people in the processes of decision-making. The Western approach often clashes with local customs and ignores traditional structures, such as medical programs and practices already in place. To raise questions regarding the "how" of population policy may seem a way of avoiding the central question of their inherent justification. But matters of implementation are hardly trivial. For it is by looking at implementation that we can tell how such policies are understood, both by those who are responsible for them and by those for whom they are designed.

This point can be illustrated by looking at the relationships, real and perceived, between policies aimed at reducing fertility in a developing country and policies aimed at improving maternal and child health. It is clear, on the one hand, that while there is no necessary connection between the two, reducing fertility is likely to improve maternal and child health. Maternal health will improve because the bearing of many children imposes enormous emotional and physiological stress, especially where women have access only to substandard nutrition and health care. Children's health will improve because, other things being equal, the fewer children a family or community has to care for, the better it can care for them.

In turn, improving maternal and child health will reduce fertility to the extent that high fertility in developing countries responds to, or is a vestige of, high rates of infant and child mortality. It is also true that when health workers win the confidence of mothers — by treating their children's illnesses and teaching them to administer unfamiliar medicines — they have greater credibility when they raise the subject of family planning. For these reasons, many agencies have decided that family planning programs can best be implemented in conjunction with programs that focus on maternal and child health.

Now to join the two in this way might seem cynical and instrumental — a deceptive repackaging of one's true goals in the place of straightforward appeals for fertility reduction. But this need not be the case. First of all, maternal and child health are centrally important values in their own right — so important, indeed, that they can hardly be overemphasized. Moreover, a commitment to these values, either as means or as ends in themselves, will have genuine practical implications: policies that incorporate these values will differ substantially in their impact on individuals from those that do not. For example, they will consider more carefully the health effects of contraceptive methods — effects that did not receive sufficient scrutiny during earlier efforts to limit population growth in developing countries.

These observations should be understood both as moral criticism of the styles, approaches, and methods of many past population programs and — not coincidentally — as an explanation of their limited effectiveness. When the argument takes this form, its gist is practical: if you want a population program to succeed, begin with an understanding of the culture for which it is designed; take its traditions seriously; treat its people with the respect you would accord your own.

DENESE SHERVINGTON

Reflections on African-American Resistance to Population Policies and Birth Control

· · · · · · · · · · · · · · · · · · · ·

P opulation policy and birth control are undeniably and inextricably linked. Birth control programs are often driven by demographic goals and other efforts to increase or reduce the size of a given population. Most recently, birth control programs have been promoted by those who blame uncontrolled population growth for current economic and environmental crises. How do African Americans enter this debate? With much mistrust and anger.

As the National Black Women's Health Project has noted, African Americans were "Brought here in chains, worked like mules, whipped one day, sold the next — for 244 years we were held in bondage. Somebody said that we were less than human and not fit for freedom. Somebody said that we were like children and could not be trusted to think for ourselves.... Somebody owned our flesh and decided if and when and with whom our bodies were to be used." [1]

African-American perspectives on population policy and birth control are deeply etched by the history of slavery and ongoing oppression. During the period of slavery, population policies directed at slaves favored high fertility, in order to increase the supply of free labor. However, after emancipation, African Americans were no longer of capital value to plantation owners. In fact, when Reconstruction briefly granted political representation to freed slaves, white Southerners took an avid interest in controlling the size of the African-American population.

Denese Shervington is the director of the Women of Color Reproductive Health Forum.

From the late 19th century until the 1970s, many social policies affecting African Americans were anti-natalist, with a strong undercurrent of eugenics and genocide. For example, in the 1920s and 1930s, in many Southern states condoms were illegal for whites, but not for African Americans. (The rationale given for this disparity was the unproven contention that African Americans suffered higher rates of venereal disease.)

Eugenics laws, which aimed to limit the number of "undesirables" — indigents, criminals, the mentally ill, and the mentally retarded — were frequently used to justify the sterilization of African-American women. Although eugenics laws were not explicitly racist, evidence suggests that they were implemented in a racist manner. For example, in the 1960s, 65 percent of all the women sterilized in North Carolina were African-American. The civil rights movement of the 1960s brought to light many instances of sterilization abuse. A typical case was that of Nial Ruth Cox, an African-American woman from North Carolina who was told that she would lose her relief benefits if she did not submit to sterilization.[2]

Since the 1960s, many have argued that population growth is responsible for poverty, food shortages, and environmental degradation. These arguments invigorated population control programs in the developing world — programs which have sometimes ignored human rights in their pursuit of demographic goals. Although the U.S. does not have an official population policy, many birth control programs aimed at African Americans bear a resemblance to the draconian population control programs in the South, particularly in their disregard for individual rights and dignity. In such programs it is possible to hear an echo of the eugenicists' refrain: "more children from the fit, less from the unfit."

For example, reproductive technologies may soon be used in a coercive and punitive manner. Several state legislatures are considering bills that would require women convicted of certain crimes to be given Norplant contraceptive implants, with or without their consent. Other states have considered legislation that would offer incentives for welfare mothers to use Norplant. Many advocates fear that such laws will be disproportionately applied to African-American women. As Julia Scott of the National Black Women's Health Project told *Newsweek* in 1993, "Because it's the closest thing to sterilization, folks have seized on this and tried to impose it on the women who have the least power in our society. They see it as social control for the women who they believe are responsible for all of our social issues."[3]

Moreover, Norplant is being aggressively promoted to adolescent African Americans without any studies to determine the long-term health effects of exposure to large doses of progestins at an early age. For example, in 1992, a program offered Norplant implants to students at a

Baltimore school for pregnant teens. The implants were given for free, but students had to pay for removal.

The emphasis on Norplant instead of condoms and other barrier methods could cost many women their lives. Of the 334,344 adult AIDS cases reported in the U.S., women account for 40,702, according to the Centers for Disease Control. More than half of those cases are found among African-American women.[4] Because low-income African-American women are currently at greatest risk for HIV infection, they are in also in greatest need of birth control methods that also provide protection from sexually transmitted diseases (STDs). But, since most women are unlikely to use *two* birth control methods, Norplant may be used instead of methods that could protect them from AIDS. A study of Norplant users in Texas showed that nearly half of those who had formerly used condoms said they would stop using condoms now that they have Norplant.[5] At the same time, many family planning programs for African-American teens emphasize contraception rather than preventive health education and/or treatment of STDs.

Family planning providers sometimes show disregard for the health and well being of low-income women of color. For example, poor women are frequently targeted for marketing of contraceptive drugs whose side-effect profiles are not fully understood, such as Depo Provera. And attempts to de-medicalize the use of oral contraceptives by selling them over-the-counter could keep many poor African-American women from receiving reproductive health care, because oral contraceptives serve, for many, as an entry point into the reproductive health care system.

A LEGACY OF DISTRUST

For African Americans, the issue of birth control/family planning has never been simply that of women's rights to choice. Rather, it is a complex issue stained by the history of racially motivated population policies. Within the African-American community, there has always been strong resistance and opposition to such policies. For example, many slave women resisted pro-natalism through the use of home-made contraceptives and abortifacients, because they did not want to bring children into a world that enslaved them. After emancipation, when fears of depopulation resulted from the policies of eugenics and genocide, many African-American leaders and organizations were pro-natalist and discouraged the use of birth control.

Recent studies show that African Americans support individual fertility regulation within the larger context of total reproductive health. However, African Americans have relatively low rates of contraceptive use and high rates of unintended pregnancies and abortions. Evidence suggests that this is not the result of inadequate access to birth control. A

1991 poll found that 59 percent of African-American women at risk for pregnancy do not use contraception; 74 percent of those women felt that they did not need birth control, and only three percent did not know where to get contraceptives or how to use them.[6]

I hypothesize that there is still a very deep unconscious and unacknowledged fear of depopulation among African Americans today. In the above-mentioned poll, 11 percent equated birth control with genocide. Could it be that high fertility rates are an unconscious response to the high mortality rates of African-American infants and young African-American men? I further hypothesize that the preferred family size among African Americans is closer to that of the developing world than that of the U.S. It may be that the two-child family norm is an alien and hence a resisted construct for African Americans.

Many African Americans do not accept the argument that population growth is responsible for environmental degradation or poverty. It is possible that African Americans still carry an archetype from the mother land: that nature will take care of you if you live in harmony with her. African Americans once strongly valued living in harmony with nature. But recent patterns of overconsumption and disrespect for nature in African-American communities seem to result from two centuries of oppression, which has led to identification with the oppressor.

TOWARDS WORKABLE SOLUTIONS

African Americans will not be able to enter the debate on population with trust, until all racist policies are eliminated from family planning programs nationally and globally. This will also entail the acknowledgement of past and present wrongs. To accomplish this goal:

- More African Americans must participate in population policymaking;
- The debate on population and the environment should focus more on overconsumption and equitable distribution of land and other resources;
- Family planning services must be integrated with a wider array of reproductive, child, and general health services including, for example, treatment of STDs, management of non-severe hypertension, and immunizations;
- More African Americans must participate in the research and development of new contraceptive technologies.

MENCER DONAHUE EDWARDS

People of Color and the Discussion of Population

....

I n the United States, the issue of population is undergoing a transformation. Once the province of demographers and a handful of activists, population is now discussed by a growing number of people who, though few in number, represent an ever-widening sampling of U.S. society. At the same time, the small community of professionals and activists that comprises the population movement has begun a serious reexamination of long-held positions and perspectives. This transformation is essential if the U.S. is to provide its leadership in addressing the problems that threaten the world's collective future.

Still, although population issues are being discussed by a broader spectrum of Americans, people of color remain underrepresented in the debate. Outreach to people of color by the population community has been limited, due to the persistence of myths and stereotypes. People of color — especially those active in environmental, economic, health and other social justice efforts — have much to contribute to a public conversation about population in the U.S. Indeed, people of color are a potentially powerful, but historically and tragically underutilized, constituency for the implementation of the progressive policies being considered now.

Perhaps the most important contribution people of color could make would be to help focus the population debate on justice issues. Many people of color understand the central role of social justice not only in addressing population growth, but in sustainable economic develop-

---

Mencer Donahue Edwards, former executive director of the Panos Institute, is a consultant on social justice and sustainable development issues.

ment, health care, and environmental protection as well. This perspective derives, in part, from the values and ethical assumptions they have drawn from their collective and individual historical experiences. The lessons of that experience have relevance for every U.S. citizen — and for the development of population policies at home and abroad.

In order to engage people of color in the debate, the population community must reexamine the ideas and presuppositions that comprise what might be called "the myth of the collective disconnect."

THE MYTH OF THE COLLECTIVE DISCONNECT

Simply put, the myth of the collective disconnect holds that people of color, as a monolithic subset of American society, cannot connect the struggles they face in their day-to-day lives with "big picture" problems like the environment, national security, and population growth. Even when the connection is made, according to the myth, people of color are simply too overwhelmed with the demands of "just trying to survive" to act in any capacity. The myth holds that struggling against racism and poverty are full-time jobs which leave no intellectual, emotional, or spiritual resources for constructive action.

Until recently, the same myth held that people of color were not concerned about the environment. Many mainstream environmentalists believed that people of color knew nothing about the fouling of the air, water, or land where they lived, worked, and played. Nor, according to the myth, did people of color connect their escalating incidence of health problems to environmental factors.

The myth was wrong. Not only are many people of color aware of environmental problems in their communities, but they connect those problems to larger themes of social and racial injustice. People of color coined the term "environmental racism" to describe the deadly duo of pollution and discrimination that plagues their communities. And in the early 1990s, a series of well-documented reports and studies served to announce not the birth, but the ascension of the environmental justice movement to national and international eminence.

The same myth has been applied to population issues as well. The myth, with respect to population, has two concepts at its hollow core: "Go Forth and Multiply" and "Family Planning Equals Genocide." The first concept rests on the stereotype of people of color as unquestioning followers of religious doctrine, mindlessly obeying the biblical injunction to "be fruitful and multiply." This formulation is patently ridiculous. Religious institutions *are* powerful forces in communities of color, but people of color are no more likely than their white counterparts to interpret scripture literally or slavishly follow its dictates.

The "Go Forth and Multiply" myth seeks to explain the fact that, broadly speaking, people of color in the U.S. have higher birth rates and lower rates of contraceptive usage than their white counterparts. But the explanation for this phenomenon lies not in immutable religious views. Instead, high birth rates reflect the persistent poverty that plagues communities of color. As experts in the field have noted, birth rates remain high in societies where poverty, high rates of infant and child mortality, and social insecurity prevail. Sadly, high birth rates can also perpetuate these problems. Many communities of color — like communities in the developing world — are caught in a cycle of poverty and high birth rates. And, just like people in developing countries, people of color often resent analyses that focus on birth rates but ignore the complex context in which childbearing occurs.

The second component of the myth holds that people of color equate family planning with genocide. This notion is partly rooted in reality: many people of color believe that there is at least an unofficial conspiracy to depopulate their communities. History and current events provide ample support for this view. The mass murder of Native Americans, the sterilization of Puerto Rican women, the Tuskegee syphilis study (in which treatment was withheld from African Americans in order to study the progress of that deadly disease) and current epidemics of AIDS and drug use all provide grist for the conspiracy theorists' mill. And there has long been a small faction that links family planning to genocide. Since the 1930s, some people of color have questioned the intent of the birth control movement and its eugenicist associations.

The "Family Planning Equals Genocide" myth contains a grain of truth — but it is only a grain. There is also a tradition of support for family planning among communities of color. In 1938, for example, W.E.B. DuBois championed birth control in an article titled, "Black Folk and Birth Control."[1] Organizations such as the National Black Women's Health Project, the National Coalition of 100 Black Women, the National Council of Negro Women, and the Hispanic Health Council, among others, vigorously support family planning within a broad context of reproductive health care. In a series of focus groups conducted for the Pew Global Stewardship Initiative in 1993, African American participants strongly rejected the assertion that family planning was tantamount to genocide. Quite the reverse: Participants associated family planning with enhanced personal choice and control over one's life. And a 1991 poll conducted by the Communications Consortium found that only 11 percent of women of color equated birth control with genocide.[2] That percentage may be unacceptably high, but it is clearly not the majority view.

The myth of the collective disconnect is wrong about people of color and population issues, just as it was wrong about people of color and the environment. Still, it would be inaccurate to say that population growth is a "front burner" issue for people of color in the U.S. The focus group research cited above found that population issues have little salience among people of color and whites alike. Population took a back seat to more compelling domestic concerns for all participants except the small number who described themselves as "internationalists" or "environmentalists."

DIFFERENT DOORS

If we are to broaden public discourse on population, we must raise the salience of the issue among all Americans. But history suggests that people of color have unique perspectives on this issue; therefore a generic outreach program cannot be guaranteed to reach whites and people of color alike.

In 1993, the author helped conduct a series of interviews with the chief executive officers and/or senior staff of fifteen highly respected national and local organizations serving people of color in the U.S. Those interviews provided key insight into how people of color frame the issues of population, consumption, and the environment. Interviewees spoke of people of color using distinctive labels or of entering these issues through different "doors."

The concept of justice — racial, gender, environmental and economic — occupied a center-stage position in the world views of the leaders interviewed. "I see the world through a justice prism," declared one interviewee. Many pointed to the inequitable distribution of resources as the leading cause of misery in their communities, as well as in developing countries. Others noted the injustice of the implicit and explicit racism driving the current wave of antiimmigrant feeling in the U.S. and energizing demands for restrictions on immigration. One complained of "environmentalists' lack of sensitivity toward immigrants," which reinforces a perception "that they are more concerned with trees and the quality of lakes than with people."

The interviewees also believed that mainstream environmentalists had failed to embrace concerns about toxic emissions in communities of color and in the developing world. For that reason, they felt that environmentalists had little basis for credible leadership in the developing world or in communities of color around demographic goals.

The interviews reveal that outreach to people of color must begin with accurate information and understanding. Accommodating these requirements would go a long way toward reducing the pernicious effects of "the myth of the collective disconnect" and increasing the potential for partnership with people of color.

TRANSFORMING POPULATION

What people of color did for environmental issues, they may do for population. Through their efforts, new constituencies are reinvigorating the environmental movement. They have brought much-needed credibility to America's role in the preservation of the global environment, from the Brazilian rain forest to the deserts of southern Africa. They have begun to "Americanize" the concept of sustainable development with examples from their own cultural traditions.

There are many encouraging signs that people of color are beginning to change the way population issues are framed in the U.S. People of color are participating in preparations for the International Conference on Population and Development at high levels of significance and within a diversity of settings. In both nongovernmental organizations and governmental agencies, people of color are directing government preparations, forming NGO networks, participating in delegations to the Preparatory Committee meetings, and publishing articles that address issues of population and development, as well as issues in related areas such as the environment, consumption, and national security.

In a 1993 speech about environmental justice, President Bill Clinton declared that "We don't have a person to waste...." Clinton's observation holds for people of color and population issues as well. We cannot afford to lose more lives in the cycle of poverty which is both a cause and result of high fertility. We cannot afford to see children's lives cut short because their mothers cannot feed them, or because they are addicted to drugs *in utero*. These are the population issues people of color recognize and consider important in their day-to-day lives.

Despite the many deprivations with which they struggle, people of color connect with and act on issues that they perceive to be in their best interests. When the issues are presented in ways that are true to their historical and current experiences, people of color do express concern about population, consumption, and the environment.

People of color are looking for authentic opportunities for leadership and partnership. They, above all, know and understand what a partnership for change can produce. They recognize honesty and accountability and believe these qualities have the power to make a difference. People of color *will* join the discussion of population. But there are realities which must be acknowledged for the conversation to achieve common ground.

RUSSEL LAWRENCE BARSH

Indigenous Peoples, Population, and Sustainability

....

The village of Karluk, Alaska, rests on a small hill overlooking the Shelikof Strait and, barely visible, the Aleutian Mountains of the Alaskan mainland. It is home to fewer than a hundred Koniagmiut,[1] the original people of Kodiak Island. The Koniagmiut were once renowned as high-seas seal and otter hunters in their characteristic canoe-like *bidarkas* and walrus-gut rain slickers. Today, the village is scarcely distinguishable from any other impoverished coastal settlement on the Gulf of Alaska: a cluster of small frame bungalows, power generators, and beached aluminum skiffs. The fishery was badly damaged by overharvesting a century ago, when one of Alaska's largest salmon canneries extended along the sandspit; the *Exxon Valdez* spill gave it the *coup de grace*. It is a vision of the fringes of the Third World — or the aftermath of the First World.

It is also a parable of the paradox of indigenous peoples, in the context of the contemporary policy debate over population, ecology, and "development." Beneath the Karluk of today, lie the remains of older Karluks stretching back continuously for an estimated 4,000 years.[2] The inconspicuous Koniagmiut village is, indeed, one of the oldest continuous human settlements on this continent, established as soon as the ice sheets exposed bare land along Shelikof Strait. It was also the trading hub for a string of even smaller settlements that once stretched far up the Karluk River, exploiting a great variety of ecosystems: beach, muskeg,

Russel Lawrence Barsh is an associate professor of Native American Studies at the University of Lethbridge and U.N. representative of the Four Directions Council.

streams, forests, and lakes. When the Russians controlled the region, they concentrated the indigenous population downstream at Karluk, where they could control them better and promote the otter-fur trade. Then came the Americans, with their canneries, their powerboats, their televised illusions, and their laws. The population of the village is smaller today than ever. It is still shrinking, despite very high fertility, feeding its youth to the unemployment lines in Kodiak City, Anchorage, and Seattle.

While Karluk's population shrinks, its ecological impact grows. As the people switched from driftwood to fuel oil, fresh fish to tinned and packaged meats, storytelling to television with its thirst for electricity, they began to accumulate the means of their own destruction. A short walk upstream, a pattern of slit trenches serves as the community landfill, brimming with styrofoam, aluminum pop cans, motor oil, batteries, and discarded electrical equipment. Already, it is deeper than the midden left by all earlier Karluk generations. The culture is dying, while the trash grows. It is a potent symbol of the power of industrial "civilization" to accumulate dead matter as a kind of substitute for life.

Like most indigenous communities worldwide, Karluk has key characteristics which distinguish it from the conventional population-growth scenario. It is in a state of population *recovery*, following a period of elevated mortality due to invasions, disease, and ecological degradation. This recovery is reflected in high fertility, relatively low mortality,[3] and a youthful age structure; it has been delayed for a generation or more by the impact of ecosystem degradation on locally available food resources and the resulting poor nutritional status of the survivors of the initial catastrophe. A significant factor in the rate of recovery has been the availability of wage-work, trade, or transfer payments, partly offsetting the loss of subsistence resources by enabling people to import food, and thereby restore their dietary quality.[4] In terms of area and productive capacity, the lands that remain in indigenous peoples' control can no longer support their original numbers — not sustainably, at least — so even a modest recovery may result in what appears to be overpopulation and population-driven poverty. This process has affected indigenous peoples everywhere, but different peoples can be found today at different stages, from initial contact and population losses to full recovery and population crises.

Indigenous peoples face a policy dilemma as a result. To restore the vitality of their cultures, indigenous nations must rebuild their families, clans, and communities. A tribal culture, based so much on clan and kinship, cannot exist without a sufficient number of people to carry out a multitude of social and ceremonial roles. Cultural renewal depends in part, at least, on population renewal. Yet the loss or ecological degradation of land generally means that population recovery will lead either to greater poverty or to greater dependence on outside sources of income.

Either of these results, in turn, threatens the integrity of culture, kinship and reciprocity, and promotes emigration.[5] Moreover, the erosion of tribal norms and practices undermines traditional means of managing family size, thus promoting a rebound beyond the limits of environmental sustainability. In brief, cultural renewal involves the risk of exceeding ecological capacity and the unintended consequence of dependence and cultural assimilation.

This is the reason why indigenous peoples are so strongly opposed to population policies that would restrict their ability to rebound to or exceed aboriginal population levels. It is also the reason why issues of land claims, restitution of land and natural resources, and ecological rehabilitation are so central to indigenous peoples' views about development, self-determination, and cultural rights. Without a redistribution of resources, freedom to recover lost population has no practical meaning.

INDIGENOUS PEOPLES AND THE ENVIRONMENT

There is much to be learned from the ways indigenous peoples originally managed both their human populations and the natural environment. The experience of indigenous peoples may help illuminate the central policy question of what "drives" population growth and what, if anything, can be done to make human societies adequately *self-regulating* in population terms.

Stephen Lansing's 1991 study of the ecological role of Balinese water temples reflects growing "scientific" recognition of the managed character of indigenous peoples' territories. Even hunter-gatherers, long regarded as relatively passive foragers, have been shown to have modified their surroundings to make useful species more abundant, and more easily accessible — for example, by burning off forests to create pasture for wild game, or by creating biodiverse enclaves within oligarchic forests. As I have suggested elsewhere, indigenous peoples devised extremely sophisticated ways of taking advantage of behavior, species diversity, and niche diversity to minimize labor and risk, while maintaining an unutilized reserve of carrying capacity as a kind of insurance against the vagaries of nature.

Indigenous fishers exploited a wide variety of marine niches, for instance, and seasonally harvested small numbers of a large variety of fish, mollusks, and seaweed, each when it became most easily accessible to capture. This minimized interference with the relative proportions of species in marine ecosystems, and left predator-prey relationships intact. In contrast, "modern" industrial fisheries target a narrow band of species (especially larger predators such as cod, tuna, and salmon), competing for them in a year-round, high-seas chase that depends on large inputs of mineral fuels. This modifies food chains, as non-targeted species opportunistically and irreversibly fill the niches of targeted ones, leading to the sudden collapse of targeted stocks such as California anchovy in the 1930s, Kodiak king crab in the 1960s, and North Atlantic cod in the 1980s. Indigenous hunters also utilize diverse niches and species, minimizing structural impacts on ecosystems.

There are significant differences between indigenous horticulture and herding and the farming and ranching systems of large-scale urban and industrial societies. Indigenous peoples organize single-species flocks in ways that mimic herbivores' own social organization, minimizing the need for supervision; or, organize mixed-species flocks that utilize a wider range of forage plants in a more balanced manner than single-species herds. Productivity has arguably fallen wherever indigenous methods have been replaced by large-scale, single-species herds on sown, monocultural grasses. Similarly, traditional methods of interspersing crop species, while labor-intensive, produce biodiverse gardens with higher sustainable yields and lower risks of failure than monocultures that depend on machines and chemical inputs. In the Mediterranean Rim, and, as recent studies suggest, by comparison, Aztec Mexico, intensive grain monocultures to feed growing urban centers may have been responsible for severe soil destruction from very early times.

Risk was also managed by organizing trade networks among villages with varied ecosystems and resources. Sharing and reciprocity among households not only involved the pooling of productive efforts, but the distribution of products, and exchanges of people among related households, adjusting local population levels to seasonal and annual variations in local resources.

At the same time, reciprocity perpetuated a relatively democratic and inclusive social order, not necessarily based on strict equality of wealth among families or households, as much as a balanced and complementary distribution of important resources, both spiritual and material. Within such a social order, there is little need to produce and hoard a surplus, because social security is provided by sharing. On the contrary, all accumulations are viewed with suspicion, because they are presumptively anti-social. Likewise, status is achieved by sharing, not accumulating resources.

These societies tend to be *ecologically* sustainable because their social order does not rely, for its reproduction and perpetuation, on the pursuit of *inequality* and a spiral of increasing consumption. Nor do they strive to maximize population, because everyone shares direct personal responsibility for the well-being of every other.[6] Tools for population management such as birth spacing, herbal contraceptives and abortives, and mechanical abortion are widely known and routinely used.

POPULATION, CONSUMPTION, AND SOCIAL TRANSFORMATION

It is also instructive to view European demographic history through the lens of indigenous peoples' experience. European society was also tribal, at one time, but its ecological character was fundamentally transformed by Roman imperialism. Peoples that Tacitus described (in the *Germania* and *Agricola)* as fundamentally egalitarian and non-materialistic were forcibly relocated, reorganized for maximum agricultural production, and divided into economic classes on the model of the Roman social order. Obedience to the state replaced kinship, while accumulation replaced reciprocity as the basis of personal status and power. The breakdown of a central Roman administrative system unleashed waves of ethnic resurgence and massive migrations (or *Völkerwanderung)*, completing the destruction of tribal cultures. The European population explosion of the 17th and 18th centuries had its roots in the days of the Caesars.

Particularly significant was the concentration of land ownership, which left the majority of Europeans with insufficient land and little or no security of tenure. This resulted in a large number of landless poor, as well as a large class of *landed* poor, as is the case today in "developing" countries such as the Philippines. The poor relied then, as they do now, on wages for their survival, so that even among farmers there is an incentive to increase family size beyond the carrying capacity of the land. Industrialization speeds this process by increasing the supply of wages and the relative value of labor. As inequalities grow, they promote migration.

Migration takes the form of urbanization and frontier settlement. On the frontier, land tenure is also insecure, and unfamiliarity with ecological conditions leads to degradation of resources. Insecurity and high mortality encourage larger families, further resource degradation and fresh migration. In the cities, labor replaces land entirely as a source of security while overcrowding, poor sanitation, and unemployment result in contagion and high mortality. Dependence on wage labor and high mortality provide strong incentives to increase family size. Hence while the growth of cities and frontiers temporarily relieve the symptoms of social inequality, the result is a cycle of greater inequality and *accelerated population growth.*

Faced with this crisis, 17th-century British economists began to argue that *population is wealth,* because people are not only producers but consumers, and consumption is growth. Thus did Sir William Petty declare in his *Treatise of Taxes and Contributions* that "Fewness of people, is real poverty." Small population levels encourage sloth, Petty concluded: "If the people be so few, as that they can live, *Ex sponte Creatis,* or with little labour,…they become wholly without Art." Long before the publication of Adam Smith's influential *Wealth of Nations,* Petty and other British economists concluded that the population growth associated with rural

FIGURE 1: Population Growth and Per Capita Consumption of Coal, Sugar, and Tea in England and Wales, 1801-1901

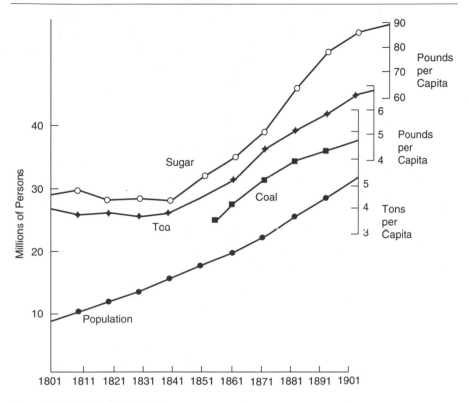

Source: Mitchell 1962: 6-7, 355-358.

landlessness, migration, and urbanization was not a curse, but a blessing, because it raises both demand and supply. In other words, population (or demand) drives growth. Malthus challenged this policy on the basis of resource constraints on population growth. But did population growth actually stimulate economic expansion in the early stages of European industrialization? What was the relationship between population and consumption?

Figure 1 compares 19th-century British population growth with per capita British consumption of tea, sugar, and coal.[7] Per capita consumption remained constant at first, then expanded, and for sugar expanded at an *accelerating* rate. A similar pattern appears in Figure 2, which shows the relationship between American population growth and the *per capita* consumption of mineral fuels, iron ore, and paper products, 1850-1950. In these two countries on the leading edge of world industrialization,

FIGURE 2: U.S. Population Growth and Per Capita Consumption of Mineral Fuels, Iron Ore, Paper, and Paperboard, 1850-1950

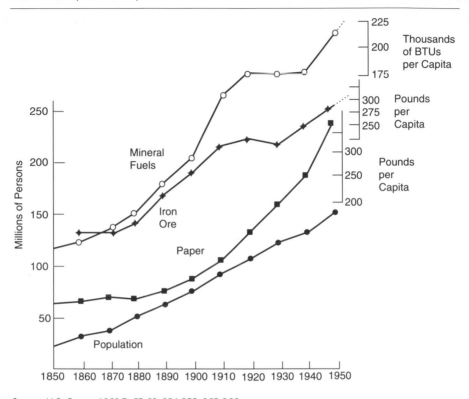

Source: U.S. Census 1960:7, 67-69, 354-355, 365-366.

then, it appears that growth was driven by consumption rather than population, *per se*, from a very early stage. What, then, drove such increases in consumption? In Europe, as in indigenous communities today, increased consumption followed the breakdown of social order.

This observation returns us to the indigenous perspective. In indigenous cultures, a sustainable relationship between human populations and the environment is maintained through social justice and reciprocity. Inequality and injustice in connection with land can stimulate population growth. But, by unleashing the need for consumption, the breakdown of social order creates an even greater force that soon outstrips population in its potential impact on the environment.

MATERIALISM REPLACES RECIPROCITY

Just as the increased consumption and environmental degradation of Europe can be seen as an echo of Roman imperialism, reports of materialism, ecological recklessness, and extensive warfare among indigenous peoples reflect the effects of invasions and displacement rather than aboriginal conditions. Like the landless and landed poor of Europe a few centuries ago, indigenous peoples have been forced to integrate wages into their economies, while traditional bonds of kinship and reciprocity have broken down. In addition, they must contend with externally imposed land-tenure regulations and other institutional constraints, which restrict ecologically sound practices such as pooling gardens, exchanging children with distant kinsmen, and redeploying flocks.

Among indigenous peoples today there is a widespread breakdown of reciprocity, *replaced by* an ethic of accumulating material and people. Some have suggested that this is a "rational" or inevitable response to markets. Participating in markets increases the relative value of resources and products which can be converted into money, because money buys imports that were scarce or absent in the past. Money can be hoarded, or even hidden indefinitely, moreover, making it easier to avoid reciprocity — should anyone wish to do so. The apparent abundance and durability of money challenges traditional beliefs in a fragile, finite, and fickle universe, which support the practices of reciprocity and conservation. The initial flush of relative prosperity suggests that accumulation is indeed more rational than sharing and saving. Hence, the early stages of market integration create a kind of "demonstration effect" favoring increased production and population growth.

However, export prices eventually fall and expectations continue to rise,[8] encouraging the newly integrated society to intensify its use of land and resources. Traditional science and ethics no longer stand in its way.[9] On the contrary, the illusory abundance and durability of money suggest that ecosystems can be *replaced* by money and, hence, are quite literally expendable. The breakdown of reciprocity has meanwhile led to a breakdown in social cohesion and cultural solidarity, so that any attempt to promote coordination, self-discipline, or self-sacrifice is certain to be resisted. After they have dismantled their preexisting system of collective security, people seek individual security in the pursuit of accumulation, domination, reproduction, and competition.

From this viewpoint, the breakdown of reciprocity is "irrational" *in the long term*, but difficult to prevent. Markets create short-term conditions that obscure long-term consequences, while discrediting the entire system of long-term ethics as practiced by indigenous peoples.

In tribal societies, as I suggested earlier, the entire community functions as a system of social security. This system is "rational," economically, only if there is enough land to go around, so that acts of generosity are likely to be reciprocated. This is especially true in the case of children who care for elderly kinfolk, since they must expect to share eventually in any inheritance, or to assume the roles or privileges of the kin-folk served, in exchange. Raising larger families does not improve the prospects for parents in a society with a fixed stock of land, because it dilutes the interest of each child. The possibility of earning cash through outside labor changes this calculation. Under expansionary economic conditions — as typically prevail on young frontiers — there is an apparent cornucopia of wages, and therefore an incentive to max-imize labor. Even in the absence of insecure tenure, ecosystem degra-dation or absolute poverty, then, integration into markets promotes population growth at the same time that it weakens cultural constraints on consumption.

LESSONS FROM THE ANASAZI

There is a growing body of evidence that our planetary climate is far more unstable than previously believed, in terms of average annual tem-perature and annual precipitation. Conditions have been *relatively* stable since the last glacial episode, however, encouraging the intensification and geographical expansion of agriculture. The earliest intensive agri-cultural systems, such as the cultivation of the Nile delta, also enjoyed a brief subsidy from the rich sediments created by retreating ice-sheets and sea-level changes. These temporary conditions favored the growth of large urban societies which depend on large, relatively stable food supplies. Another period of climatic instability — possibly hastened by greenhouse gases and ozone depletion — would find the human species vastly overextended.

Archaeology offers a preview of the danger of overestimating the long-term stability of climatic factors. The Colorado Plateau appears to have been drying out gradually for about a millennium, with periodic episodes of severe drought. The Anasazi, maize farmers indigenous to the region, responded to the early drought episodes by intensifying cultivation and storing larger quantities of food. This led to illusory prosperity and an *increase in population*. Subsequent droughts were truly catastrophic because all new productive capacity had been consumed rather than con-served. In *Muqaddimah,* his 15th-century attempt to explain world histo-ry, the Islamic scholar Ibn Khaldun recorded a number of parallel exam-ples from North Africa. The planetary environment has been sufficiently unstable, locally, to draw some societies into a snare of overexpansion. Global interdependence, far from mitigating this danger, has increased it.

Indigenous peoples have long carried with them the memory of that earlier age, before the last glaciation, when the earth was a far less pre-dictable habitat and it was basic wisdom for societies to be small, con-serve large reserves of biodiversity, and remain highly adaptable. For the past several centuries, this old wisdom seemed more like myth, contradicted by the apparent triumph of technology over environmental limits. Are we about to find that one of our most basic assumptions — that climatic changes are gradual — is just as erroneous as the belief, shaken by recent studies in paleontology and genetics, that biological evolution is gradual and incremental?

A technological *deus ex machina* cannot prevent the human species from succumbing to sudden changes in planetary conditions on the order of what happened 10,000 to 50,000 years ago. There simply would be no time to respond before the global food system, and economy, collapsed. Population control will not solve this problem, not only because world population is probably already overextended, but because growing rates of consumption will continue to deepen our predicament even if global population stabilizes. Family planning does not alleviate insecurity, inequality, or empty materialism, the shared root causes of population growth and consumption growth. The only lasting solution is restoring the security, reciprocity, social values, and ecological caution of the indigenous civilizations, which themselves are rapidly losing faith in their own validity as they witness the continued profligacy of Western and Westernizing societies.

Population and Religion

· ·

L. ANATHEA BROOKS AND TERESA CHANDLER

American Religious Groups and Population Policy

·····················

In recent years, American religious attitudes have influenced government policy toward international family planning programs. Religious
conservatives were instrumental in persuading the Reagan and Bush
administrations to withhold funding from two global agencies, the
U.N. Population Fund and the International Planned Parenthood
Federation, because they believed that these organizations were major
promoters of abortion around the world. They held to this view even
though the U.N. Population Fund does not directly finance abortion services, and even though the London-based International Planned
Parenthood Federation reportedly devotes only one percent of its budget to abortion-related activities. The decision to withdraw support from
these organizations did not signal a large-scale reconsideration of U.S.
assistance to population programs in individual countries. But it did
come to symbolize the American government's retreat from a visible
leadership role in addressing global population issues—a role that the
Clinton administration is now determined to reclaim.

As the United States becomes a more energetic advocate of population programs, it will find that some religious groups are prepared to
support its efforts. Such groups are an important force in environmental and humanitarian movements which are concerned about rapid population growth. Moreover, almost every American religious group takes
seriously the biblical obligation to be stewards of the earth, and faces

*L. Anathea Brooks is an environmental research assistant and Teresa Chandler is
a graduate assistant with the Institute for Philosophy and Public Policy at the
University of Maryland. This chapter is reprinted from the* Report from the
Institute for Philosophy and Public Policy, *Fall 1993.*

the challenge of spelling out this obligation concretely and in contemporary terms.

THE AMERICAN CONTEXT

Most Americans consider themselves religious. They look to religion for moral guidance; they understand ethical issues ultimately in religious terms. According to a recent Gallup poll, approximately 56 percent of Americans identify themselves as Protestant, 28 percent as Catholic, two percent as Jewish, and four percent as other (Buddhist, Muslim, Hindu, Orthodox, etc.). Only 10 percent claim no religious affiliation at all. Sixty-nine percent describe themselves as church or synagogue members, and 43 percent attend religious services regularly. Their religious institutions hold justice and compassion as ideals and struggle to realize these ideals in circumstances that tradition may not explicitly address.

Rapid population growth has become a critical issue for some religious groups because of their long-standing involvement in humanitarian and relief work. Today approximately 75 percent of congregations report contributing to some form of international assistance; these include 88 percent of mainline congregations and 62 percent of very conservative ones. According to the Institute for Development Training, religious groups, taken together, are the second-largest health care provider in the world. At the same time, however, it is important to recognize that the denominations most deeply involved in foreign aid and development do not focus primarily on population. Rather, they emphasize maternal and child health, the education of women, the importance of improved economic opportunity, an equitable distribution of resources, and the need for more moderate consumption in developed nations.

Moreover, church members often object to the "crisis rhetoric" that pervades discussions of rapid population growth. Some, citing Julian Simon, simply do not believe there is a population crisis, while most of the rest agree that to the extent there is a problem, it results largely from an unjust distribution of wealth. They argue that the "lifeboat" analogy—which offers developed nations a place inside the boat, while less developed nations are clambering to get aboard—encourages those in the wealthier countries to think that their survival depends on the reproductive restrictions they place on families in the developing world. Such an attitude clashes with religious teachings which emphasize self-sacrifice, equity, and reverence for life.

In a stance characteristic of mainline Protestant denominations, the United Methodist Church argues that rapid population growth must be addressed in the context of development; any attempt to bypass this slow and difficult process is likely to be ineffective or even harmful.

Specifically, the church argues that a reasonable response to population growth must include the following:

> ...better education, and the opportunity for people to participate in decisions that shape their lives; the provision of basic economic security, including old-age security; upgrading the status of women; improved maternal and child health care; and finally, a strong birth control program, including the right to abortion and sterilization procedures which are both legally obtainable and voluntary.

The Lutherans have similar views and policies, as do the Presbyterians. In sum, mainline Protestants insist that treating population outside the context of social justice and empowerment is like treating the symptoms of a disease rather than its causative agents.

Some ecumenical organizations today do make population policy an explicit priority. These include the Ministry for Justice in Population Concerns, which supports "a voluntary limit of global population growth through increases in economic and racial justice, contraceptive availability and women's status worldwide," and the Institute for Development Training [IDT], whose Save the Mothers program specifically aims to improve the health, education, and status of women. The IDT emphasizes the link between improvements in women's status and lower birth rates. Several religiously affiliated environmental awareness groups, of which there are a growing number, also focus attention on the population problem.

POPULATION AND ENVIRONMENTAL JUSTICE

Religious groups have become increasingly committed to protecting nature and have re-examined the relationship between God, man, and creation. Some religious traditions have long considered nature to have intrinsic value, independent of its utility to mankind. Buddhism, Hinduism, and Native American religions regard human beings as but one among innumerable species and forms of life, all of which inhabit the earth on equal terms. Judaism and Christianity have sustained a tradition of reverence for creation as a manifestation of God's perfection, as exemplified by the Song of Solomon and the writings of St. Francis of Assisi and Abbess Hildegarde von Bingen. On the other hand, the Judeo-Christian tradition has more often portrayed nature as having chiefly instrumental value, or as serving merely as a backdrop to the personal drama of salvation.

Recent Protestant commentaries on the book of Genesis emphasize God's love and concern for all creation, and our responsibilities towards it. In Gen. 1:28 God appoints man to "rule over" the animals, and in Gen. 2:15 Adam is instructed in the Garden of Eden "to till it and to care for it." Modern Christian theology has also been enriched by exchanges with both North American and Far Eastern religions, which hold that people have no right to destroy a natural order that they did not create. This idea lies at the heart of the eco-spirituality movement, which emphasizes the integrity of all creation.

However, many religious groups influenced by eco-spirituality have been careful to say that a theology which excludes human needs and social concerns is both shortsighted and unchristian. This position is reflected in the 1983 statement of the World Council of Churches, which proposes three values for contemporary Christian emphasis: peace, justice, and the integrity of creation. A commitment to these values has given rise to the environmental justice movement, which holds that socioeconomic equity is inseparable from ecological integrity.

The link between social justice and the environment was initially made in the early 1970s by the American Baptist Churches' National Ministries, which insisted that the emergence of environmental activism must not displace concern for civil rights. Today the environmental justice movement has strong interdenominational support, and the ideas at its heart constitute a major theme in contemporary theology. According to the Evangelical Lutheran Church of America, "Justice in the environmental context is defined by right relationships in creation: fairness within the human family and the honoring of the integrity of creation."

Pope John Paul II, in his 1990 message "Peace With All Creation," declared that the people of every nation, collectively and individually, share moral responsibility for the current ecological crisis, since humankind has upset the harmony of a creation that God entrusted to it. But by November 1991, the Conference of Catholic Bishops was assigning responsibility more narrowly: Western societies, the bishops said, were consuming an inordinate share of the world's resources and exacerbating the problems caused by rapid population growth. The Vatican has endorsed this view and has cited it as one reason for questioning the necessity and fairness of efforts to limit population growth in developing countries. On June 1, 1992, in a position paper released in anticipation of the Earth Summit, the Vatican asserted that family planning programs promoted by wealthy nations can easily become "a substitute for justice and development." The "goods of the earth," the Vatican declared, "are for the benefit of all, and all peoples have a right to fundamental access to those goods." Thus, "aid programs should not be conditioned on acceptance of programs of contraception, sterilization or abortion."

Other conservative religious movements, finding common ground with a wide range of denominations on the issue of environmental justice, have voiced support for international family planning efforts. Several such movements, along with Jewish and mainline Christian groups, appeared as signatories to the 1992 "Joint Appeal by Religion and Science in Partnership for the Environment: Declaration of the Mission to Washington," which conceives of environmental problems as moral problems, resulting from greed and selfishness. The declaration "affirm[s], in the strongest possible terms, the indivisibility of social justice and the preservation of the environment," and suggests that population growth is related to the low status of women in many cultures. It explicitly states that "there is a need for concerted efforts to stabilize world population with humane, responsible and voluntary means consistent with our different values."

Although the largest conservative Protestant group, the Southern Baptist Convention, was a signatory to the Washington declaration, some conservative Protestants, like their liberal counterparts, worry about coercion in population policies. At the same time, they tend to be more skeptical of claims that increased population will cause dangerous environmental degradation. For example, the Family Research Council questions the assumption that rapid population growth interferes with or precludes economic development. Conservatives also worry that family planning programs may engage in over-zealous promotion of birth control, and thereby undermine cultural traditions that value the formation of families.

FAMILY PLANNING AND VIEWS OF SEXUALITY

Although many denominations may be prepared to recognize the importance of limiting population growth, a mere 32 percent of congregations offer financial support for family planning programs. The National Council of Churches dropped its family planning program several years ago after it became a source of controversy. Church World Services has been similarly cautious.

Such caution stems in part from the continuing conflict surrounding the issue of abortion. Some people see "family planning" as code for abortion-on-demand—a policy from which nearly all religious groups distance themselves. In an effort to gain as much distance as possible, some humanitarian or missionary programs may understate the amount of family planning counseling they actually provide. The danger, though, is that by seeking to dissociate themselves from abortion altogether, these programs may actually neglect to provide services crucial to the well-being of families.

For some religious groups, one of the moral obstacles to using birth control is that it usurps divine authority by interfering with the genera-tion of new souls. The Vatican has called on believers to display the "courage" necessary to "accept" new children and the "stout hearts" required "to cooperate with the love of the Creator and the Savior, who through them will enlarge and enrich His own family day by day." Such pronouncements emphasize the passive, vessel-like role humans play in procreation. On the other hand, in the interest of the "responsible trans-mission of life," the Vatican also stipulates that in making reproductive decisions, parents must consider both their own welfare and that of their children, those already born and those whose birth may be foreseen. For this purpose, they must take into account both the material and the spiri-tual circumstances of their lives, weighing the interests of the family group, of temporal society, and of the Church itself.

Issues of sexual morality also provide a powerful source of resistance to family planning. Contraception introduces a separation between sex and the power of procreation, and many Christians believe that without this power, sex is sinful. This view has been propounded by Christian thinkers (most notably St. Augustine) for centuries, and the contemporary Catholic Church has affirmed it by asserting that "each and every mar-riage act must remain open to the transmission of life."

Some Protestant theologians have developed the view that sexual rela-tions in marriage are a sacramental means of expressing love, and thus have a holy purpose in addition to procreation. These theologians hold that if a husband and wife believe that God does not wish them to bring a child into the world, they are not only permitted, but have a positive duty, to use the most effective contraception available. Similarly, the Jewish view presents sex within marriage not as a sin, but as a mitzvah, or a posi-tive deed and obligation. This obligation overrides that of procreation itself, and contraception is therefore more acceptable in Jewish law than in many Christian traditions. Moreover, because Jewish law places the woman's well-being before that of a fetus, it does not categorically forbid abortion. This is not to say that Judaism actively advocates contraception and abortion, but rather that the tradition acknowledges circumstances in which such measures are necessary.

DOCTRINE AND PRACTICE

In any religious group, however, the beliefs and practices of the laity can fail to correspond with official doctrine. Catholics provide one con-spicuous example of this phenomenon. Certainly their low birth rates in many parts of the developed world suggest that they have been using effective means of birth control. In developing countries, where the

Church has exerted its influence to block legal reform and government initiatives with regard to family planning, the picture may be equally complex. Ms. Guadelupe de la Vega of FEMAP, a program providing family planning services in Mexico for the past 20 years, recalls telling a Roman Catholic bishop that as long as couples were denied access to contraception, it was inevitable that some women would resort to abortion in order to limit the size of their families. The bishop blessed her and told her to continue her work. Survey data reported by Catholics for a Free Choice show that in spite of the Vatican's teachings on family life and sexuality, Catholic women in the United States "use contraception and abortion in the same numbers as the population as a whole." Moreover, Catholic attitudes toward birth control, abortion, and sex tend to correspond with those of the public at large. In some matters, Catholics even prove to be more tolerant. For example, significantly more Catholics than non-Catholics favor public funding for abortion (62 percent versus 42 percent).

A similar split between doctrine and popular attitudes appears in Joseph Chamie's 1981 study of Christians and Muslims in Lebanon. "It is likely," Professor Chamie writes,

> ...that a religious sect may maintain an official position on fertility control or procreation that is quite different from the attitudes and practices existing among its followers or clergy. . . . Groups with the strongest sanctions against contraception and abortion have the highest rates of birth control approval among their followers; groups with relatively liberal and flexible positions have followers who are the least approving.

In addition, Professor Chamie found that "local orientations"—the "current attitudes, views, and positions on fertility and fertility control prevailing among the particular religious communities"—varied depending on whether members of a community had lived abroad for a time or maintained contacts with "outside nations, groups, and cultures."

In a society as diverse as our own, we must acknowledge that attitudes toward family planning vary considerably. But we may also expect them to evolve. Among certain religious communities, there does seem to be an emerging consensus affirming environmental justice as part of a larger commitment to social justice and human flourishing. And it is this commitment, conjoined with the obligation of global stewardship, that is most likely to prompt American religious groups to support international family planning efforts.

REVEREND JAMES B. MARTIN-SCHRAMM

Population Policies and Christian Ethics

• • • • • • • • • • • • • • • • • • •

P opulation policy has undergone some significant changes over the last twenty years.[1] Since the failure of ill-conceived and invasive policies in the 1960s, a narrow focus on *population control* and contraceptive technologies has been increasingly replaced by a wider emphasis on various social, political, and economic factors that contribute to *fertility reduction.* Most countries and non-governmental agencies now recognize that population policies must be part of broader public health, education, and economic development initiatives, although there remains a substantial gap between rhetoric and action on these fronts. As numbers continue to grow, however, it is likely that proponents of population control will rejuvenate their appeals for specific measures that go beyond voluntary family planning. For example, while China's population policy is criticized by some for its coercive aspects, it is hailed by many as a model of effective population control.

In the past, ethicists have evaluated population policies that contain various incentives, disincentives, and forms of coercion by considering their impact on four primary human values: freedom, justice, general welfare, and security or survival. These terms have been defined in various ways. Freedom has been primarily defined as the capacity to make reflective choices in life and the power to act upon those choices. Justice has been understood primarily as the distribution of burdens and bene-

---

Reverend James B. Martin-Schramm teaches theological ethics at Luther College. This article is partly adapted from "Population Growth, Poverty, and Environmental Degradation," Theology and Public Policy, *Vol. 4, no. 1, Summer 1992, 26-38.*

fits in a society. Some ethicists have limited the value of general welfare to the overall well-being of individuals and families with regard to having adequate food, clothing, shelter, and health care. Others have broadened the concept to extend it to the general welfare of other species and ecosystems. Similarly, some ethicists have considered the value of security/survival solely in terms of an individual's right to freedom from risk and endangerment to one's life, while others have expanded the scope of this value to encompass the security and survival of other species and Earth's ecological systems.[2]

While responsible moral deliberation involves reflection on all four of these primary values, ethicists have arranged these values in different orders of priority. For example, some have made the values of general welfare and security/survival subordinate to the more fundamental values of freedom and justice. They argue that efforts to maximize freedom and equality are more effective means of securing the common good than coercive means that violate human dignity.[3] Others have taken the opposite approach and have emphasized that without a fundamental measure of general welfare and security it is impossible to experience the values of freedom and justice. They argue that it may be necessary and justifiable to limit certain individual rights and regulate human fertility out of a concern for the common good of present and future generations.[4]

JUSTICE AS THE "TELOS" OF CHRISTIAN ETHICS

It is important to recognize, however, that the relationship and ordering of these four primary values is largely determined by what end these values serve. If the end is the preservation of a certain quality of life, then the values of welfare and security/survival will be elevated to a higher normative status. Similarly, if the end is the preservation of human dignity in the face of calamity, then the values of justice and freedom are given more moral weight.

I want to identify the end which these values serve specifically within the context of the Christian tradition. For Christians, the "telos" of all life is the in-breaking of the Reign of God and the experience of Shalom which God is bringing. The content of this Shalom and the nature of God's Reign are described variously in scripture as Peace, Reconciliation, Liberation, and Community. I believe that the theological image which best describes this end or goal of Christian existence is the metaphor of Justice as right relationship with Creator and creation.[5] Justice conceived as a quest for the restoration of right relationship with God, neighbor, and all creation is much more than an approximation of love, or simply one virtue among many. It is "the animating passion of the moral life"[6] which seeks to do "whatever is required for the fullest possible flourishing of creation."[7] The content of the Christian covenant with God is to "do justice" in creation just as God has done justice through the life, death, and resurrection of Jesus Christ.

There are some, however, who are less than enthusiastic about positing justice as the central theological metaphor and "telos" for Christian social ethics. Stanley Hauerwas has recently argued in *After Christendom?* that "appeals to justice have simply gotten out of hand."[8] He means that Christians uncritically appeal to liberal conceptions of justice which have very little to do with biblical understandings of the term. Hauerwas contends that liberalism's over-emphasis upon the autonomy of the individual has warped Christian understandings of justice and freedom rooted in community and servanthood, and he suspects that what lures Christians to use the term justice is the familiar desire of Christendom to wield power in the world. He does not oppose the Church wielding power but thinks it ought to be redirected to another end. Hauerwas writes: "We forget that the first thing as Christians we have to hold before any society is not justice but God."[9]

Hauerwas' disassociation of God and justice must not go unchallenged. Justification and justice are coherent in God; they are not separable even though they can be distinguished. The God revealed in the scriptures is a lover of justice. To hold up before society any conception of God other than the God of justice is to hold up a different god. But what does it mean to say that God is a God of justice? How is this justice defined?

The first thing that must be said is that this justice is rooted in God's love for the world. That is to say, God's love for all of creation confers an intrinsic value to all of God's creatures and creations. It is this fundamental measure of worth which serves as the foundation of justice.[10] God calls the faithful to love what God loves, to value what God values, and to join with God in the process of redemption to restore the right relationships of a good creation in a fallen world. Justice understood in this light is not simply "rendering to each their due," but is more profoundly understood as "rendering to each their dignity as a creation of God."

Throughout the Bible God acts on behalf of those who are poor and oppressed. It is precisely because God loves all creation that God shows special attention to those who do not live with the dignity that they deserve. In turn, Christians are called by God to restore dignity by remedying the injustices that violate the dignity of God's creatures. The heart of justice in Hebrew and Christian thought is the meeting of fundamental human needs.[11] This description of Christian responsibility helps us see that the heart of a Judeo-Christian understanding of justice lies not in retribution but in restoration.[12] While justice as the restoration of right relation does not provide a comprehensive theory of justice, it does offer some fundamental criteria for judging all theories of justice and responses to moral issues: 1) Is priority attention given to the needs of the poor and disadvantaged? 2) Are oppressive structures which perpetuate injustice challenged and are more just alternatives proposed? 3) Is there a provision for reparations to the victimized? 4) Is the goal the restoration of a fundamental level of human dignity corresponding to basic human needs?

With this framework and these criteria in mind, I want to propose an ethical foundation for the development and evaluation of population policies.

POPULATION GROWTH AND JUSTICE

Faced with the problems created by rapid population growth, Christians rooted in God's justice of right relation are empowered by the vision of a world where all pregnancies will be desired and welcome, where children and their parents will have all they need to live full and productive lives, and where the human species will live in harmony with each other and with the rest of God's creations. But this vision remains elusive. Instead of being the blessing that God intended (Genesis 1:28), the multiplication of human life is endangering God's creation. Continued population growth contributes to enormous levels of human suffering through its relationship to poverty and environmental degradation.

Any effort to redress this suffering means seeing the reciprocal relationship between ecological integrity and social justice. The ecological jeopardy which the world currently faces is grounded in an unjust distribution of wealth and power between the affluent few and the numerous poor. This robs nearly half the world's population of fundamental sustenance and basic human dignity and creates the prospect for ecological peril the likes of which the world has never seen.

Differently said, population growth today is inextricably intertwined with poverty, which has its roots in injustice. Therefore, the primary Christian response should constitute an attack upon poverty and injustice. At the forefront of this effort must be the attempt to provide for basic human needs. This will necessitate substantial social, economic, and political reforms, and not just the spread of Western economic development, since most conventional courses of development have only served the middle and upper classes in many developing countries. Alan Durning, citing Gandhi, rightly reminds us that "true development puts first those that society puts last."[13] The areas in which reform is desperately needed include more equitable distribution of land and income, improvement in access to education and employment, the elimination of discrimination based on race or sex, and substantial improvement in access to affordable housing, food, and health care.

For moral and practical reasons, however, the most important area of social reform involves improvement in the lives of women. A wide variety of studies indicate that when the status of women's lives improves, fertility declines. As women have received adequate nutrition, proper sanitation, access to basic health care, increased educational opportunities, and equal rights, the fertility rate has dropped markedly.[14]

Christian communities must defend and strive to improve the lives of women and all people in this process of social transformation. High on the list must be the fundamental right to voluntary family planning. Currently, approximately $4.5 billion is spent per year by governmental and non-governmental organizations for family planning efforts.[15] These funds provide family planning services to only 30 percent of men and women of reproductive age in the developing world outside of China. Christian churches should join others who are calling for universal access to family planning by the end of the decade.[16] The provision of such services is estimated to cost $10 billion — a sheer bargain compared with third world debt payments of $125 billion and annual global military expenditures of $880 billion.[17]

In this process, pressure must be brought to bear on the U.S. government to resume its leadership through a substantial increase in support for family planning programs. In addition, Vatican policies which proscribe artificial means of contraception must be critically challenged in the light of the effects of population growth on poverty and environmental degradation. These "pro-life" policies are ambiguous at best since they certainly contribute to an increase in unwanted pregnancies which have a deleterious effect on the lives of poor women, their families, and the environment.

This commitment to universal and voluntary family planning must serve as the cornerstone of any population policy. Moreover, because women disproportionately bear the costs and burdens of reproduction, women must ultimately judge whether the programs of any population policy serve their needs. Ideally, such programs will include access for women and men to an increasing variety of contraceptive and birth control technologies, including legalized, voluntary abortion. While contraception is certainly the morally preferred means of birth control, the unjust treatment and exploitation of women makes legal recourse to voluntary abortion necessary. In addition, legalizing abortion would also make the procedure safer for the estimated 20 million women who undergo an illegal abortion each year.[18]

Considering measures which seek to go beyond voluntary family planning, the range of options between incentives, disincentives, and coercion should be viewed on a spectrum between suspicion and derision.[19] Of paramount concern must be the potential impact these measures would have upon the lives of poor women and their families. The use of incentives like cash or consumer goods to promote family planning may be justifiable, but only if the incentive offers a significant gain in social or economic welfare and only if the recipient believes he or she benefits in a substantial way. Disincentives, like taxation schemes or the restriction of health, education, or medical benefits to limit births, are highly unjustifiable because of the unfair impact these measures would have upon poor families and their children. The use of coercive measures like compulsory abortion, sterilization, or adoption is morally abhorrent. One coercive measure which seems to be gaining support within and outside the United States is the forcible implantation of the contraceptive Norplant into women who wish to continue receiving welfare payments. This unjustifiable and coercive violation of reproductive rights must be vigorously opposed.

Other measures to reduce fertility involve the alteration of social institutions and the use of government-administered programs affecting the distribution of the right to bear children. Some have proposed altering the institution of marriage by raising the minimum legal age. When substituted for reforms aimed at improving the lives and choices of women, this proposal would have a disastrous effect upon young women who currently have few other options in life than marriage. Herman Daly and John Cobb have proposed the governmental implementation of "transferable birth quota plans" which would issue birth rights certificates to parents to sell or use as they deem fit on an open market. One of the major flaws with their proposal, however, is that if no attempt is made to level an unequal economic playing field, the poor are left with the terrible option of having to sell their fundamental right to bear children in order to purchase fundamental necessities like food, clothing, and shelter.[20]

REALITY CHECK

The reality is that, apart from my emphasis on improving the lives and moral agency of women, much of what I have proposed is not new. Ethicists like Ronald Green and Daniel Callahan were saying much the same thing nearly twenty years ago. Still, population growth has continued largely unabated. While the demographic transition has lowered the rate of population growth, demographic momentum has steadily increased the numbers of people born each year.

This failure of what Garrett Hardin describes as a "laissez-faire approach" leads him to the following conclusion: "If the proposal (for population control) might work, it isn't acceptable; if it is acceptable, it won't work."[21] In this pithy but highly dangerous maxim, Hardin has defined the key tension between effectiveness and ethical acceptability. This will only become more acute in the near future. In all likelihood, most nations facing continued high rates of population growth will either respond by bombarding their people with contraceptives or by forcing their people to participate in government-managed population control programs. In both cases, the fundamental need for comprehensive social reform and substantial improvement in the lives of women will be ignored—for such changes will require a shift in the balance of social, economic, and political power. As a result, the bulk of the consequences of population growth will fall upon poor women and their families.

Christian communities, always mandated to side with the poor and disempowered, must speak out on their behalf. Delegates to the 1994 United Nations International Conference on Population and Development must be challenged to follow this alternative approach which addresses the needs of women and the poor in its attempt to redistribute wealth and power, end discrimination, provide for basic human needs, and open doors to education, employment, and health care. While it is unreasonable to expect nations, or the Church for that matter, to accomplish perfect justice, it is clear that a greater measure of justice can be achieved, and that with its increase population growth will decline.

I will also argue that this commitment to justice for the poor and improvement in the lives of women represents the best way to insure the protection of the common good. There are many who point to the increasing level of environmental degradation caused by the sheer growth in human numbers and the threat human population growth poses to other species and the basic ecological systems of Earth. We must resist all attempts to pit the welfare of countless poor human beings against the welfare of the planet. The reality is that individual well-being and the common good cannot be separated. In the words of one of my teachers, Beverly Harrison, we "live and breathe or die together."[22]

Legitimate concern for the common good must be re-focused upon the other two variables which contribute to environmental degradation: harmful technology and the level of affluence. The greatest threat to the common good continues to be posed by the destructive consumption of the rich and not the meager consumption of the poor. The growing ecological threat which population growth does pose can only be resolved by redistributing wealth and power and by providing for basic human needs. In Lester Brown's words:

> We can no longer separate the future habitability of the planet from the current distribution of wealth.... [A] meaningful sustainable development strategy anywhere must now embrace the satisfaction of *basic human needs everywhere*.[23]

What we know for certain is that the entire world's population is going to nearly double during the course of the lives of most people alive today, with almost all of that growth occurring in the less developed world. This means the exponential increase of suffering and misery among people who are already hungry, ill, and poverty-stricken. The only question is whether the international community will care enough about this suffering to prevent it from reaching unprecedented levels. The consensus among policy-makers and demographers is that decisions made during this decade will significantly determine the rate and consequences of population growth. I believe that Christian communities must find their voices and join with others in offering this kind of moral leadership. In a world driven by the values of wealth and security, Christians must lift up the value of justice and champion the cause of the poor and disenfranchised in an attempt to meet the ethical challenges posed by global population growth.

Why Some Christians Object to Abortion

Susan Power Bratton

Among Christians interested in population and demographic change, abortion is, and will remain, a sensitive issue. Christians are divided in their views on abortion, and the positions of both individual Christians and of the major denominations do not entirely follow conservative/liberal political alignments. For example, some Christians who oppose the development of nuclear arsenals (usually considered a "liberal" political stance) consider abortion morally wrong unless the mother's life is in danger.

Although Christian arguments against abortion often center on the question of whether abortion is murder, there are several other theological and ethical objections to abortion. First, the New Testament writings, particularly the four gospels, contain motifs that suggest a Christian world view results in a reversal of the social order. The last shall be first, the weak rather than the strong are blessed, the meek shall inherit the earth, and God has a special love and concern for the poor and helpless. New Testament ethics emphasize the spiritual importance of caring for the "least" among human kind, including the politically oppressed and the socially abandoned. Although Christian teachings would encourage a strong concern for the poor woman with child, the child itself is also weak and worthy of care and attention. The fetus, in this view, is completely vulnerable and has no voice of her own. The fetus can be viewed as the least of the least, which clouds the question of whether the mother's rights and needs should be the primary concern.

A second major theme is that of other-centered neighbor love. To treat the rights or needs of the fetus as irrelevant can be seen as violation of concern for the "other." Christianity often makes a conscious attempt to extend the concept of "neighbor" to those who might be excluded, ignored, or considered marginally human. Interestingly, the Bible does not directly address the morality of abortion, even though abortion and infanticide were widely practiced in the Greco-Roman world. However, Christian writings dating from as early as the second century do speak against abortion, primarily out of concern for the fetus. The literature of the early church also indicates that Christians were involved in rescuing and raising abandoned children. "Child rescue" is thus seen as an heroic and virtuous occupation for Christians. To sacrifice one's immediate economic or social goals for the sake of one's own or someone else's child is an ultimate expression of unselfish love.

A third theme is avoidance of bloodshed, or, conversely, honoring life as absolutely sacred. Early Christians refused to serve in the Roman army on the basis that taking any life was morally wrong. Today, some churches speak out against the acceleration of arms development or use of military force. Christians still register as conscientious objectors and risk imprisonment for refusing military service or protesting military activities. Ronald J. Sider and other Christian ethicists have developed "completely pro-life" ethics, which encourage Christians to avoid any actions which threaten or destroy human life — including building nuclear weapons, forwarding structural violence that leads to poverty and starvation, and performing abortions in non life-threatening circumstances. Other Christians consider "just war" to be moral, yet would permit the taking of human life only when the individual or the individual's family or neighbors are placed in life-threatening danger.

A fourth Christian theme is the welcoming of the child into the human family as a potential member of the coming Kingdom of God. This gives even the poor or unwanted child value in Christian eyes. Every child should be received as Christ would receive her. Both the family and the sexual act have sacred meaning, implying the integrity of the former and the purity of the latter should be maintained. Christian worship devoted to Mary as the mother of God and to the infant Jesus centers on the wonders of birth itself and on a child-king. The attention given to the baby Jesus focuses Christian spirituality on the potential of a child in a context of the birth and renewal of the entire human race.

A last theme that influences Christian opposition to abortion is that of forgiveness. Abortion may be perceived as rejecting or denying a wrong or improper sexual act, rather than seeking divine forgiveness and lovingly bearing the consequences. Although many a "traditional" Christian community has cast unwed mothers out, there are also Christians who try to assist women who are pregnant and lack social or financial support. Contemporary pro-life groups may sponsor programs that aid unwed mothers or find adoptive parents for newborns. Although Christians can be tediously self-righteous about such activities, the basic thought behind them is that forgiveness prevents additional injury, to the family, to the mother, to the child, and to the community.

Christians who consider abortion ethically permissible in some circumstances may identify termination of a pregnancy as the least of several evils, rather than as an ethically neutral act. Although these Christians will not vote for legislation banning abortion, they may resist the inclusion of abortion as an element in population programs. This view holds that abortion is a last option to be used in cases of serious social, health, or economic difficulties, thus it is not a woman's "right" per se. Christians subscribing to variants of this position, including many "main-line" Protestants, are often strongly in favor of contraception as a means of reducing the risk of problematic pregnancies.

In summary, it is likely that Christians will remain split over the abortion issue, and that some Christians will continue to oppose abortion as a component of population/family planning programs. However, most of those who oppose abortion support family planning and provision of better medical services to women, as well as programs to alleviate poverty.

Susan Power Bratton is a professor of theology at the University of North Texas.

FRANCES KISSLING

Theo-Politics: The Roman Catholic Church and Population Policy

....

Among the many institutions that seek to influence population policies, the Roman Catholic church is arguably the most controversial and least understood. The controversies associated with the positions taken by the church as well as its style of intervention are bound to continue. Indeed, they may even increase as we enter a new era in which the links among population, environment, development, and women's well being are better understood.

Even as Catholics themselves largely ignore its teachings on matters of sexuality, the church has emerged as an effective player in both international forums and national debates about family planning, sometimes providing the determinant voice in policy that affects the lives and well being of millions. The church has either directly or indirectly caused the closing of in vitro fertilization (IVF) services from Poland to Uruguay; kept condom distribution and education out of AIDS prevention programs; influenced U.S. policy regarding international financial assistance for family planning; and eliminated specific references to family planning methods in U.N. documents. Any policy advocate as influential and articulate as the church clearly merits thorough scrutiny.

It is difficult for policy analysts to understand the positions taken by the church in the public arena not only because those analysts are usually not theologians, but also because the church's articulation of its theology for policy purposes is often oversimplified, even where it is not disin-

Frances Kissling is the president of Catholics for a Free Choice.

genuously presented. The uninitiated observer may also have difficulty discerning the extent to which the Vatican's policy positions are taken to serve internal church purposes or, alternatively, to benefit the public good. Much of what the church has said on population issues falls unambiguously into the latter category. For example, since the 1974 U.N. population conference in Bucharest, the Vatican has consistently spoken for greater responsibility by the North in reducing wasteful consumption, for more equitable distribution of resources, for vigilance against coercion and respect for the rights and dignity of couples. It has also condemned population policies that flow from demographic and political goals. However, history is filled with examples of ways in which church policy on sexuality, contraception, abortion, and population weave a tapestry of self interest emerging from demographic trends, religious rivalry and theology that is rooted in part in ancient misconceptions about biology and hostility toward women and sex.

Throughout history, church leaders have held a variety of views on family planning. St. Thomas Aquinas, following Aristotle, suggested his approval of laws limiting population size; he believed it was not possible for a country to allow unlimited growth.[1] But the church has also invoked pronatalism when it was expedient to do so. For example, some scholars believe that the Protestant Reformation led the church to strengthen its rigid anti-contraceptive posture partly in order to build membership in that period of religious rivalry.[2] And when the French birth rate fell by 17 percent during the last decade of the 18th century and France lost the 1870-71 war with Prussia, a Swiss cardinal cried, "You have rejected God and God has struck you. You have…made tombs instead of filling cradles with children."[3] The 1930 encyclical on marriage (Casti Connubii) told couples they "have a duty to raise up…members of God's household that the worshipers of God and Our Savior may daily increase."[4] John Ryan, a prominent Catholic thinker and university professor, sounded that note again when he declared that U.S. Catholic couples had a duty "to maintain at least the previously existing proportion between Catholics and non-Catholics."[5] Recently, in 1991, Pope John Paul II encouraged Brazilian couples to help solve the shortage of priests by having more children.[6]

Since the 1950s, the church has acknowledged the problems associated with rapid population growth. In 1951, the "rhythm method" of family planning was approved for "serious motives," including "medical, eugenic, economic and social reasons," and couples were told they could even use rhythm throughout their married lives, though not "habitually" or for less than "grave" reasons.[7] Beginning in 1958, Cardinal Suenens of Brussels convened regular meetings on population growth. Suenens is believed to have persuaded Pope John XXIII to convene the 1963

Commission on Population and Birth Control, which ultimately recommended that the Vatican permit the use of artificial contraception — only to be ignored by John's successor, Paul VI, who recondemned contraception in 1968.

Yet even as the church solidified its opposition to contraception, concerns about population growth persisted. In the 1961 encyclical entitled *Mater et Magistra,* Pope John XXIII asked, "How can economic development and the supply of food keep pace with the continual rise in population?"[8] And in 1967, Pope Paul VI noted in *Populorum Progressio,* that "the accelerated rate of population growth brings many added difficulties to the problems of development."[9] At the 1992 Earth Summit, the church's dilemma on population and fertility control was brought into sharp relief. Criticized for its apparent resistance to population policy, the Vatican insisted that the church "does not propose procreation at any cost" — a tacit recognition of the importance of population stabilization. But the Vatican resisted remedies that "are contrary to the objective moral order," i.e., contraception and abortion.[10]

THE ABORTION CONTROVERSY

While both contraception and abortion have been condemned by the church as tantamount to murder, it is the abortion controversy that is front and center these days. Church officials proclaim that Catholic teaching on abortion is clear and has remained unchanged over time. Various statements imply that church opposition to abortion is based on religious doctrine that considers the fetus to be a person from the moment of conception, and that requires absolute respect for the "right to life" of all existing and potential human beings.[11] The church contends that this position is not just "Catholic," but a moral insight that should form the basis of ethical policy.

The 1974 *Declaration on Procured Abortion,* issued by the Vatican Congregation for the Doctrine of the Faith (formerly the Inquisition), called on Catholics to work for legislation that would make abortion illegal in all countries. Perhaps the most dramatic statement in this regard was voiced by New York's Cardinal John O'Connor on the occasion of Earth Day 1990. O'Connor expressed suspicion of the event and its link to population concerns by declaring that, "one of the most dangerous environments in the world today is the mother's womb. Millions of babies are killed there each year."[12]

In fact, Catholic teaching on abortion and the status of the fetus is far more complex than such directives and rhetoric imply. While church law and discipline have always treated the act of abortion severely (although even here the history is by no means constant), theological positions on the issue have been more nuanced and ambivalent. The *Declaration on*

Procured Abortion acknowledges that the question of the personhood of the fetus cannot be determined by science or medicine, but is properly a philosophical and theological question. It also notes that church leaders are not in agreement on the moment along the continuum of pregnancy when personhood begins. Over time, different hypotheses have been put forward by theologians and favored by the church, but none has been accepted as doctrine.

Ambivalence about abortion extends back to the early church. In the fourth century, Augustine held that while abortion was always sinful, church law could not consider the act homicide for "there cannot yet be said to be a live soul in a body that lacks sensation when it is not formed in flesh, and so not yet endowed with sense."[13] In the Middle period, theologians including Thomas Aquinas, who was influenced by Greek philosophy, held that fetal life went through various stages of development before achieving a clearly human "anima" or spirit. Aquinas held that in the case of male fetuses this occurred at 40 or 45 days after conception, while for females ensoulment took twice as long.[14] It was not until 1869 that church law eliminated gestation-based differences in the penalties attached to abortion, declaring all abortion to be subject to the penalty of excommunication.

This is where Catholic theology remains today. The teaching office of the church admits that it does not know when the fetus becomes a person, but insists that the fetus's potential for personhood and doubt about its personhood are sufficient to call both for absolute church sanctions and for attempts to achieve parity between church law and secular law.

In practice, most Catholics have rejected the church teaching on abortion, just as they have rejected church teaching on contraception. In the U.S., Catholic women are as likely to undergo abortion as women of other faiths.[15] Nor does the incidence of abortion in predominantly Catholic countries seem to be influenced by religious beliefs. Abortion is common in the predominantly Catholic countries of Latin America. For example, a Colombian government health minister recently estimated that more than 250,000 abortions are performed on teenagers annually in that country alone.[16] In Latin America as a whole, experts estimate that there are roughly four million clandestine abortions each year.[17]

In theological circles, the ban on abortion is widely criticized. One can, however, see the influence of current Vatican demands for orthodoxy on the issue. A review of scholarly articles shows extensive writing in the 1960s and early 1970s, with a precipitous decline in intellectual inquiry following the 1978 installation of Pope John Paul II.[18] Most criticism focuses on the following points:

- In the absence of an ability to speak definitively on the facts of a moral problem, the church has traditionally granted individuals the freedom of conscience to determine an appropriate course of action in their specific circumstances. This freedom is not granted in the case of abortion in spite of the clear doubt about the personhood of the fetus.

• The prohibition on abortion, which involves the taking of potential rather than actual life, is more stringent than the standards applied to areas of moral decisionmaking that involve taking the life of actual persons. For example, "just war" theory permits the taking of life when one's physical life is in jeopardy (self-defense); to protect a nation's integrity (the Gulf War, the Falklands, etc.); and to protect values judged to be proportionally more important than the anticipated loss of life (religious and political freedom, freedom from slavery). The church allows no equivalent guidelines to be used by women in judging the licitness of an act of abortion.

In light of the institutional church's recognition of the link between population growth and poverty and the absence of a definitive position on the status of the fetus, one wonders about the source of the vehemence and passion invested in fighting against the very reproductive health measures (contraception and abortion) that would help alleviate poverty and certainly expand human rights. Feminist theologians, among others, have concluded that the heart of the church's objection to abortion can be found in its deeply embedded hostility toward sexuality and women.

That hostility, which extends back to the first centuries of the Christian church, is not unique to Roman Catholicism. It is present in Greek and Roman philosophy, as well as in the early Judaic tradition. Ancient mythology is full of tales portraying women as temptresses and sexuality as the downfall of man. Unfortunately, this mythology persists in contemporary Roman Catholicism. Powerful examples from each period of history illuminate the development of a rigid theology — and anthropology — of women and sex.

Rosemary Radford Ruether, a leading Catholic feminist theologian, notes that women's status was compromised by Christ's immediate successors. An example from the writings of Paul establishes the pattern:

Let a woman learn silence with all submissiveness. I permit no woman to teach or to have authority over men; she is to keep silent. For Adam was formed first, then Eve; and Adam was not deceived but the woman was deceived and became a transgressor. Yet woman will be saved by bearing children....[19]

Thus, writes Ruether, "Women are denied the attributes of speech, of self-articulation, of autonomous personhood. Instead, women are defined as subordinate in the very nature of things, yet prone to insubordination....Childbearing becomes her way of atoning for this sin."[20] Emblematic of their innate sinfulness, women were also presumed to be physically defective. As Aquinas wrote in *Summa Theologica:*

> As regards individual nature, woman is defective and misbegotten, for the active force in the male seed tends to the production of a perfect likeness in the masculine sex; while the production of woman comes from defect in the active force or from some material indisposition, or even from some external influence, such as that of a south wind...

Faulty notions of biology reinforced the primacy of men over women and contributed to the condemnation of contraception. Tertullian, an influential third-century theologian, held that semen contained individuals waiting to be born: "He is a human being who will be one; the whole fruit is actually in the seed."[21] Tertullian's mistake begot a second one — the belief that ejaculation outside the "vessel" (woman) murdered the "seed" individuals. This view was taught in the fourth century by St. Jerome.

Negative views of women were translated into church policies, which survive to this day. Women are barred from serving as priests or deacons (much less bishops, cardinals, or popes), and they are thus excluded from power and authority. In fact, it was not until the middle of the twentieth century that women could even sing in church choirs, because all liturgical functions were closed to females.[22] In any other context, such barriers certainly would be considered sexist discrimination, if not overt hostility.

Contempt for women and for sexuality are are closely linked in the writings of Catholic authorities. Jesus Christ was dismissive of the family, urging his followers to devote themselves to God, and Christ's disciples translated this urgency into sexual asceticism as a form of religious devotion in itself.[23] This value, married to the Greek idea of the dichotomy of body and soul, was in turn corrupted into disgust with the body and with sex. Female bodies were especially disparaged: "To embrace a woman is to embrace a sack of manure," wrote one twelfth-century authority, Odo of Cluny.[24]

St. Augustine defined the church's position on sexuality for a millennium and remains a leading influence. Augustine reinforced the early Christian identification of women with sex and the negativity about both by reducing women to procreative servants — sex objects. In *De genesi ad litteram,* he wrote:

> I don't see what sort of help woman was created to provide man with, if one excludes the purpose of procreation. If woman is not

given to man for help in bearing children, for what help could she be? To till the earth together? If help were needed for that, man would have been a better help for man. The same goes for comfort in solitude. How much more pleasure is it for life and conversation when two friends live together than when a man and a woman cohabitate.

Church leaders continued to see sexuality as a distraction from the spiritual life, but were able to redeem the sex act in the context of pro-creation. Sex was not to be seen as good in and of itself, certainly not when pursued for pleasure, but could be tolerated if married couples approached it out of a desire to procreate.

The church's profound disapproval of sexual activity is evident in the rules set forth by bishops and popes throughout the Middle Ages stating the times when it was unacceptable for married couples to have sex. Sex was proscribed when the woman was pregnant, menstruating, or lactating; during the forty days before Easter (Lent) or the four weeks before Christmas (Advent); on Ember Days (almost no one remembers what they are); on Fridays, because Good Friday was the day Jesus died; and on Sunday, the Lord's Day.[25] Well into the nineteenth century, the church debated whether it was possible for a couple to ever have sex without committing a sin, for it was believed that experiencing sexual pleasure, even in sex directed to procreation, was sinful. Was there a difference in sinfulness, church leaders asked, if pleasure was sought or if it just happened in spite of one's good intentions? Women continued to be seen as the source of sexual pleasure, thus sin. The development of a celibate, all male priestly class contributed to the continuation of these theories long after they had been discarded by most other religions and cultures.

With the 1930 encyclical *Casti Connubii*, the church began to acknowledge that sex in marriage had a purpose beyond procreation, in that it contributes to the growth and development of marital love. While this view has not been explicitly rejected by John Paul II, it has not been at the core of his pronouncements on sexuality and procreation. In fact, many observers feel that church statements on a variety of issues of sexuality and reproduction are more influenced by the thinking of the medieval rather than the modern church.

While church leaders are somewhat circumspect in discussing church attitudes and beliefs about women and sexuality (perhaps understanding the extent to which they would lose credibility in the larger society), the fact remains that medieval attitudes towards women and sexuality survive among church leaders. The Vatican does not accept human sexuality as a positive good if it is separated from the intent to procreate or is conduct-ed outside the confines of a lifelong monogamous union. Nor has the church reached the point where women are trusted as competent, capable moral agents to be granted autonomy in making sexual and reproductive decisions.

A continuing belief that all sexual acts must be open to procreation informs both the moral teaching and policy positions taken by the Roman Catholic church. This belief serves as the guiding principle for church positions not only on abortion, but also on contraception and contraceptive research, sterilization, assisted reproduction such as IVF, sex education and AIDS education, as well as population policy.

CHURCH AND STATE

Many may still wonder why non-Catholics should bother to consider sectarian Catholic views. The answer is that Catholic bishops believe they have the right and responsibility to participate in matters of state and they pursue their agenda with vigor. It is important to remember that it was only 30 years ago, at the Second Vatican Council, that church leaders acknowledged for the first time the legitimacy of the separation of church and state. For most of European history — from Constantine to the Protestant Reformation — church and state were inextricably linked; indeed for much of that time temporal leaders ruled only with church permission. It was assumed that laws would reflect the moral wisdom of the church.

Vestiges of this system remain present in the modern world. Roman Catholicism is the only religion that is simultaneously a nation state — the Vatican — with an ambassadorial court and permanent observer status at the U.N. Examples of church/state symbiosis abound: in Argentina, the Constitution requires that the president be a Roman Catholic (President Carlos Menem, who took office in 1989, converted to Catholicism to meet this test); the Polish bishops have sought a constitutional amendment that would declare Poland a Catholic country; and a number of countries, from Ireland to Chile, retain laws against divorce to satisfy the demands of the church.

The political power of the church, combined with passionate opposition to the principle of personal choice in sexual and reproductive health, has hindered efforts to legalize abortion and increase access to reproductive health care worldwide. On the international level, the Vatican has used its permanent observer status at the U.N. to influence multilateral policies and documents on these issues. At the 1984 U.N. population conference in Mexico City, the Vatican was instrumental in developing the U.S. "Mexico City policy," which denied family planning funds to any agency that included abortion information or services in its program.[26] And in 1992, the Vatican launched a successful effort to remove language that referred specifically to family planning from docu-

ments of the U.N. Conference on Environment and Development
(UNCED). While that effort was framed in the most moral of concerns
— a desire to respect the rights of couples, to ensure that population
policies respect the dignity of the human person and cultural norms —
the intent was clearly understood. Family planning methods prohibited
by the church were not to be officially approved by the U.N. and its member countries.[27]

The Vatican is heavily involved in the planning process for the 1994
International Conference on Population and Development (ICPD). At
the recent Asian and Pacific Population Conference (a regional preparatory conference for ICPD), the Vatican participated as both a nation state
and as an NGO (non-governmental organization) observer.
Representatives of the Holy See, unlike other NGOs, were seated at the
main conference table. The Vatican representative objected to the use of
the term "contraceptive" instead of "family planning," because contraceptive implies artificial methods, and the document that came out of the
meeting was worded accordingly.

The church has been even more successful in influencing national policies. In the Philippines, church officials who had particularly close links to
the Aquino administration were able to insert a Human Life Amendment
in the new Constitution. They also managed to chill the distribution of
modern contraceptives under the national family planning program and
divert broad-based family planning funds to the promotion of periodic
abstinence — the only method approved by the church.[28] In Ireland,
church leaders were a major force behind a 1983 constitutional amendment which effectively outlawed abortion in the Republic of Ireland. That
amendment was so strictly interpreted that, until 1992 when new referen-
da softened the law's impact, it was illegal to tell an Irish woman where
she could get a legal abortion or for Irish women to travel abroad to
obtain the service.[29] In Mexico, the bishop of Chiapas threatened legislators with excommunication if they did not rescind a law making abortion
legal; the law was tabled.[30] Similar threats were made to legislators in
Guam, and the same type of pressure has been exerted in Puerto Rico.[31]
In Poland, following the transition to a democratic government, bishops
convinced the health department to tighten eligibility for abortion by
requiring women to see several doctors and a state- (church-) approved
psychologist before having an abortion. Reportedly, abortions dropped by
50 percent in the six months that followed. Several maternal mortalities
due to complications of self-induced abortions, as well as cases of infanticide, were also reliably reported. On the extralegal level, Catholic pharma-
cists and priests in Poland regularly buy up birth control pill supplies and
destroy them. Catholic women on line for confession are told to go home
and not come back until they have their IUDs removed.[32]

In the U.S., Catholic bishops have testified successfully before Congress against funding for contraceptive research, against reauthorization of federal family planning programs, against fetal tissue research, and against lifting the import ban on RU-486, the French abortion pill. At the state level, they are major opponents of school-based clinics that provide contraceptive education or services.[33] The bishops are so single-minded in their objection to all "artificial" methods of contraception, including condoms, that they routinely oppose the inclusion of condom education in AIDS prevention programs.[34] At the theological level, bishops have declared that married heterosexual couples in which one partner has AIDS may not use condoms to prevent the spread of the disease.[35] This judgment is meant to apply to Catholics worldwide.

Assisted reproduction is also opposed by the church. In response to a Vatican document (Donum Vitae), which declared almost all methods of assisted reproduction illicit, a number of IVF programs have been closed. In Donum Vitae, the Vatican declared that only through genital intercourse could a couple create a child that was truly the product of love.[36]

The list goes on. However, there is some evidence that policymakers are becoming reluctant to treat the church as a privileged player in the policy debate and to challenge its interventions on their merits. Even the Archbishop of Canterbury criticized the Vatican's role at the UNCED meeting in Brazil. One can only hope that public officials, policy analysts, and the news media will follow suit.

MAURA ANNE RYAN

Reflections on Population Policy from the Roman Catholic Tradition

....

"Don't squeeze so," said the Dormouse to Alice.
"I can't help it. I'm growing," she replied.
"You've no right to grow here."
"Don't talk nonsense; you know you're growing too."
"Yes, but I grow at a reasonable pace and not in that ridiculous fashion." [1]

The debate over international population policy often seems to resemble this spat between Lewis Carroll's famous Alice and her inhospitable neighbor. Alarmed by accelerating population growth in the developing world—in areas such as sub-Saharan Africa, the Middle East, Southeast Asia, Latin America and the Indian subcontinent—those in favor of global population control cry that the earth is groaning under the weight of too many people, its life-sustaining capacity "squeezed" toward the limits. No longer can unrestricted, "irrational" population growth go unquestioned, they argue, nor the "right to grow *wherever*," willy-nilly, be protected. Those opposed to birth regulation programs frequently protest, with an Alice-like air of indignation, that restrictions on procreative liberty are unfairly levied. First-world observers are willing to propose a level of government interference in reproduction for others (for those in developing nations) which they would not (and need not) tolerate for themselves. Global resources are taxed indeed, and people suffer; but, from this side of the debate, the problem is not that various peoples are growing too large or too rapidly,

Maura Anne Ryan is an assistant professor of Christian ethics at the University of Notre Dame.

but that they are given too little space in which to grow, too small a share in the earth's bounty.

The Roman Catholic church is one of the most vocal opponents of birth regulation programs. Bishop Jan Schotte, speaking as head of the Vatican delegation at the 1984 U.N. International Conference on Population in Mexico City, reiterated the longstanding position of the church that "population policies should be part of overall policies of socioeconomic development and not a substitute for them."[2] Socioeconomic development must be promoted, he argued, "not simply [as] a matter of economic philosophy or strategies, but [as] an ongoing process that respects the value and individuality of every person and in which each person is free to take responsibility for his or her own destiny and growth."[3] "Scientific and technological progress should find new ways to make it possible for increasing numbers of people not simply to survive, but to live together in dignity, in social unity, harmony and in peace;...[g]overnments have a duty to create conditions that enable couples to exercise responsibly their fundamental right to form families, to bear and rear their children, without coercion or pressure to conform to the small family model or limit their childbearing to one or two children per family."[4] While Schotte acknowledged that couples have a moral and religious duty to procreate responsibly, he expressed the familiar Roman Catholic ecclesial position that sterilization, abortion, and contraception are unacceptable means for exercising responsibility—individual or collective.

We are unlikely to hear any new or different message when the head of the Vatican delegation addresses the UN Conference on Population and Development in Cairo this September. A great many people disagree with the official position of the church on contraception, including many faithful Roman Catholics. But the magisterium's insistence on approaches to global population growth which honor human dignity and the capacity for responsible choice, which acknowledge the significance of reproduction and reproductive liberty in human experience, and which favor noncoercion and economic development over government-imposed fertility regulation is not without a sympathetic audience. Indeed, many feminists, who are generally committed to guaranteeing reproductive autonomy for all women through safe and available contraception (and often safe and legal abortion) also reject "top-down" or mandatory population control policies as contrary to human dignity.[5] Moreover, the fierce debates which erupted between "North" and "South" at the 1992 Earth Summit in Rio suggest that the church is not alone in being concerned that efforts to limit population growth will substitute for soul searching in the developed world, for a sincere will to address "first world" patterns of producing, distributing, and consuming goods.

Is it possible that feminists and the Vatican can truly share common ground on the question of population policy, when they are often in such disagreement on fundamental matters of sexual ethics? Put another way, is it possible for the Catholic tradition to provide genuine resources for those who seek compassionate and human solutions to the vast political, social, and environmental problems posed by accelerating birth rates, despite the controversial and unpopular stand of the institutional church on contraception? There are obstacles to be overcome, but I believe that a Catholic theological vision can indeed enrich the debate — in particular, that it can point the way beyond *population control* to *just reproduction*. To illustrate this, however, we need to situate the questions of population and contraception within the broad tradition of Roman Catholic social teaching. We need to show how the meaning of procreative liberty for Catholic social ethics is discovered within a rich account of human dignity and sociality, and why, ultimately, the problem of overpopulation can only be engaged as part of the deeper Christian mandate to guarantee a just social order.

THE DEEP SETTING: ROMAN CATHOLIC SOCIAL TEACHING

The Roman Catholic church has a long tradition of taking up the pressing social and economic problems of its time. From Pope Leo XIII's passionate defense of exploited workers (voiced in the encyclical *Rerum Novarum* in 1891) to Pope John Paul II's stinging critique of consumerism in *Centesimus Annus*, we can trace a centuries-old effort to view the social order through the lens of Christian faith. Four broad commitments (or themes) in Catholic social teaching are of particular importance for the questions of population, environmental quality, and economic relations which are of such concern today: (1) human personhood is essentially communal; (2) rights imply duties; (3) the goods of the earth are meant for all; and (4) justice for the poor has moral priority.

Human personhood is essentially communal: At its core, Roman Catholic social teaching resists both the "liberal" model of society (where the common life is principally an arena for competing individual interests) and socialist/marxist or romantic communitarianism (where the individual is more or less absorbed into the state). Rather, the individual is "necessarily the foundation, cause and end of all social institutions."[6] Thus, social institutions exist to serve the essential dignity of each human being, a dignity which can be realized and protected only in solidarity with others. To paraphrase Reinhold Niebuhr, it is the human individual's *capacity for union* with others which makes society possible; it is the human individual's *need for assistance from and communion with others* which makes society necessary.

This account of personhood as essentially communal provides the theological foundation for the meaning of "justice" and "the common good" in Roman Catholic social teaching, as well as the warrant for taking up either as a moral concern. Advanced in the Catholic tradition is a concept of justice as *relational* and *mutual*.[7] What is required of individuals, of institutions, or of the social order is specified by the concrete needs of individual persons as they seek to achieve fully human membership in various communities. Basic justice, in the words of the United States Catholic Bishops, is a matter of establishing "minimum levels of participation in the life of the human community for all persons."[8] A fundamental concern for persons as they develop or fail to develop human potentialities *in relation* results, therefore, in a preoccupation with marginalization in all its forms (political, economic, and social) and a criticism of social arrangements which serve to deny equity of access to the means for human fulfillment.

Since the human person is "being-in-relation," individual good and collective good are intrinsically intertwined in Roman Catholic social teaching; individual freedom is to be oriented toward the common good, although it is never collapsed into it. As it was defined in the Second Vatican Council, the *common good* means "the sum of those conditions of social life which allow social groups and their members relatively thorough and ready access to their own fulfillment;" it is, thus, "the comprehensive human good of all who make up society."[9] A concern for the common good embraces the social conditions for intellectual, social, and spiritual development (strong educational institutions, healthy family life, rich cultural and artistic activity) and the conditions for the maintenance of physical well-being (the provision of material goods sufficient to meet basic needs of all and to allow participation in the civic community). From the point of view of Roman Catholic social teaching, therefore, it is always a mistake to construe social problems (such as accelerating population growth or environmental decline) as fundamental conflicts between individual good and the common good. Rather, since individual good and the common good imply one another, access to the means for pursuing important individual goods is a community problem; contributing to the conditions of the common life is an individual obligation.

Rights imply duties: In Roman Catholic social teaching, "Respect for human rights and a strong sense of both personal and community responsibility are linked, not opposed."[10] A wide range of basic human rights are named and defended in the tradition, including: the right to life, bodily integrity, employment security; the right to reproduce and rear children, and to worship without interference; the right to a just wage, to free initiative, and to own private property; and the right to take an active part in public affairs and to contribute to the common good.[11] Such fundamental rights are to be protected in the just society not as "negative liberties" (rights of non-interference), but as conditions for empowering the individual to contribute to the common life. Precisely because individual flourishing requires a rich and protective common life, it does not make sense, from the point of view of the Roman Catholic tradition, to defend individual rights except as they imply responsibilities, or to argue for an isolated sphere of individual free initiative abstracted from the context in which it is both protected and exercised. Thus, individual rights are treated as relative (as modified by commitments to the common life) and as reciprocal (arising in a social field involving correlative duties and counter-claims).

In order for the common good to be realized, individual liberty must be respected at the same time as individuals are held accountable to their responsibilities for the work of community, for the maintenance of a good human life in common. Once again, in the Catholic tradition, the common good is not achieved at the expense of personal liberty or without a call to personal accountability; rather, it becomes possible precisely "when personal rights and duties are maintained."[12] What emerges from Catholic social teaching, therefore, is a strong defense of inalienable human rights within a comprehensive theory of social responsibility. Thus, a fundamental human right to participate in the political life is asserted for all because each one has a duty to participate according to his/her own abilities; a right to work follows from one's duty to care for one's family and to contribute to the development of conditions for realizing universal well-being, and so on. Securing the conditions for the protection and exercise of human rights is a duty incumbent on the state itself, on various institutions and associations within the state, and upon individuals themselves (who act properly both on their own behalf and for the sake of those who lack power or resources).

The concerns for securing access and for guaranteeing the "minimum levels of participation in the life of the community for all people" which shape the meanings of justice and the common good in Catholic social teaching are given flesh in two further commitments: to the *universal destination of goods* and to a *preferential option for the poor.*

The goods of the earth are meant for all: In reflecting on the realities of global poverty and hunger in the twentieth century, John Paul II observed in *Sollicitudo Rei Socialis* (issued December 30, 1987):

> It is necessary to state once more the characteristic principle of Christian social doctrine: The goods of this world are originally meant for all. The right to private property is valid and necessary,

but it does not nullify the value of this principle. Private property, in fact, is under a 'social mortgage' which means that it has an intrinsically social function based upon and justified precisely by the principle of the universal destination of goods.[13]

In light of this doctrine, social and economic structures (or conditions) which permit the concentration of vast amounts of wealth and resources in the hands of the few are to be denounced. This is so, not only because under such conditions persons have inequitable access to the goods necessary for human flourishing (that is, not only because a failure of community responsibility occurs) but because such structures reflect a violation of God's intentions for creation. The equal dignity of all human beings under God gives rise to an equal claim (at least prima facie) to everything that issues from God's gracious and creative energy. Therefore, the exercise of human rights—for example, the right to private property—has as its boundary condition the universal claim of all human beings to a share in the earth's goods. In application to issues of economic justice, the principle has functioned to underscore a right of development for all peoples, to justify preferential treatment for those who have been unfairly denied a decent share in the earth's goods, and to challenge those with power and resources to act in solidarity with those who have neither. The moral and theological force of this principle shifts the question of social action from charity to justice: Christians are not simply called to share generously "of their surplus," but to respond to systemic violations of equity obligations.[14] Although the Roman Catholic church has been slow to incorporate a genuine concern for the quality of the physical environment, environmental justice is also implied in this principle. As we have come to understand in this century, it will be impossible to honor the claims of successor generations to a decent share in the goods of creation unless we alter environmentally toxic patterns of behavior.

Justice for the poor has moral priority: All that has been said thus far about human dignity, access, the right of all to participate in society, the common good, the universal destiny of God's generosity, and so on, is crystallized in the doctrine of the *preferential option for the poor.* "Love for others, and in the first place love for the poor, in whom the church sees Christ himself, is made concrete in the promotion of justice."[15] To love justly is not simply to be charitable, but to help those people (and peoples) "who are presently excluded or marginalized to enter into the sphere of human development."[16] Justice involves not only giving persons what they need to live decent human lives, but enabling them to exercise their rights and fulfill their duties within various communities. Today, addressing marginalization in all its forms calls for the establishment of a "floor of material well-being on which all can stand,"[17] and for a critical examination of systemic privilege (political, economic, racial, sexual).

In the words of the U.S. Catholic Bishops, the poor have the "single most urgent economic claim on the conscience of the nation."[18] This claim issues in societal obligations to "evaluate social and economic activity from the viewpoint of the poor and powerless" and personal obligations on the part of the fortunate "to renounce some of their rights so as to place their goods more generously at the service of others."[19] A growing consciousness of economic and political interdependence in the tradition presses the moral admonition to secure justice for the poor to its global dimensions. Thus, a call is sounded throughout the contemporary encyclicals to those in developed nations: the reality of massive world poverty calls for a commitment to sustainable development, an end to the systemic economic exploitation of more vulnerable nations, and a critical examination of those values and practices which feed and support the economic and political domination of others (e.g, consumerism and/or materialism).

In summary, in the principle of a preferential option for the poor, the goal of securing a minimum level of participation in the life of the community for all takes on the urgency of immediate obligation through the real suffering and privation of the most marginalized. The Christian is challenged to go beyond alms-giving, beyond charity, to the practice of social, political, and economic justice. It is a challenge which truly undertaken involves self-critique, sacrifice, and the renunciation of privilege in all its forms.

IMPLICATIONS FOR POPULATION POLICY

It should not be difficult to see why international development and the redistribution of goods is favored over birth restriction in official Roman Catholic statements on the population question. Quite apart from theological objections to the use of contraceptives (based in a natural law reading of the proper ends of the sexual act), the impetus in Roman Catholic social ethics is toward enabling persons to assume control of their lives and to pursue the interests and goals constitutive of human flourishing (one of which is reproduction). The central question for Catholic social ethics is: How are the conditions under which access to the goods of human community and participation in its life are to be guaranteed for all? As important as it is for the community to meet the material needs of its members, it is just as important to make possible the satisfaction of *intangible* human needs and desires (expressed as generativity, creativity, self-actualization, relationality, and spirituality). Therefore, the deep "Catholic" resistance to resolving the population problem by overriding procreative autonomy is not just the result of a particular perspective on sexual ethics. It is also based on the belief that proposals for alleviating the problems of poverty, environmental decline, and unsustainable development which rest on the violation of human rights (or the denial of self-determination) are not finally *human* solutions.

The Catholic tradition affirms the deep significance of reproduction in individual human flourishing (and in the life of the human community); reproduction is recognized as an inalienable human right, and governments are argued to have a duty to make it possible for all couples to bear and rear the children they desire. At the same time, even here, the right to reproduce is not treated as an entirely unlimited right. From the obligation to pursue the *common* good, to assure a "minimum level of participation in the life of the human community for all persons," it follows that reflection on the impact of reproductive decisions on the communal conditions of possibility is a moral responsibility. Although the Vatican consistently argues that a married couple cannot be morally required to violate the procreative dimension of their sexual relationship, taking account of the demographic situation, of available personal and social resources, and of one's capacities to care for children, is part of reproductive responsibility.[20] Indeed, the Vatican delegation in 1984 renewed its support for the proposal in the 1974 "Recommendations for the Further Implementation of the World Population Plan of Action," which stated that governments should provide couples with accurate demographic information and education in and access to the means of natural family planning—in other words, that governments should provide assistance for couples in properly "assess[ing] their duties and responsibilities."[21] While continuing to believe that the goods of the earth can be harvested and distributed broadly enough to make decisions to limit or "space" births on demographic grounds unnecessary, the Vatican nonetheless recognizes that demographic and economic considerations appropriately enter into responsible or just reproduction.

The Catholic tradition does not evade the problems of reproductive accountability, poverty, wasteful consumption, or environmental decline, as it is often accused of doing. Rather, it could be said to redescribe them in light of a commitment to individual freedom and responsibility. Still, the question of how far one can go with a Roman Catholic social ethic in debates over today's global problems remains. The Catholic church's long standing confidence that God's bounty is plentiful enough to provide a decent life for all (and for all who might come into being) is just one of the "sticking points" or obstacles we encounter in an effort to make the theological and ethical resources of the Roman Catholic tradi-

tion accessible to others who are looking for a way beyond birth regulation policies. Today's social teachings do not display the same level of
confidence as those of the Vatican II-era in the power of coordinated
international development to eliminate global poverty; still, an optimism
about the essential fecundity of the earth pervades Roman Catholic social
teaching, an optimism which many experts would call unfounded. The
present pope seems to place his confidence for the earth's future in the
unlimited power of human persons to respond to the challenge of solidarity—to transcend sin and greed in a free and voluntary redistribution
of goods—rather than in the capacity of the earth to produce all that is
needed for persons to live humanely. [22] But, even here, it is assumed that
good will (which is plentiful) can always substitute for coercion in the
work of justice (i.e., a minimum level of participation in the community
for all can always be achieved through cooperative social transformation). Those with a sharper sense of urgency about the earth's present
feeding and sustaining capacity might well be uncomfortable with this
assumption, as might those whose reading of structural sin has left them
deeply pessimistic about human capacities for other-centeredness.

Moreover, the Catholic church's failure to incorporate gender analysis
into its social critique, and its continued practice of dichotomizing the
public and private realms hampers the translation of its insights into
prophetic social policy.[23] Recent church teaching continues to reaffirm
the natural complementarity of the sexes and to support a gendered division of domestic labor.[24] As a result, the importance of guaranteeing equitable and universal access to opportunities for human flourishing which
underscores rights to health care, a just wage, private property, etc. in the
public realm does not carry over into the realm of the family. An awareness of the reality that disproportionate responsibility for the care of children is a serious obstacle to women's full participation in the goods of
human community is not absent from Roman Catholic social teaching;
indeed, the magisterium recognizes that the world is a hostile place for
women and children. However, the answer offered is unsatisfying for
most people with feminist sensibilities: Rather than acknowledging that
women need, therefore, to have control over their own fertility, the official church argues for better social support for women's domestic role.
Thus, the complex and all-important question of the relationship
between the status of women and demographic trends, a central issue in
the population debate, is not posed within the Catholic tradition as help-
fully as it needs to be.

Failure to incorporate gender analysis into its social ethic renders the
conclusions of Roman Catholic social teaching on population problematic in another sense as well. The magisterium advocates reproductive

responsibility via the use of natural family planning methods. Prescinding from the important questions of effectiveness and reliability, this response does not acknowledge sufficiently the failure of male cooperation in birth control, particularly in cultures where status attaches to fertility. Advocates for natural family planning methods as safe and effective means for exercising responsibility to self and others often assume a level of cooperation between men and women (or in the alternative a level of power in women) which does not often exist. Many people, therefore, who share the Catholic church's bias for personal autonomy and responsibility in procreative matters, depart company at the insistence that artificial means of birth control are never morally permissible.

Finally, the official position of the Roman Catholic church on birth control is, obviously, a significant stumbling block to giving its social ethic a voice in contemporary debate. Despite the reasoned objections of many loyal members of the church that the magisterium's position on contraception rests on an overly physicalist account of the sexual act,[25] the official teaching of the church remains unchanged. One consequence of retaining a prohibition against birth control, which many people (inside and outside of the church) hold to be untenable, is that the Catholic tradition's rich account of just reproduction is dismissed. Another consequence of arguing that it is never permissible to use artificial means of birth control (and that the frustration of the sexual act's procreative potential is always wrong) is that the tradition's best insights on the communal character of all human rights are never really translated into the private realm. Therefore, for all intents and purposes, the principle that the exercise of rights (e.g., to private property) must sometimes be limited by the survival needs of others bypasses the domestic sphere. For many observers, the distinctions between the exercise of procreative liberty and other sorts of liberties relative to human capacities are simply not so great as to warrant isolating procreation from the scrutiny of justice.

In sum, Roman Catholic social teaching provides a very rich context in which to reflect on the questions of procreation and population. By posing the problems of hunger, poverty, and demographics as ultimately issues of the common good, the Catholic tradition avoids the temptation to ask only "who has a right to grow where?" From the outset, population is bound up with the problem of providing all people with the means for a decent life. Moreover, a deep appreciation of human dignity, for the importance of honoring self-determination, squares with the inner convictions of many people that however we go about transforming social arrangements, it is a mistake to secure material well-being at the price of liberty. It is fair to argue that, indeed, the Catholic tradition has something valuable to offer to the debate. At the same time, the difficulties to be resolved in mining Catholic social teaching ought to be appreciated. The Church's teaching on contraception continues to be a matter of theological dispute, as does its interpretation of women's "natural roles." The magisterium's failure to undertake gender analysis poses serious issues of credibility for contemporary Roman Catholic sexual ethics. Whether these "sticking points" render Catholic social teaching finally impotent on the problem of population is a serious question which cannot be resolved here. What I have tried to suggest is that whatever one makes of Roman Catholic sexual ethics, there are riches beyond the contraception teaching which should not be missed.

CONCLUSION

In *Alice and Wonderland,* the Dormouse solved his space problem by getting up and moving away. And, eventually, Alice shrunk back to normal size anyway. It is obvious today that our population problems will not be solved by "someone" moving away or by "someone" just shrinking back to size. The global problems we face will require political commitments, economic transformations, and, most importantly, changes in habits and behaviors. It will take no less than the ability to begin thinking in terms of a global common good. The Roman Catholic church has the theological and philosophical resources to offer a prophetic voice to this endeavor. Whether the church *will* be prophetic depends upon many things, not the least of which will be the magisterium's willingness to take seriously the voices of environmental and demographic experts, of theologians, of women, and of the poor. But for those who are searching for an ethic beyond birth control, the church's long thinking on social justice is too valuable a resource to miss completely, whatever serious theological issues may still need to be resolved.

Population Distribution: Urbanization and International Migration

....

NANCY YU-PING CHEN AND HANIA ZLOTNIK

Urbanization Prospects for the 21st Century

······················

The late twentieth century has witnessed unprecedented changes in the dynamics of population growth and distribution. By the year 2005 — for the first time in human history — more than half of the world's people will live in urban areas.[1] The change is most striking in the developing countries, where urban populations have grown fivefold since 1950. Of the world's 20 largest cities in the year 2000, 17 will be in developing countries.

The burgeoning cities of the 21st century will certainly test the ability of social institutions to provide for their citizens. Some analysts believe those institutions will not withstand the strain; dire predictions abound in both the popular media and in the literature on urbanization. Conditions in today's cities appear to offer support for this view. Poor housing conditions, overcrowding and homelessness, lack of access to clean water, problems with garbage collection and disposal, inadequate sewage systems, environmental hazards, and growing crime rates are just some of the problems facing urban managers the world over. Some predict that these problems will be vastly magnified by population growth in the years to come.

Are dire predictions warranted? Perhaps not. Cities are plagued with social and health problems, but those same problems are often more acute in adjoining rural areas. And urban areas are usually the locus of a

Nancy Yu-Ping Chen and Hania Zlotnik are population affairs officer and chief of the mortality and migration section, respectively, at the United Nations Population Division. The views and opinions expressed in this paper are those of the authors and do not necessarily reflect those of the United Nations.

nation's most vigorous economic activity. Today's urban ills are not without precedent; many of the problems faced by cities in the developing world today are no different than those faced by cities in the now-developed countries over the last century. Undeniably, the magnitude of urban problems in the developing world today is greater than in Europe a century ago, but just as in Europe, large cities in developing countries have considerable potential to muster the forces of modernization and development for the benefit of their people. The challenge is therefore to devise ways of maximizing the benefits of population concentration while avoiding or minimizing its negative consequences. In order to do so, it is important to understand the magnitude and pace of the changes taking place. Below is an overview of urbanization trends; a detailed region-by-region report follows on page 353.

URBANIZATION TRENDS AND PROSPECTS

Over the past forty years, the number of urban dwellers has more than tripled, growing from 737 million in 1950 to about 2.5 billion in 1993. About 60 million people — the equivalent of four cities the size of New York — join the world's urban population every year. As noted above, most of that increase is taking place in developing countries. Indeed, whereas in 1950 only 39 percent of the urban population lived in developing countries, by 1990 that proportion had risen to 61 percent. By 2025, developing countries will be home to nearly four times as many city dwellers as developed countries. Although the rate of urban growth worldwide is expected to decline by the year 2015, cities in developing countries are expected to gain some 1.6 billion new inhabitants, while those in developed countries gain 200 million. To put those figures in perspective, note that between 1970 and 1993 the urban populations of developed countries grew by 208 million and those in developing countries grew by 910 million (See Table 1). Thus, in the next 22 years, the urban areas of developing countries are expected to accommodate about 72 percent more people than they absorbed between 1970 and 1993.

There are three principal causes of urban population growth: the excess of births over deaths (natural increase); the net gain of migrants from rural areas; and the reclassification of areas from rural to urban as they grow or gain services. It is not possible to derive global estimates of the extent to which each of those components contributes to the growth of the urban population. Estimates of the components of urban growth are available for a limited number of countries and refer mostly to the 1960s and 1970s. On the basis of data for 29 developing countries during the 1960s, the U.N. concluded that about 61 percent of the growth of urban areas was attributable to natural increase and only 39 percent to a combination of reclassification and net rural-urban migration.[2]

TABLE 1: Population and Average Annual Rate of Change of the Total, Urban and Rural Population of the Major World Regions

Major area and region	Population (thousands)									Average annual growth rate (percentage)					
	Total			Urban			Rural			Total		Urban		Rural	
	1970	1993	2015	1970	1993	2015	1970	1993	2015	1970-1993	1993-2015	1970-1993	1993-2015	1970-1993	1993-2015
World total	3697	5572	7609	1352	2470	4232	2345	3102	3376	1.8	1.4	2.6	2.4	1.2	0.4
Developed countries	1049	1231	1366	698	906	1105	350	325	262	0.7	0.5	1.1	0.9	-0.3	-1.0
Developing countries	2648	4341	6242	654	1564	3128	1994	2777	3115	2.1	1.7	3.8	3.2	1.4	0.5
Africa	363	702	1265	83	235	601	280	467	663	2.9	2.7	4.5	4.3	2.2	1.6
Eastern Africa	110	214	402	11	44	137	98	170	265	2.9	2.9	5.9	5.2	2.4	2.0
Middle Africa	40	77	146	10	26	71	30	52	76	2.9	2.9	4.1	4.6	2.3	1.7
Northern Africa	83	151	242	30	68	139	53	83	102	2.6	2.1	3.6	3.2	2.0	0.9
Southern Africa	25	46	74	11	22	44	14	24	29	2.6	2.1	3.0	3.2	2.3	0.8
Western Africa	105	213	401	21	75	210	85	137	191	3.1	2.9	5.6	4.7	2.1	1.5
Asia	2102	3291	4461	482	1083	2110	1620	2209	2351	1.9	1.4	3.5	3.0	1.3	0.3
Developing Asia	1998	3166	4331	407	986	2002	1590	2181	2329	2.0	1.4	3.8	3.2	1.4	0.3
Eastern Asia	987	1406	1680	244	497	868	743	910	812	1.5	0.8	3.1	2.5	0.9	-0.5
China	831	1205	1458	145	345	685	686	861	774	1.6	0.9	3.8	3.1	1.0	-0.5
Japan	104	125	130	74	97	108	30	28	22	0.8	0.2	1.2	0.5	-0.3	-1.1
Southeastern Asia	287	470	648	58	143	295	229	327	353	2.2	1.5	3.9	3.3	1.6	0.3
Southern Asia	754	1271	1894	148	349	761	607	923	1134	2.3	1.8	3.7	3.5	1.8	0.9
Bangladesh	67	122	194	5	22	66	62	100	128	2.6	2.1	6.4	4.9	2.1	1.1
India	555	897	1265	110	235	471	445	661	794	2.1	1.6	3.3	3.2	1.7	0.8
Pakistan	66	128	220	16	43	108	49	85	111	2.9	2.5	4.2	4.2	2.4	1.2
Western Asia	74	143	239	32	94	186	42	49	52	2.9	2.3	4.7	3.1	0.7	0.3
Latin America	283	466	637	162	341	521	121	125	116	2.2	1.4	3.2	1.9	0.1	-0.3
Caribbean	25	35	46	12	21	32	13	14	13	1.5	1.2	2.6	1.9	0.2	-0.1
Central America	67	121	177	36	82	136	31	40	41	2.6	1.7	3.5	2.3	1.1	0.2
South America	191	310	414	115	238	353	76	71	61	2.1	1.3	3.2	1.8	-0.3	-0.7

Equivalent data for developed countries indicate that natural increase contributed less to urban population growth (40 percent) than reclassification and net migration combined (60 percent). Since the 1970s, net migration and reclassification have become more important as components of urban growth in many countries.[3] This trend is likely to continue in the future as fertility declines in developing countries. Unfortunately, data in this area are still too limited to permit solid conclusions.

URBANIZATION, MIGRATION, AND DEVELOPMENT

For the past two decades, the governments of developing countries have consistently singled out the spatial distribution of population as a

TABLE 2: Percentage of the Population Residing in Urban Areas and Rate of Urbanization by Major Area and Region

Major area and region	PERCENTAGE URBAN			RATE OF URBANIZATION	
	1970	1993	2015	1970-1993	1993-2015
World total	37	44	56	0.8	1.0
Developed countries	67	74	81	0.4	0.4
Developing countries	25	36	50	1.6	1.5
Africa	23	34	48	1.7	1.6
Eastern Africa	10	21	34	3.0	2.3
Middle Africa	25	33	48	1.3	1.7
Northern Africa	36	45	58	1.0	1.1
Southern Africa	44	47	60	0.3	1.1
Western Africa	20	35	52	2.6	1.8
Asia	23	33	47	1.6	1.7
Developing Asia	20	31	46	1.8	1.8
Eastern Asia	25	35	52	1.5	1.7
China	17	29	47	2.1	2.3
Japan	71	78	83	0.4	0.3
South-eastern Asia	20	30	46	1.8	1.8
Southern Asia	20	27	40	1.5	1.7
Bangladesh	8	18	34	3.8	2.8
India	20	26	37	1.2	1.6
Pakistan	25	34	49	1.3	1.8
Western Asia	43	66	78	1.8	0.8
Latin America	57	73	82	1.1	0.5
Caribbean	47	61	71	1.1	0.7
Central America	54	67	77	1.0	0.6
South America	60	77	85	1.1	0.5

major source of concern.[4] Thus, they often deem urban population growth unmanageable and try to curb "excessive" migration from rural to urban areas. But considerable evidence suggests that it is wrong to regard the urbanization of developing countries in purely negative terms. Among the countries of Asia, for instance, there is a strong positive correlation between the level of urbanization and per capita gross national product (GNP). Similarly, a positive correlation has been found between the level of net rural-urban migration experienced by Asian, Latin American, and Northern African countries in recent decades and GNP growth.[5] Countries experiencing relatively high rural-urban migration levels tend to record major gains in GNP. Not surprisingly, these countries generally have more favorable social and health indicators than those experiencing slow or negative economic growth because improved economic performance can both provide a higher proportion of the population with adequate incomes and permit larger government expenditures in social services. It would therefore seem counterproductive to assume that rural-urban migration per se is an economic drain or a social problem.

TABLE 3: Annual Growth of the Urban Population and Percentage of Total Population Growth Occurring in Urban Areas, 1970-1993 and 1993-2015

Region	Annual increment in urban areas		Percentage of total population growth occurring in urban areas (in millions)	
	1970-1993	1993-2015	1970-1993	1993-2015
World total	49	80	60	87
Developed countries	9	9	114	147
Developing countries	40	71	54	82
Africa	7	17	45	65
Eastern Africa	1	4	31	49
Middle Africa	1	2	42	65
Northern Africa	2	3	56	79
Southern Africa	0	1	52	83
Western Africa	2	6	51	71
Asia	26	47	51	88
Developing Asia	25	46	49	87
Eastern Asia	11	17	60	136
South-eastern Asia	4	7	47	86
Southern Asia	9	19	39	66
Western Asia	3	4	90	97
Latin America	8	8	98	105
Caribbean	0	0	95	104
Central America	2	2	84	97
South America	5	5	104	110

Urbanization is often associated with improvements in public health. In Latin America, for example, mortality in large cities is usually lower than that at the national level, especially among infants and children.[6] Better access to maternal and child health programs, better immunization coverage, higher educational levels among the population of larger cities and better sanitation than in smaller urban centers or in rural areas contribute to this outcome. There remain, however, considerable differences between socioeconomic groups within cities, with the poor suffering considerably higher rates of morbidity and mortality than those who are better off.

Contrary to common belief, the poorest urban dwellers are not necessarily migrants from rural areas. In fact, studies of migrants to cities in Latin America and other regions usually indicate that their adaptation is relatively successful and that most differences between migrants and non-migrants — in terms of labor force participation, occupation, or income — are attributable to differences in age, sex, educational attainment, and length of stay in the city rather than to migrant status per se.[7] Indeed, in many contexts, migrants exhibit higher rates of labor force participation and lower unemployment rates than non-migrants, and their presence makes possible the development of certain economic activities that could not flourish without the labor they supply.

There is a long tradition of ascribing urban problems to recent migrants.[8] Yet, in urban areas experiencing a significant influx from rural areas, a high proportion of migrants are young people willing to work for low wages. This allows employers to save considerable sums of money, especially if, as is often the case, the employers do not meet existing occupational and social-security regulations. It is unfair and erroneous to characterize as a drain on the economy the very people who contribute to a city's economic base by providing cheap labor and services.

But what about the migrants themselves? Do they benefit as much as those who employ their labor? Research indicates that migrants seek — and find — more opportunity in cities than in the rural areas they left behind. Indeed, most micro-level studies indicate that migration is a rational response to changing economic circumstances. At the macro level as well, migration and population distribution can best be explained in terms of the economic strengths and weaknesses of different locations. Latin America provides an example of the influence of economic growth in shaping migration. During the 1980s, the region experienced declines in migration to large cities, which appears to reflect the poor economic performance of the region in general and of urban areas in particular.[9] In Asia, the newly industrializing economies (NIEs) have generally experienced rapid urbanization and sizable rural-urban migration as industrialization and manufacturing have grown and the agricultural sector has been modernized. Indeed, the demand for cheap labor in the manufacturing sector has spurred a large-scale movement of women to the region's cities. In the NIEs, migration has functioned as an effective mechanism to allocate labor between sectors and alleviate poverty in rural areas.[10]

The other component of urban population growth, natural increase, also responds to socioeconomic differences between urban and rural areas. In developing countries, urban areas generally have lower birth rates, and therefore lower levels of natural increase, than rural areas. Lower fertility in urban areas has been attributed to greater availability of family planning services and increased educational and employment opportunities for women, among other factors. During the past 30 years, significant reductions in fertility have been recorded in a growing number of developing countries, although those reductions have not always been associated with economic growth. Still, many poor countries have yet to show signs of the onset of fertility decline. Therefore, in coming years, much of the urban population growth in poor countries will be attributable to high levels of natural increase among urban dwellers.

The problem facing many developing countries is that urbanization has not been anchored on sustained industrialization and the concomitant development of an urban infrastructure. Because of colonization, the urbanization process in the developing world was part of its incorporation to the global economy, whereby a few urban centers grew by concentrating administrative activities and acting as conduits for the flow of raw materials to the developed world. This mode of incorporation led to the development of dual labor markets, in which a small elite prospered while most of the population remained engaged in low-paying work in agriculture or the informal sector.[11] During the post-colonial period, the efforts of developing countries to expand their industrial base and their urban infrastructure have often been outpaced by rapid population growth. Many countries have had to resort to deficit financing to provide needed services to their growing urban populations, with deleterious effects for the economy in general.

Yet, despite deficiencies in their basic infrastructure and social services, urban centers in much of the developing world are the core of modern economic activity and offer better wage-earning opportunities than rural areas. The rates of economic growth in the main urban centers of developing countries are usually higher than elsewhere and they are the main sources of innovation, training, and funds for development. The challenge is to harness the dynamism of urban centers to fuel development more generally.

MAKING CITIES WORK

For over thirty years, developing country governments have taken measures to slow urban growth. Policies have been directed at helping rural economies support and retain more people, at controlling in-migration to large cities, and at redirecting migration to medium-sized and small urban centers.[12] There is ample evidence, however, that few of the policies and programs implemented to control or redirect migration have had a major impact. In part, the poor performance of those policies has been ascribed to their failure to take into account the economic and social forces that propel and sustain migration.

The effectiveness of those policies has also been limited by their lack of consistency with other national policies regarding trade, industrialization, or investment in infrastructure. For example, the practice of subsidizing basic food products or the cost of public services in urban areas favors city dwellers over their rural counterparts, creating an incentive to migrate to urban areas. To improve the efficiency of cities, those subsidies should be eliminated or reduced considerably, and equitable cost-recovery schemes must be instituted to help raise funds for the further improvement of infrastructure. Such schemes would involve charging well-off urban dwellers realistic fees for the services provided by municipal authorities. Subsidies on water, sanitation, garbage collection or electricity should be phased out and replaced by programs designed to improve the access of marginalized urban groups to those services.

Cities must invest more in their people — particularly their poorest people. The urban poor claim only a tiny share of material and natural resources. On a per capita basis, poorer groups draw less on non-renewable resources and on the capacities of local ecosystems to supply fresh water or biomass than their wealthier counterparts. Moreover, because governments of developing countries spend relatively little on social services for urban areas, it is the urban poor themselves who bear the costs of rapid urbanization — costs that include disease, illiteracy, and environmental deterioration.[13] And, while low-income households in cities of developing countries have contributed substantially to increase urban housing stock through their own self-help and mutual help efforts, they often face the hostility of public authorities, who often use force to destroy illegal settlements or to evict the poor from makeshift housing. In most of the cities of the developing world today, environmental problems are largely the result of inadequate provision of piped water, sanitation, drainage, garbage collection, and health services, particularly for the urban poor. Under these conditions, rapid population growth is generally less a cause of those inadequacies than the failure of government to ensure a more equitable access to services, to implement pollution controls, or to tax fairly those who benefit from publicly funded infrastructure.

A major priority for developing countries is therefore to improve the effectiveness of local government, particularly by increasing the capacity and competence of municipal authorities to respond to the needs of urban dwellers and especially of the urban poor. This capacity can be enhanced by granting local authorities the right to raise revenue and to expand the tax base. To enhance the situation of the poor, the barriers that reduce their economic efficiency and prevent them from improving their own status must be removed. The provision of credit, the protection of those working in the informal sector, and the granting of property rights to those who have built their own housing are some measures that deserve consideration.

THE IMPORTANCE OF RURAL DEVELOPMENT

In addition, given that many developing countries face rapid population growth in urban *and* rural areas, the importance of rural development programs cannot be overstressed. Yet, such programs should not be seen only as a means of retaining population in rural areas. When development is pursued by providing the rural population with better educational facilities or by fostering better integration of rural areas with adjacent urban centers, the rising expectations of rural dwellers often prompts migration to cities. Partly for that reason, it is crucial to combine rural development strategies with measures directed at the economic development of small- and medium-sized urban centers. In many instances, such urban centers have spurred the growth of the rural economy by becoming agricultural processing sites, attracting small-scale industry, and providing marketing facilities for agricultural products. Such efforts will yield the best results when implemented simultaneously with effective measures to increase agricultural productivity, raise rural incomes, and generate employment opportunities in rural areas.

Another important dimension of rural development is the provision of adequate social services. The expansion of education, health and family planning services in rural areas is sorely needed, particularly in those regions where the rural population is still growing at relatively high rates. Indeed, the expected urbanization trends discussed below are based on projections that assume the continuing reduction of fertility levels in the developing world. Thus, the likely growth of both the urban and the rural populations of regions such as Africa or Southern Asia would be considerably greater if fertility were to remain constant at current levels. Even in regions where fertility has already declined considerably, the contribution of natural increase to urban growth is substantial.

TABLE 4: Number of Urban Agglomerations and Population in Those Agglomerations by Size, Class and Region — 1970, 1990, and 2010

Class size	Africa			Latin America			Developing Asia		
	1970	1990	2010	1970	1990	2010	1970	1990	2010
10 million or more									
Number of agglomerations	0	0	2	0	4	5	1	5	14
Population in agglomerations (000)	0	0	34 507	0	55 600	80 054	11 154	58 257	229 021
As percentage of urban population	0	0	7	0	18	17	3	7	13
5 to 10 million									
Number of agglomerations	1	2	6	4	1	3	5	11	17
Population in agglomerations (000)	5 333	16 375	34 792	32 585	6 475	19 151	31 345	80 934	110 881
As percentage of urban population	6	8	7	20	2	4	8	9	6
1 to 5 million									
Number of agglomerations	7	23	56	14	35	63	48	92	202
Population in agglomerations (000)	10 754	42 785	123 122	23 907	68 305	126 140	91 543	174 827	401 017
As percentage of urban population	13	21	25	15	22	26	23	20	23
500 000 to 1 million									
Number of agglomerations	11	34	45	17	49	57	59	120	168
Population in agglomerations (000)	7 592	23 772	29 682	11 388	32 343	43 549	39 697	83 498	118 482
As percentage of urban population	9	12	6	7	10	9	10	10	7
Under 500 000									
Population in agglomerations (000)	70 290	122 570	270 948	94 564	152 755	213 521	233 693	481 414	879 631
As percentage of urban population	68	60	55	58	48	44	57	55	51

Consequently, a population policy aimed at satisfying the unmet family planning needs of the population in general and of the lower socioeconomic groups in particular is also one of the major means of reducing the pressures associated with rapid urbanization.

CONCLUSION

There is no doubt that the world of the future will be mostly urban nor that most urban dwellers will live in what are now developing countries. Already, the cities of the developing world are functioning under extraordinary pressures. Yet, since the dawn of civilization, cities have provided the impetus for change and innovation. Despite their many problems, cities remain the most dynamic components of the modern world.

Urbanization: A Report From Around the World[1]

URBANIZATION AND COUNTERURBANIZATION IN THE DEVELOPED WORLD

Developed countries as a whole are highly urbanized: in 1993, 74 percent of their population lived in urban areas. Currently, the annual rate of growth of the urban population in developed countries is low (1.1 percent) but it is still higher that the rate of growth of the population as a whole (0.7 percent). Consequently, the urban population in the developed world, which grew from nearly 700 million in 1970 to 900 million in 1993, is expected to increase to 1.1 billion by 2015. In contrast, the rural population of developed countries has been decreasing at a rate of -0.3 percent per year since 1970.

In the first half of the twentieth century, urban population growth was concentrated in the largest cities. But in the 1960s, 1970s, and 1980s, many of those large cities lost population, at least in relative terms, to smaller urban regions. The term "counterurbanization" was coined to describe this trend.[2] A widespread shift towards counterurbanization was first detected in the U.S.[3] During the 1970s, the population in large U.S. cities grew at half the rate of medium-sized and smaller cities.[4] Other studies confirm that similar developments occurred in other countries, including Australia, Denmark, France, Germany, Italy, Japan, Norway, and the United Kingdom, though the scale and timing of the phenomenon, and to some extent its nature, differed from country to country.[5]

The counterurbanization trend was expected to accelerate during the 1980s, but recent evidence suggests that the largest cities are making a comeback. The most significant reversal seems to have taken place in the U.S. during the 1980s, though there is evidence that Paris and London have also regained population.[6]

These trends remain poorly understood. Counterurbanization has been attributed to, among other factors, changes in corporate organization which facilitated the relocation of manufacturing away from major industrial centers. Government investments in infrastructure, including the expansion of transportation networks or the improvement of health and educational services in smaller communities; support provided for agriculture, forestry, and rural development in general; and the adoption of decentralization policies and "new town" development, all contributed to counterurbanization. However, the economic downturn that affected much of the developed world during the late 1970s and early 1980s weakened government commitment to those policies, and their subsequent deregulation of various elements of economic activity and service provision are likely to have renewed the advantages associated with the concentration of business and, consequently, of population in large metropolitan areas.

LATIN AMERICA: THE MOST URBANIZED REGION
OF THE DEVELOPING WORLD

In 1993, 73 percent of the population of Latin America lived in urban areas, a proportion comparable to that of the developed world. This represents a substantial increase since 1970, when only 57 percent of the population lived in cities. The change in the level of urbanization experienced by Latin America in 25 years is comparable to that experienced by developed countries over the course of the last 40 years.

As in other developing regions, there is considerable variation in the level of urbanization among the countries of Latin America. In Antigua and Barbuda, Haiti, Montserrat, St. Vincent and the Grenadines, and Guyana, over 60 percent of the population still lives in rural areas. At the other end of the spectrum, over 85 percent of the population of Argentina, Chile, Uruguay, and Venezuela lives in urban areas.

Latin America's urban population tends to be more concentrated in large cities than in other developing regions. In 1990, 19.7 percent of Latin America's urban population lived in cities of at least five million people (see Table 4). In comparison, only 15.8 percent of the urban population of Asia was concentrated in cities of that size and in Africa the equivalent proportion was only eight percent. The tendency of the urban population to be concentrated in a single city was particularly marked in countries such as Argentina, Chile, Cuba, Mexico, Peru, and Puerto Rico. Indeed, in 1990, 53 percent of the urban population of Puerto Rico lived in San Juan, the capital; Santiago was home to 44 percent of Chile's urban population; and Buenos Aires accounted for 41 percent of Argentine city dwellers.

Since the 1970s, however, many large Latin American cities have attracted fewer migrants from rural areas. As a result, some major cities — including Buenos Aires, Havana, and Montevideo — have seen their share of the urban population decline in recent years.[7] A similar trend is expected for other Latin American cities during the 1990s.[8] Indeed, recent data indicate that steep reductions in migration have been experienced by Mexico City, Rio de Janeiro, and Santiago (Chile).[9] At the same time, medium-sized and small urban centers have seen their population growth increase.

ASIA: A REGION OF CONTRASTS

Roughly 3.2 billion people — 57 percent of the world's total population — live in the developing countries of Asia. The region as a whole is characterized by a great deal of heterogeneity, comprising as it does some of the most urbanized, as well as the most rural, countries in the world. In Hong Kong, Israel, Kuwait, Qatar, and Singapore, for example, 90 percent or more of the population lives in urban areas — while in Bhutan, Nepal, and Cambodia, only 12 percent of the population was classified as urban in 1990. Asia also includes some of the most populous countries in the world: China, India, Indonesia, Pakistan, and Bangladesh. All of these countries have relatively low levels of urbanization (under 32 percent) and their statistical weight is largely responsible for producing a level of urbanization of only 33 percent for the region as a whole. In fact, excluding Japan, the proportion of the Asian population living in urban areas stood at only 31 percent in 1993 (see Table 1).

Nonetheless, Asia accounts for 40 percent of the world's urban population. The cities of Asia were home to 986 million people in 1993, a number higher than the current urban population of the developed world. By 2015, more than two billion Asians will live in urban areas, thus accounting for 47 percent of the world's urban population.

High levels of urban population growth in Asia have been accompanied by relatively high levels of growth among the rural population, despite the considerable rates of rural-urban migration experienced in some countries. Between 1970 and 1993, the urban population of developing Asia grew at an annual rate of 3.8 percent and is expected to continue growing at a level of 3.2 percent annually during 1993-2015. The equivalent rates of growth of the rural population are 1.4 and 0.3 percent, respectively. Thus, although Asia's rural population is expected to keep on growing well into the 21st century, a significant decline in its rate of growth is expected. Nevertheless, by 2015, the rural population of developing Asia is still expected to exceed that in urban areas by about 16 percent or more than 300 million persons (see Table 1).

Because Asia's urban population is growing much faster than its rural population, developing Asia is expected to record the world's highest rate of urbanization for the years 1970-2015. Indeed, the proportion of the Asian population living in urban areas is doubling every 38 years, and is expected to increase by 130 percent during 1970-2015, rising from 20 to 46 percent.

The process of urbanization in Asia's most populous countries has given rise to some of the largest cities in the developing world. In 1970, six Asian cities had populations of at least five million, compared to only four in Latin America and none in Africa. In fact, Asia was the only developing region with a city of more than 10 million inhabitants in 1970 (Shanghai). By 1990, there were five such cities in the developing countries of Asia (Shanghai, Bombay, Seoul, Beijing, Calcutta) and that number is expected to increase to 14 by the year 2010. Yet, many Asians live in smaller cities; in 1990 about 55 percent of all urban residents in Asia lived in cities of less than a half million persons. By 2010, those cities are expected to account for 51 percent of all urban dwellers in the region.

Consequences of Rapid Urban Growth: The Example of Mexico City

Robert W. Fox

Many of the consequences of rapid urban population growth are no different from the stresses and problems that have faced cities for centuries. However, the scale of population growth in the large cities of the developing world is without precedent in history. In some places, that growth has had staggering effects on natural and social systems, which has convinced some analysts that we are now in dangerous, uncharted waters.

Consider the example of Mexico City. Mexico City had 3.5 million inhabitants in 1950, five million in 1960, 8.5 million in 1970, 13 million in 1980, and roughly 16- 20 million today. In the 1970s, it was charted on a course to reach 32 million by year 2000. This expectation has now been lowered to around 24-26 million, a change that can be attributed to the overall reduction in the national rate of population growth — a top priority of the Mexican government since the mid-1970s.

The city's natural support systems are under severe stress. The city rests on a lake bed (the lakes were drained in the 19th century) spongy enough to cause buildings to slowly sink. Fresh water from the lake-aquifer is taken out for commercial and residential use, but now at a rate faster than it is replenished by the afternoon storms that sweep through the valley in the rainy season. To remedy the problem, the city spent billion of dollars to pipe in water from afar and to build "infusion" wells to inject water back into the ground. But why infuse the water into the ground? Why not use it directly from the pipe? Because by doing so, the city would sink further into the soil. It is also needed there to dilute the settling sewerage from cracked pipes leaking into the aquifer. (Earthquakes regularly hit Mexico City and sewerage pipes are broken in many places. Enough are now leaking to threaten the safety of the entire water supply.) This program has been enormously costly. Water taken from the Cutzamala region (the supply source) is no longer available for use there in agriculture and hydro power generation. Moreover, pumping water over a distance of several hundred kilometers into the Mexico City valley consumes a significant amount of Mexico's total energy generation capacity. Yet, all in all, it was estimated that the entire project would "buy" only five to seven additional years in population growth terms before additional water supplies were needed.

The topography of Mexico City is such that unflushed sewage and overflow drains onto the site of former Lake Texcoco, a district previously on the city's eastern outskirts. The site now hosts the settlement of Nezahualcoyotl, a vast slum with over two million inhabitants. During the dry, windy season (January and February) dry fecal matter blows from there throughout the entire metropolitan area, where it combines with auto and factory exhausts. Mexico City's polluted air is trapped in the valley through thermal inversion, and tens of thousands suffer from respiratory system illnesses and infectious disease as a result.

Most of Mexico City's untreated sewage flows slowly out of the metropolitan area via the Grand Canal, passing through numerous slum districts before some of it is intercepted and used to irrigate agricultural fields. Vegetable crops must be soaked in iodine solution before being consumed.

Adjusco, a nearby mountainous region, is known as the "lungs" of Mexico City. It is dying, stressed out by pollution. The result — apart from elimination of this oxygen producing and air cleansing agent — is erosion, as heavy rains cause mud slides.

Social stresses include two- to four-hour daily commutes in packed buses and subway cars as the residents of Mexico City, spatially segregated by social class, crisscross the city to work. Traffic problems have become so serious that the city has adopted an "alternate day" driving system, in which cars are licensed to drive only every other day. Wealthy residents have circumvented the rule by buying second cars with the desired alternate day license plate number. The city "beltway," or *periferico* is normally a near-parking lot during the morning and evening rush hours. Thousands who have jobs at the western and southern edges of Mexico City have moved to cities some 30-70 kilometers distance such as Cuernavaca, Tepoztlan, Toluca, and Cuautla from which they commute by bus and car over the mountains and into the city.

Newspapers and technical journals are filled with articles on the severe stress from life in the city. People do not leave their neighborhoods for months at a time because of traffic congestion and a general feeling of insecurity. Robberies and kidnappings are up, and gangs control territory. Add to this the inflation of the 1970s and 80s, which reduced purchasing power by more than 50 percent, and you can appreciate the difficulties faced by the average Mexico City dweller.

Perhaps the most disturbing statistics are those on Mexico City's employment outlook. Mexico is now struggling unsuccessfully to provide employment for the large generation born in the 1970s; roughly one-half of the Mexican labor force is now unemployed or severely underemployed. Mexico's total labor force now adds one million new workers every year and is projected to increase from 30 million in 1990 to 40 million in the year 2000, and to 61 million by 2025. The U.S., with a population four times the size of Mexico's and an economy 25 times as large, adds two million workers yearly.

Robert W. Fox is a consultant on population and natural resources.

AFRICA: THE FASTEST-GROWING URBAN POPULATION

Africa, with 34 percent of its population living in urban areas in 1993, had a slightly higher level of urbanization than developing Asia. By 2015, that proportion is expected to increase to 48 percent. Although there is considerable variation in the level of urbanization of the sub-regions of Africa, their range of variation is narrower than among the Asian sub-regions. Southern Africa is and is expected to remain the most urbanized sub-region of the continent, followed closely by Northern Africa (47 and 45 percent in 1993, respectively). The least urbanized sub-region is Eastern Africa (21 percent), with Middle and Western Africa having urbanization levels close to the continental average (see Table 1). A number of countries in Africa have very low urbanization levels (below 20 percent), including Burkina Faso, Burundi, Ethiopia, Malawi, Rwanda, and Uganda — most of which are among the least developed countries in the world.

The urban population of Africa has been characterized by very high rates of growth, averaging 4.5 percent annually between 1970 and 1993. Rapid growth is expected to continue between 1993 and 2015, implying that Africa's urban population will be the fastest growing in the world. Growth will be particularly brisk in Eastern Africa, where it is expected to surpass five percent per year. Relatively high rates of growth are also expected for the urban populations of Middle and Western Africa (4.6 and 4.7 percent annually, respectively). This means that the urban population of Africa is doubling every 15 or 16 years, so that it will increase seven-fold between 1970 and 2015, from 83 million to over 600 million (see Table 1).

In comparison with other regions, the urban population in Africa tends to be more concentrated in smaller towns and urban centers. In 1990, 60 percent of all African urban dwellers lived in cities of a half million people or less, and only two cities on the continent — Lagos and Cairo — had more than five million inhabitants. Although the number of large cities is projected to increase during the coming decades, Africa will remain a continent of small- and medium-sized cities. Even by 2010, over three-fifths of the African urban population will still live in cities of one million inhabitants or less (see Table 4).

Africa's rural population is also growing at a rapid rate. Indeed, Africa is the only major region whose rural population is expected to keep on growing at moderate rates beyond 2015. In 1993, 467 million Africans — two-thirds of the continent's population — lived in rural areas, up from 280 million in 1970. By 2015, 663 million Africans will live in rural areas and will account for one fifth of the total rural population of the world. Thus, most African countries are faced with the necessity of absorbing considerable increases in the rural population even as they experience unprecedented urban growth.

HANIA ZLOTNIK

International Migration: Causes and Effects

························

I nternational migration is a source of growing concern. Many people —
particularly in countries that are on the receiving end of migration
flows — believe that the numbers of migrants are increasing too rapidly
and that mechanisms to control them are failing. Anti-migration movements are gaining momentum in many countries, including the U.S.

In Europe, the frequently-used term "fortress Europe" conjures up the
image of a region under attack, threatened by hordes approaching from
both East and South. The population of native Europeans is expected to
decrease within the next 30 years, while that of neighboring developing
countries is likely to continue growing at a robust pace. For this reason,
some Europeans fear that migration from developing countries will necessarily increase to unacceptable levels. It is as if population, like nature,
abhorred a void and were compelled to move from high-growth to low-
growth areas.

Does rapid population growth spur international migration? As we will
see, population growth is but one factor among many that drive migration flows. It is well known that migration is often a response to differences in economic opportunities between areas of origin and those of
destination. In international migration, however, governments play a
crucial role in determining to what extent and under what conditions
economic forces impel migration. Furthermore, at least during this century, large international migration flows have generally resulted from

*Hania Zlotnik is chief of the mortality and migration section of the United
Nations Population Division. The views and opinions expressed in this paper are
those of the author and do not necessarily reflect those of the United Nations.*

changes in the configuration of nation states and have therefore been driven by political rather than economic factors. Clearly, the end of the bipolar era, by modifying the world order and giving rise to new nation states, has had major impacts on international migration that have yet to run their course. Changes in the world economy — including the shift from manufacturing to service industries in the developed countries and the pursuit of freer terms of trade at the global level — are also expected to have a substantial impact. Although some of these developments suggest that international migration may indeed continue to rise in the foreseeable future, an analysis of past experience is needed to establish both the likelihood and nature of such increases.

THE EXTENT OF INTERNATIONAL MIGRATION

One of the major problems encountered in assessing the likely course of international migration is the relatively poor information available on the subject. In some regions, data on international migration are virtually nonexistent. In others, the data are hard to interpret because they are not comparable between countries and because there are so many different types of migrants. In general, governments gather better information on migrants from abroad[1] than on their own citizens who emigrate to other countries. To understand the international migration process in its entirety, one would need information on immigrating and emigrating citizens, as well as on immigrating and emigrating foreigners. Data on the migration of citizens would be essential to ascertain to what extent population growth influences migration, yet very few countries gather and publish such information. One must therefore resort to information obtained from the main countries of destination to assess the level of emigration or net emigration experienced by a country.

Information regarding migration between developing countries is particularly deficient. Population censuses are perhaps the most comprehensive source of comparable data on the international migration experienced by the developing world. Unfortunately, censuses are carried out at lengthy intervals and their timing does not coincide for all countries. Furthermore, census information is often lacking for poor countries, especially those known to be hosting relatively large numbers of refugees. Nonetheless, because of their universal coverage, censuses may be the best instruments available to measure the number of migrants present in a country.

It is estimated that in 1985 there were 100 million international migrants — people living in countries other than those in which they were born — throughout the world. That estimate was made on the basis of census data from 1970 and 1980 for about three quarters of the world's countries. It includes refugees as reported by United Nations

High Commissioner for Refugees (UNHCR) and the United Nations Relief and Works Agency for Palestine Refugees in the Near East (UNRWA). It does not, however, include any migrants present in the former Soviet Union, for which data are unavailable. For China, the estimate includes only resettled refugees from Vietnam. Except in the receiving countries of the developed world, the estimate was based on crude extrapolations from available information.

In 1985 almost half of all migrants, 48.2 percent, lived in developed countries.[2] The traditional countries of immigration (Australia, Canada, New Zealand, and the U.S.) accounted for about half of the migrants in developed countries[3] and Europe largely accounted for the rest. Most of the migrants in the developing world were concentrated in Asia. Southern Asia, the region that includes the Indian subcontinent, hosted slightly over 16 percent of all migrants. India was the second major migrant-receiving country in the world, largely as a result of the population movements that took place at the time of partition. The developing countries of Eastern and Southeastern Asia accounted for 5.3 percent of all migrants in 1985; Western Asia and Northern Africa, about 14 percent; sub-Saharan Africa, 10 percent; and Latin America and the Caribbean, six percent.[3] The remaining migrants lived in the developing countries of Oceania. Most of the migrants in developing countries, particularly in Asia and Africa, originated within the same region.

Given that in 1985 world population was estimated at 4.8 billion, international migrants represented only slightly more than two percent of the world's people. In some regions, however, migrants constituted a much larger share of the population. In Oceania and North America, international migrants accounted for 16 and 7.8 percent of the population, respectively. In Europe, they accounted for 4.7 percent. In the developing regions migrants constituted fairly low proportions of the population: 2.3 percent in sub-Saharan Africa; 1.2 percent in Asia and Northern Africa; and 1.6 percent in Latin America. In relative terms, therefore, international migration has had a smaller impact in the developing world than in developed countries.

INTERNATIONAL MIGRATION AND POPULATION GROWTH

To determine whether population growth spurs international migration, one would need to compare demographic data with a time series of estimates of net migration from particular countries, net migration being defined as the difference between the number of migrants to, and emigrants from, a given country. Unfortunately, such estimates do not exist. However, some inferences about the impact of population growth on migration can be made on the basis of estimates obtained from the migration statistics of the main migrant-receiving

TABLE 1: Average Annual Number of Immigrants Originating in the Developing Regions and Average Annual Net Migration by Region of Destination and Period, 1960-1991

	1960-64	1965-69	1970-74	1975-79	1980-84	1985-89	1990-91
Sub-Saharan Africa							
Emigrants to:							
Northern America	1,900	4,105	10,094	15,817	17,506	20,942	16,191
Oceania	2,072	2,722	3,832	3,124	4,454	5,708	3,960
Western Europe	13,287	37,688	50,998	42,254	37,368	51,644	—
Total	17,259	44,515	64,925	61,194	59,328	78,293	20,151
Net migration							
Oceania	1,906	2,337	3,401	2,852	4,146	5,548	3,865
Western Europe	9,710	- 524	7,426	7,338	-1445	25,659	—
Maximum net emigration							
to developed countries	13,517	5,918	20,921	26,007	20,207	52,149	20,056
Northern Africa and Western Asia Emigrants to:							
Northern America	8,494	16,185	19,319	26,038	26,983	31,482	26,521
Oceania	2,171	4,792	9,087	7,461	2,625	6,217	5,701
France	40,869	48,245	79,159	32,474	43,841	22,623	—
Western Europe	45,079	98,391	221,785	168,205	124,303	114,751	—
Total	96,613	167,613	329,349	234,177	197,752	175,073	32,222
Net migration to:							
Oceania	2,119	4,677	8,886	7,281	2,455	6,039	5,621
Western Europe	34,716	53,110	137,517	28,883	25477	38,614	—
Maximum net emigration							
to developed countries	86,198	122,218	244,881	94,675	47,802	98,757	32,142
Southern Asia Emigrants to:							
Northern America	2,753	11,208	29,048	36,188	56,632	71,530	64,553
Oceania	1,371	2,798	4,890	1,909	2,209	5,687	8,280
Western Europe	4,234	35,758	25,995	35,430	36,916	45,344	—
Total	8,358	49,763	59,933	73,527	95,757	122,562	72,833
Net migration							
Oceania	1,206	2,558	4,673	1,793	2,106	5,607	8,235
Western Europe	2,528	24,466	14,096	25,519	23,965	29,506	—
Maximum net emigration							
to developed countries	6,486	38,232	47,818	63,500	82,703	106,644	72,788
East and Southeastern Asia Emigrants to:							
Northern America	15,088	46,450	102,970	162,272	248,670	236,643	213,775
Oceania	2,205	3,039	4,853	12,624	25,436	35,909	47,759
Western Europe	6,522	32,991	42,683	58,162	70,463	101,828	—
Total	23,814	82,481	150,506	233,057	344,569	374,380	261,534
Net migration							
Oceania	1,912	2,513	4,417	12,079	24,795	35,347	47,354
Western Europe	3,883	9,972	11,855	18,986	23,245	51,041	—
Maximum net emigration							
to developed countries	20,882	58,935	119,243	193,338	296,710	323,031	261,129

Table 1 Continued

	1960-64	1965-69	1970-74	1975-79	1980-84	1985-89	1990-91
Latin America and the Caribbean Emigrants to:							
Northern America	100,416	159,011	211,433	364,845	432,364	343,709	224,933
Oceania	216	687	4,577	3,946	1,721	4,197	3,945
Western Europe	4,433	21,200	20,805	30,071	23,945	25,486	—
Total	105,065	180,898	236,815	398,862	458,030	373,392	228,877
Net migration							
Oceania	191	617	4,405	3,721	1,493	4,019	3,840
Western Europe	2,327	6,921	5,958	16,864	8,785	11,308	—
Maximum net emigration to developed countries	102,934	166,548	221,796	385,430	442,642	359,036	228,772
Developing countries Emigrants to:							
Northern America	128,650	236,959	372,864	605,159	782,155	704,306	545,972
Oceania	8,035	14,038	27,239	29,064	36,444	57,718	69,645
France	40,869	48,245	79,159	32,474	43,841	22,623	—
Western Europe	73,555	226,028	362,266	334,121	292,996	339,053	—
Total	251,109	525,270	841,527	1,000,817	1,155,436	1,123,700	615,617
Net migration							
Oceania	7,334	12,702	25,783	27,726	34,995	56,560	68,915
Western Europe	53,164	93,944	176,852	97,591	29,073	156,127	—
Maximum net emigration to developed countries	230,017	391,850	654,658	762,950	890,064	939,616	614,887
Developed countries Emigrants to:							
Northern America	243,161	303,965	208,199	153,364	133,090	125,854	134,086
Oceania	138,278	164,177	146,063	55,244	69,248	56,767	51,813
France	133,416	155,490	113,426	37,924	35,048	14,037	—
Western Europe	655,598	874,469	935,176	577,898	549,873	890,435	—
Total	1,170,453	1,498,101	1,402,864	824,429	787,258	1,087,094	185,899
Net migration							
Oceania	115,771	121,124	103,670	24,022	42,016	47,305	46,493
Western Europe	214,575	88,565	146,857	-44297	-38830	303,315	—
Maximum net emigration to developed countries	706,923	669,144	572,153	171,013	171,324	490,511	180,579
All regions Emigrants to:							
Northern America	371,811	540,924	581,063	758,523	915,244	830,160	680,058
Oceania	146,313	178,215	173,302	84,307	105,692	114,485	121,458
France	174,285	203,735	192,585	70,398	78,889	36,660	—
Western Europe	729,153	1,100,497	1,297,442	912,018	842,868	1,229,489	—
Total	1,421,562	2,023,370	2,244,392	1,825,247	1,942,694	2,210,794	801,516
Net migration							
Oceania	123,105	133,826	129,453	51,748	77,011	103,865	115,408
Western Europe	267,738	182,509	323,710	53,294	-9757	459,442	
Maximum net emigration to developed countries	936,939	1,060,994	1,226,811	933,964	1,061,388	1,430,127	795,466

countries in the developed world. Table 1 presents estimates of the gross annual number of migrants admitted by selected receiving countries during each five-year period since 1960, classified by region of origin. However, these data are not fully representative of migration flows for three reasons. First, each of the countries listed defines migrants differently. For example, New Zealand lists arriving and departing long-term migrants, while the United Kingdom lists anyone arriving or departing who plans to stay for more than a year. Second, the developed countries listed here do not exhaust the possible destinations of migrants in the developed world. And third, there are substantial migrant flows among the developing regions which are not represented here.

Despite the limitations of the available data, it is still instructive to consider the story they have to tell. The data permit the calculation of the average annual number of persons originating in a given developing region who were admitted as migrants during each five-year period by the ten receiving countries considered. Those numbers can be interpreted as rough indicators of the total number of interregional emigrants leaving each developing region. Table 2 presents the estimated growth rates of the total population of the different regions of origin, together with emigration data from Table 1.

Notice that during the late 1960s, fertility was still high in most of the developing world and growth was rapid. Yet, during that time, *developed countries* were by far the major source of migrants to other developed countries. Among the developing regions, Northern Africa, Western Asia, and Latin America generated similar numbers of emigrants settling in developed countries, and their growth rates were high. However, the region having the highest population growth rate, sub-Saharan Africa, sent the lowest number of emigrants to developed countries.

By the late 1980s, sharp fertility declines were experienced by some developing regions. Viewing the data from this period, it becomes clear that population growth tends to be inversely related to the level of emigration, an outcome that contradicts the view that high population growth spurs migration.[4] Developed countries continued to be the main sources of emigrants to other developed countries, although considerable fluctuations were recorded, particularly between the early and the late 1980s. Latin America and Eastern and Southeastern Asia, regions with population growth rates below 2.0 percent, became the main sources of emigrants to the developed world. Moreover, if data were available on migration to developing regions, Eastern and Southeastern Asia would have probably surpassed Latin America as a source of emigrants. Yet Eastern and Southeastern Asia had the lowest population growth rates of any developing region during the late 1980s. Again, sub-

TABLE 2: Comparison of Estimated Inter-Regional Migration Levels and Rates of Population Growth for Major World Regions

	Growth rate		Annual number of emigrants		
	1965-1969	1985-1990	1965-1970	1980-1985	1985-1990
Sub-Saharan Africa	2.69	3.05	44,515	59,328	78,293
Northern Africa and Western Asia	2.64	2.69	167,613	197,752	175,073
Eastern and Southeastern Asia	2.58	1.63	82,481	344,569	374,380
Southern Asia	2.38	2.22	49,763	95,757	122,562
Latin America	2.58	1.96	180,898	458,030	373,392
Developed	0.90	0.64	1,498,100	787,258	1,087,094
Total			2,023,370	1,942,694	2,210,794
	1965-1969	1985-1990	1965-1970	1980-1985	1985-1990
Sub-Saharan Africa	2.69	3.05	5,918	20,207	52,149
Northern Africa and Western Asia	2.64	2.69	122,218	47,802	98,757
Eastern and South-eastern Asia	2.58	1.63	58,935	296,710	323,031
Southern Asia	2.38	2.22	38,232	82,703	106,644
Latin America	2.58	1.96	166,548	442,642	359,036
Developed	0.90	0.64	465,408	92,435	453,850
Total			857,259	982,499	1,393,467

Saharan Africa, the region with the highest population growth rate, sent the smallest number of migrants to developed countries.

Despite their many limitations, the data in Table 2 indicate that, at least during the past thirty years, high levels of population growth have not been associated with high levels of emigration from developing regions to the developed countries of the West. Indeed, the regions experiencing the sharpest reductions of population growth have been among the major sources of emigrants to developed countries. The reasons for that outcome will be explored below.

POLICIES SHAPING INTERNATIONAL MIGRATION IN THE MAIN COUNTRIES OF DESTINATION

Western Europe — The Need for Imported Labor

Throughout history, population mobility has been one of the tools used to satisfy the demand for labor. On countless occasions, the equivalent of today's foreign labor has been secured by force. Perhaps the most recent use of such practices took place during the Second World War, when the German Reich recruited thousands of foreign laborers to work in factories whose output was essential to the war effort. It is estimated that by 1944, there were over 6.7 million foreign men and women working in the Reich, about 1.5 million of whom were prisoners of war.[5] In fact, Germany has a long tradition of importing foreign labor. During a period of rapid industrialization in the 1870s and 1880s, German agricultural workers migrated en masse to urban areas, and landowners hired seasonal foreign workers to replace them. It was therefore not surprising that, when labor shortages arose during reconstruction in the 1950s, West Germany again resorted to the recruitment of foreign workers.

Population losses caused by the War had left several European countries with reduced labor forces and, as their economies began to recover, they too imported workers from abroad. Although practices varied from country to country, most operated under the assumption that the need for foreign labor was temporary and that foreign workers would leave when no longer needed. Complex systems of work and residence permits were adopted to ensure the "rotation" of labor; that is, to compel workers to leave at the end of their contracts. However, despite such measures, a significant proportion of workers stayed for lengthy periods and often managed to bring in their families. The data in Table 1 indicate that the scale of the population inflows experienced by the main labor-importing countries of Europe was large. Thus, during 1965-1969, some 1.1 million incoming migrants were admitted annually by Belgium, Germany, the Netherlands, Sweden, and the United Kingdom combined. However, outflows were almost as high, so that net migration amounted to only 183,000 persons annually.

The data in Table 1 show that net migration from developing countries to Western Europe increased steadily, surpassing that from developed countries by six percent during 1965-1969 and by some 20 percent during 1970-1974. In the decade after 1975, a major change took place. The countries of Western Europe gained population from the developing world while losing population from developed countries (including their own citizens). That is, whereas net migration from developing countries remained positive, that from developed countries remained negative.

There was, however, a substantial decline in net migration from developing countries between 1970-1974 and 1980-1984. This was due to the fact that the oil crisis of 1973 and the ensuing economic slowdown brought about a major change in policy among the labor-importing countries of Europe. By 1974, most governments had stopped actively recruiting labor but, recognizing that it was neither possible nor desirable to expel all foreign workers, they allowed migrant workers to stay and bring in their families under certain conditions. Some governments, however, instituted incentives to promote the return of migrants to their home countries. As the data in Table 1 indicate, the measures taken were successful in reducing both the total migrant inflow and the net migration gain registered by Western European countries. Poor economic conditions also had an effect, particularly during 1980-1984 when net migration became negative.

The year 1985 marked yet another turning point. During the late 1980s, net migration from developing countries increased to levels not seen since the height of labor migration, and that from developed countries surpassed all levels registered since 1960. This time, however, the countries of Western Europe were not experiencing a shortage of labor. In fact, for over a decade their unemployment levels had remained high despite improving economic growth. The changes that triggered increased migration to Western Europe during the late 1980s occurred therefore in a climate that was far from propitious. Foreigners were increasingly seen as a threat because they were thought to compete for scarce employment opportunities and were deemed to make an excessive use of social services. Consequently, the growing presence of migrants from developing countries, though by no means excessive, was increasingly perceived as unacceptable.

THE COUNTRIES OF IMMIGRATION   FROM QUOTAS TO FAMILY REUNIFICATION

In contrast with Western Europe, the migration policies of the traditional countries of immigration have generally not been geared to the recruitment of foreign workers. Indeed, at the beginning of the twentieth century, immigrants to the U.S. could be denied admission at the port of entry if they declared that a job awaited them. (That policy was intended to protect unionized labor from the use of immigrants to break strikes or otherwise undermine the unions.) Although the need for labor was always a subtext of the policy debate on immigration to the U.S., actual policies usually reflected other concerns. Thus, the immigration acts of 1921 and 1924 restricted immigration on the basis of national origins, establishing a "quota system" whereby fixed proportions of the total immigrant visas allotted were reserved for the exclusive use of natives of specific countries. In addition, immigration from most Asian countries was practically barred. Similarly, in Australia and Canada immigration policies were largely shaped by the perceived need to populate those countries with "acceptable" people, and during most of this century, the immigration policies of both countries favored Caucasian immigrants.[6]

Yet, whatever the acknowledged or unacknowledged justification for immigration policies, during the first part of this century immigration levels to all three countries corresponded closely to general economic conditions. Thus, the number of immigrants to the U.S., which amounted to 5.7 million during 1911-1920 and remained at a relatively high level during the 1920s (4.1 million), dropped precipitously during the Depression years to scarcely more than a half million annually during 1931-1940.[7] Immigration levels remained low during the Second World War; so that over the decade 1941-1950 the number of immigrants barely surpassed one million. Immigrant admissions by Australia and Canada reflected similar trends, though at considerably lower levels.

Since the War, immigration to the U.S. has increased steadily, particularly after 1965 when the national origins quota system was replaced by legislation favoring family ties with U.S. citizens or residents as the basis for admission.[8] In the other countries of immigration, the number of immigrants has fluctuated considerably, with average migration since 1960 reaching a high in 1965-1969 of about 150,000 annual admissions in Australia and 180,000 in Canada.[9]

All three countries have experienced a steady increase of the proportion of immigrants from developing countries. In the U.S., that proportion already amounted to 41.9 percent in 1960-1964, compared to 12.3 percent in Canada and 7.8 percent in Australia. By 1985-1989, the proportion of immigrants from the developing world had risen to 88 percent in the U.S., over 70 percent in Canada and about 53 percent in Australia.[10] In both Australia and Canada those changes were the result of the "universalist" admission policies adopted during the 1970s, when both countries began selecting immigrants on the basis of their skills, their likelihood to adapt to the host society, and their ties with citizens or former immigrants. In practice, universalist policies give the most weight to family ties as a criterion for admission.

It must be noted that all countries of immigration also allow the temporary admission of foreigners for a variety of purposes, one of which is to exercise an economic activity. In the U.S., in particular, the number of persons admitted annually as temporary workers or trainees has increased from 48,000 during 1974-1979 to 124,000 during 1987-1990. In contrast to the trend observed among permanent immigrants, the proportion of temporary workers and trainees originating in developed countries has been high and rising, passing from 43.8 percent in 1974-1979 to 52 percent during the 1980s. All other categories of temporary migrants who are allowed to work when admitted to the U.S. have also been growing, so that by 1987-1990 the country recorded, on average, the entry of 487,000 such migrants annually, accompanied by about 99,000 dependents.[11] These numbers represent the number of entries and are therefore likely to overestimate the number of persons involved. Nevertheless, they indicate that temporary migration is an important source of needed labor and particularly of skilled labor for the U.S.

THE DEVELOPING COUNTRIES AND JAPAN — AN EXPANDING LIST OF DESTINATIONS

Among the developing countries, for the last 20 years the most popular destination for international migrants has been the oil-producing countries of Western Asia, particularly the member states of the Gulf

Cooperation Council (GCC): Bahrain, Kuwait, Oman, Qatar, Saudi Arabia, and the United Arab Emirates. Because their small native populations largely lacked the skills needed to carry out the development projects that became possible thanks to their large oil revenues, the GCC countries imported foreign workers. As in Western Europe, temporary work or residence permits were used to prevent migrants from settling in for the long term. But here again, those measures achieved only limited success. Many workers, especially those who were recruited from other Arab countries, stayed for relatively lengthy periods. Although reliable data on the migration experience of most GCC countries are lacking, data for Kuwait show that in 1980, out of the 792,000 foreigners present in the country, 32 percent had been living there for at least ten years.[12]

To reduce their reliance on imported Arab labor, GCC countries began recruiting workers from other Asian countries in the late 1970s. During the 1980s, the sending countries in Asia granted over one million permits annually to persons wishing to work abroad, a majority of whom had job offers in GCC countries.[13] Nevertheless, even during that period, demand for labor in GCC countries slackened as major infrastructure projects were completed and oil revenues declined. Furthermore, the 1990 invasion of Kuwait by Iraq and the ensuing war led to a rapid repatriation of many of the foreigners in Kuwait and neighboring countries (Yemenis in Saudi Arabia, for instance, were forced to return to their country).

Although the migration of Southeastern Asian workers to Kuwait and other GCC countries resumed after the war's end in 1991, Asian migrant workers were already choosing from a wider range of destinations. Thus, whereas 74 percent of all work permits granted by Indonesia, the Philippines, the Republic of Korea, and Thailand during 1980-1984 were to persons intending to work in Western Asia, by 1985-1988 that proportion had dropped to 65 percent.

Increasingly, workers from Eastern and Southeastern Asia are migrating to other countries within the same region. Japan, Malaysia, and Taiwan began attracting migrant workers during the 1980s when their rapid economic growth produced labor shortages, particularly in the construction and service sectors. The number of foreigners legally present in Japan, for example, rose from 851,000 in 1985 to 1,219,000 in 1991.[14] Changes in Japan's immigration law that allow the admission of a wider range of skilled workers and that grant residence rights to the descendants of Japanese are partly responsible for that increase. However, the governments of Japan, Malaysia, and Taiwan have, in general, been unwilling to acknowledge their need for unskilled foreign workers or permit their legal admission. Consequently, undocumented migration is on the rise. In Japan, the number of foreigners charged with violating the

Immigration Control Act increased from 5,600 in 1985 to 31,900 in 1991, and Japanese authorities estimate that there were about 160,000 foreigners residing illegally in the country as of mid-1991.

Most developing countries either lack explicit policies regarding the admission of international migrants or maintain policies that are at odds with current circumstances. In some Latin American countries, for instance, immigration policies are still geared towards the admission of immigrants from European countries, particularly those from Eastern Europe,[15] but there are scarcely any mechanisms to regulate the intraregional flows that are common. Consequently, current international migration in the region tends to be dominated by flows of undocumented migrants, and governments have resorted to amnesties and regularization drives to introduce a certain control in the migration process. In recent years, efforts to create viable trading blocs among certain Latin American countries have put the migration issue on the policymaking agenda, but definite strategies for action have not yet emerged.

In Africa, the policy vacuum regarding international migration is probably even greater, given that many African states are still in the first phase of the nation-building process and are therefore reluctant to articulate their position with respect to the admission of people who are technically foreigners but who often have close ethnic if not familial ties with citizens. Perhaps the most serious attempt to establish a regional migration policy within Africa was that undertaken by the Economic Community of West African States (ECOWAS), the aim of which was to reduce barriers to intraregional travel and migration. However, the ECOWAS plan has not yet reached its goal, partly because the governments of some countries are reluctant to open their doors to the citizens of the others. Indeed, during the early 1980s, Nigeria expelled many citizens of other ECOWAS countries who had violated the terms of the agreement by overstaying or working illegally in its territory.[16]

During the colonial period, African authorities often used forced migration to secure the labor needed in the plantations or the mines. Currently, only the Republic of South Africa still has provisions for the recruitment of foreign workers from neighboring countries. Most of those workers are employed by South Africa's Chamber of Mines on a temporary basis and the rotation of labor is strictly enforced. In 1974 Malawi, one of the major suppliers of labor to South Africa, unilaterally acted to restrain labor migration to South Africa in order to protest the treatment of its workers. This action prompted South Africa to reduce its dependence on foreign labor. As a result, the number of foreign workers employed by the Chamber of Mines declined from 308,000 in 1971 to 207,000 in 1985.[17]

THE ECONOMIC ASPECTS OF INTERNATIONAL MIGRATION

The preceding section indicates that economic considerations often influence government policy on the admission of migrants. However, although the political debate on migration generally involves some assessment of the effect migrants may have on economic variables, including wage levels, unemployment, or the cost of welfare services, there is little hard data available to confirm or refute those assessments. Still, as economist George Borjas notes:

> The fear that 'immigrant hordes' displace natives from their jobs and reduce the earnings of those lucky enough still to have jobs has a long (and not so honorable) history in the policy debate. The presumption that immigrants have an adverse impact on the labor market continues to be the main justification for policies designed to restrict the size and composition of immigrant flows to the U.S.[18]

Borjas studied the impact of immigration on the employment opportunities and earnings of domestic workers in the U.S. during the late 1970s. He concluded that immigration barely affected the earnings and employment opportunities of natives. From an economic standpoint, that finding implies that native and immigrant workers are generally poor substitutes for one another in production. In other words, immigrants generally take jobs that domestic workers do not want. Moreover, no study found evidence supporting the claim that minority groups, such as African Americans, were more likely to see their wages negatively affected by the presence of immigrants. On the contrary, some studies reported that African Americans residing in cities with relatively large numbers of immigrants had slightly higher wages than African Americans residing in other urban centers. All the studies reviewed controlled for differences in the local labor markets as well as for differences in the characteristics of migrants and natives.[19]

Regarding the use of welfare services by immigrant households, the evidence reviewed by Borjas does not support the conjecture that immigrant households are more likely to use welfare than native households with demographically comparable characteristics. However, in the U.S. there is a strong negative correlation between an immigrant group's use of welfare services and the gross national product (GNP) of the country of origin. Therefore, because recent immigrants are more likely to come from poor countries, they may be more likely than their predecessors to use welfare.[20]

Unskilled migrant workers often earn wages that keep them below the poverty line. Although American consumers benefit from the low wages of migrants through reduced prices of some goods and services, their presence is likely to aggravate the economic and social problems associated with poverty. However, it is important to stress that it is not migration itself that causes such problems, but rather the fact that there is a real demand for persons willing to perform relatively unappealing tasks for low wages. For a variety of reasons, it has been easier in many developed countries to satisfy that demand by importing workers than by taking other measures. In the U.S., the agricultural lobby helped over a million agricultural workers to obtain legal residence as part of the provisions of the Immigration Reform and Control Act of 1986 (IRCA) — an implicit acknowledgement of the need for cheap foreign labor. Clearly, it is unfair and unreasonable to blame migrants for responding to economic opportunities.

Evidence suggests that a substantial number of unskilled migrants work abroad only for limited periods and end up returning to their home countries[21] where the wages earned abroad afford them an acceptable standard of living. Indeed, there is general agreement that one of the major economic impacts of international migration is the flow of remittances from receiving countries to the countries of origin. It is estimated that in 1989 remittances from migrants amounted to US$61 billion worldwide, a figure that compares favorably with the US$51 billion provided as official development assistance (ODA) by OECD member states to developing countries in 1988.[22] Although there is no consensus about whether or not remittances are used in a manner that accelerates development, they undoubtedly contribute to improving the living standards of family members left behind and are generally used rationally by migrants and their families, provided viable investment opportunities exist. For some countries, the economic impact of remittances has been estimated to be considerably larger than that of foreign direct investment. Thus, the $US1.7 billion received as remittances by the Philippines in 1989 would generate $US3.9 billion of economic growth, assuming a multiplier effect of 2.3. Since every dollar of remittance-generated economic activity is estimated to be equivalent to that produced by $5.40 of capital investment, the remittance level for 1989 would be equivalent to about $US21 billion of investment, a figure that dwarfs the $US850 million that the Philippines received in actual foreign direct investment that year.[23]

The importance of remittances for many developing countries exemplifies the linkages that increasingly characterize the global economy. Another instance of growing ties between developed and developing countries is the growth of transnational corporations. These corporations, which are based mostly in developed countries, have for at least two decades exported capital rather than imported labor as a means of reducing labor costs and penetrating new markets. Japan, in particular, has promoted the establishment of Japanese subsidiaries abroad, especially in the newly industrializing economies (NIEs) of Eastern and Southeastern Asia. Yet, that strategy has not stemmed migrant inflows to Japan. Indeed, it may have even facilitated migration by creating or strengthening linkages between Japan and other Asian countries.

In addition, some of the beneficiaries of foreign direct investment have also been important sources of migrants. The Republic of Korea and Taiwan, for instance, rank among the main sources of emigrants to the U.S.; Malaysia is an important source of migrant workers for Singapore and of emigrants to Australia; and Indonesia, Thailand, and the Republic of Korea have all sent considerable numbers of workers to Western Asia. Interestingly, those countries have sent large numbers of migrants abroad even as they experienced rapid economic growth. The Asian experience thus appears to confirm the findings of the Commission for the Study of International Migration and Cooperative Economic Development, established by the U.S. Congress in 1986. Among other things, the Commission concluded that the development process itself is destabilizing and that, in the short run, it probably enhances rather than reduces emigration pressures.[24]

While international investment can foster migration, so can trade barriers that cost jobs in developing countries. For example, the U.S. imposes import quotas in order to keep the price of domestic sugar above world prices. It is estimated that those quotas impose an annual implicit cost on American consumers of $US76,000 per sugar worker, and that they have helped reduce export revenues from Caribbean countries from $US544 million in 1981 to only $US97 million in 1988.[25] Consequently, the region has lost some 400,000 jobs, a development that is unlikely to reduce migration pressures. The Commission stressed that it was essential for receiving countries to consider carefully the implications of all such policies on migration pressures abroad.

The Commission concluded that expanded trade between countries of emigration and the U.S. was the single most important strategy to reduce migration pressures in the long run. Similar suggestions have been made in other fora,[26] but developed market economies have found it difficult to reduce trade barriers for those goods that developing countries can produce competitively, especially farm products and textiles. However, the signing of the North American Free Trade Agreement (NAFTA) and the successful conclusion of the Uruguay Round of multilateral trade negotiations under the General Agreement on Tariffs and Trade (GATT) suggest that many trade barriers may soon be dismantled.

The Commission's conclusions imply that international migration cannot be considered in isolation from other processes that affect the functioning of the world economy. Improvements in transport and communications, the growing role of transnational corporations, exchange and consumption processes that extend beyond national boundaries, the interdependence fostered by trade, the flows of technology and capital, and the movement of management and other trained personnel — all have implications for migration. Perhaps the development with the greatest impact is the emergence of a new development paradigm based on free-market strategies. As that paradigm is implemented, it is causing major structural changes in some developing countries, and, as the Asian NIEs illustrate, economic transformation is often accompanied by emigration because successful development means that more people are capable of covering the substantial costs of international migration. Yet, the potential for emigration will become manifest only if conditions in receiving countries warrant it, that is, if viable economic opportunities exist and migration is condoned even if not fully approved. Today, as in the past, the prevalence of poor economic conditions in receiving countries, coupled with strict migration controls, is almost certain to reduce migration considerably, at least as measured in terms of net gains.

THE INSTABILITY OF NATION STATES AND INTERNATIONAL MIGRATION

The late 1980s witnessed a series of disruptions of the nation state system which have major implications for migration. In particular, as sociologist W. Rogers Brubaker notes, "the breakup of the Soviet Union has transformed yesterday's internal migrants, secure in their Soviet citizenship, into today's international migrants of contested legitimacy and uncertain membership."[27] Thus, most of the 25 million Russians who in 1989 lived in non-Russian republics of the former Soviet Union are now minorities whose right to citizenship in the successor states may be in question. The antagonism they face in some states makes it likely that many will choose to return to Russia. In some cases, voluntary return movements have already begun.

Unfortunately, the "unmixing of nationalities" has not always been achieved by peaceful means. The disintegration of Yugoslavia provides the grimmest contemporary example of how forced displacement can be used for political ends. By June 1993 there were 3.6 million people in need of international assistance in the former Yugoslavia. Most had been displaced by the conflict, generally as a result of brutal "ethnic cleansing" aimed at dislodging unwanted ethnic groups from a specific territory. Most European countries are reluctant to admit ex-Yugoslavs as refugees, claiming that their admission would abet those practicing ethnic cleansing. Consequently, most would-be refugees remain within the boundaries of the former Yugoslavia.

As the world moves away from the bipolar Cold War era, developing country governments operate with fewer constraints, and expansionary pursuits and internal disputes are more likely to be acted out. Military intervention by the main powers occurs only if their vital interests are threatened, as by the Iraqi invasion of Kuwait. Otherwise, they are likely to remain on the sidelines, as in the civil wars of Liberia, Sudan, and

Ethiopia. Therefore, in contrast with most of the refugee-producing conflicts of the 1970s and 1980s, those of the 1990s are occurring largely without the backing of major world powers. As a consequence, today's refugees do not have sponsors whose leadership could muster the assistance they so sorely need.

The 1980s saw a dramatic increase in the number of refugees seeking asylum in Western-bloc countries — from about 70,000 in 1983 to over 690,000 in 1992.[28] Countries that previously admitted only small numbers of pre-screened refugees for resettlement were suddenly confronted with a continuous flow of asylum-seekers. Because many of those asylum seekers come from countries that traditionally supplied labor to Western European countries, some believe that asylum is being used as a "back door" to labor migration. However, about a third of all asylum seekers come from Eastern-bloc countries (the traditional sources of refugees during the Cold War) and the rest tend to come from developing countries affected by civil strife, war, or violent inter-ethnic conflicts.

Fearing the consequences of uncontrolled migration, the member states of the European Community have instituted more restrictive rules for the granting of asylum. The 1990 Schengen Agreement and the 1990 Dublin Convention typify the new approach. Together they narrow the eligibility criteria for asylum, shorten the determination process leaving few avenues for appeal, and facilitate the deportation of those persons who are not granted asylum.[26] The European Community is working to extend the application of such rules, by establishing enforcement agreements with the Eastern-bloc countries, Australia, Canada, and the U.S.

The traditional countries of immigration are also receiving increasing numbers of asylum-seekers and seeking ways to control that unwanted inflow. However, the U.S., which is a major destination for asylum-seekers, has failed to exercise leadership in this area mainly because its own asylum policies, like those of other industrialized countries, are dictated largely by foreign policy concerns rather than by humanitarian considerations. Thus, whereas Cubans and Nicaraguans fleeing from leftist regimes have generally been granted asylum, Salvadorans have generally been denied it,[30] and Haitians have in most cases not even been allowed to reach U.S. territory to lodge asylum requests.[31]

The growing presence of asylum seekers has generated considerable resentment in receiving countries. Indeed, the costs of feeding, housing, and evaluating the legal claims of asylum seekers can be high. Yet, because few asylum-seekers are granted refugee status, the number of refugees in most developed countries remains small.[32] Indeed, developed countries shoulder a relatively small share of the global refugee burden. By early 1991, before the effects of the war in Yugoslavia were felt, the Western-bloc countries of Europe were hosting 826,000 refugees, up from 575,000 in early 1982.[33] Although by early 1993 the number of refugees in those countries had more than doubled, to 1.7 million, that number represented only nine percent of the 19 million refugees in the world at the time.[30] In contrast, 71 percent of the world's refugees were living in developing countries, a proportion that would be even greater if the 2.5 million Palestinian refugees under the mandate of UNRWA were included. Furthermore, four million refugees (30 percent of the total) had found asylum in the world's least developed countries and were therefore exacerbating the many strains on those very poor countries. From the perspective of those countries, the plight of developed countries was to be envied.

The end of the Cold War brought hope of reducing the number of refugees in the world by establishing conditions for a lasting peace that would encourage voluntary repatriation. In many instances those hopes have proved justified. During 1992 alone, an estimated 2.4 million refugees returned to their countries of origin.[35] The number of refugees in developing countries declined from 14.8 million in late 1991 to 13.5 million in late 1992, an encouraging development. However, because of new outflows, the number of refugees in the world today is about 10 million higher than in 1985. Meanwhile, the prospects for resolution of many ongoing conflicts still seem dim, and other conflicts are looming on the horizon. The implications of those prospects are moving the international community to address the complex problems at the root of refugee movements.

Recognizing that human rights violations are a major cause of refugee movements, multilateral organizations are reconsidering the responsibility of states towards their citizens and the boundaries of national sovereignty. According to the U.N., "the concept of the state's responsibility towards its citizens is being extended to encompass a responsibility towards the international community for the way those citizens are treated."[36] The creation of a safety zone to protect the Kurds in northern Iraq by means of military intervention and the establishment of "safe areas" in Bosnia and Herzegovina to protect the displaced population are instances in which the international community has put humanitarian concerns above national sovereignty. However, there is much ambivalence about such measures and it remains to be seen whether the international community will intervene in situations of marginal interest to powerful states.

CONCLUSION

As the twentieth century draws to a close there are strong reasons to believe that international migration will increase. However, the causes of recent migration flows suggest that migration will not arise mainly because of rapid population growth or deepening poverty. Rather, migration will arise in two very different contexts. On the one hand, countries that are politically stable and economically vibrant will generate migration flows as their citizens acquire the resources needed to pursue better training or employment opportunities abroad. Such flows would be responsive to economic conditions both in the country of origin and in that of destination, and though they may not always be sanctioned by the latter, they are unlikely to have negative effects on the host country and will probably have positive effects on the country of origin. Furthermore, such migration will wane as the development process runs its course.

On the other hand, the outflow of refugees and asylum-seekers will continue from countries whose governments engage in gross violations of human rights, or where there is political instability or outright conflict. In many instances, those situations will coexist with economic stagnation or environmental deterioration, but the latter are not expected to be the primary or even the fundamental causes of most refugee movements. Ethnic tensions are more likely to lead to refugee-producing conflicts, particularly when they are exploited for political purposes either by governments or external forces. During the 1990s, tensions that are simmering in various parts of the world could explode in violent confrontations. The challenge for the international community is to find means of alleviating ethnic tensions through mediation and to prevent their exploitation by unscrupulous political leaders. Although differences in population growth and in economic endowments can certainly aggravate ethnic tensions, they are not simply the product of poverty, underdevelopment, or high fertility.

Unfortunately, some of the likely foci of instability are uncomfortably near to Western-bloc countries. As the cases of Iraq and Yugoslavia illustrate, it is likely that the international community will take measures to maintain would-be refugees within the affected region, but such measures will be costly both in terms of lives and of the need for intervention.

It is very disturbing that, while Europe faces a major humanitarian crisis at its doorstep, hate crimes against migrants and asylum-seekers by racist and xenophobic groups are proliferating. In a period of recession, the general distrust of the alien is exacerbated by reports of ever-growing migration flows, which often portray migrants as poor, backward, and undesirable. European leaders are failing to stress that migration is far from out of control; indeed, in France the foreign-born population declined between 1982 and 1990. Many have apparently forgotten that, for most of this century, Europeans themselves accounted for the major share of migration flows in the world. States have the power to effect changes in migration flows, and it is their responsibility to exercise that power in a manner that does not endanger the lives and well-being of migrants.

SHARON STANTON RUSSELL AND MICHAEL S. TEITELBAUM

Migration: What Policymakers Can Do

....

Given the complex causes and consequences of human population movements, making sense of them can seem to be a daunting task, particularly in this period of a rapidly changing world order. Nonetheless, a few general points can be made about how migration is being viewed today, about the nature of the challenges posed by population movements, and about some of the things policymakers can do.

There is growing recognition that voluntary migration is a rational response to interregional economic differences, and a logical strategy for individuals and families to enhance their opportunities, to ensure their survival, and to minimize risks. Migration thus has a major role to play in furthering human development.

At the same time, there are very deep and real concerns about the immediate, practical implications of population movements for both sending and receiving areas and for the individuals and families involved. With respect to internal migration, governments worry about how to keep pace with growing demands for infrastructure, services, and jobs in urban areas. With respect to international migration, governments worry about how to control their borders and how to balance the needs and rights of migrants against those of their citizens, especially in a climate of racial and ethnic tensions, high unemployment, and economic stagnation. Individuals, families, governments, and international

Sharon Stanton Russell is a research scholar at the Center for International Studies, Massachusetts Institute of Technology, and Michael S. Teitelbaum is a program officer at the Alfred P. Sloan Foundation. This chapter is adapted from the authors' presentation to the International Parliamentary Workshop on "Selected Issues of the ICPD Debate: How to Achieve Success in Cairo."

organizations alike worry about how to ensure the protection of migrants whose circumstances make them vulnerable to abuse or exploitation.

At the international and intergovernmental level, a number of factors have converged to demand new international responses. These include the end of the Cold War; rapidly changing national boundaries; dramatic increases in the numbers of displaced people, asylum seekers, and refugees; and the proliferation of internal crises (such as those in northern Iraq, Somalia, and former Yugoslavia).

There is now a widely recognized need to rethink the global policy framework regarding population movements, and important new debates are bringing to light significant underlying tensions. To mention only a few examples: There are tensions between the rights of states to determine who enters their borders and the rights of individuals to seek asylum; there are tensions between the principles of national sovereignty and national security, and the desires of the international community to preserve human rights and human security through new measures such as "humanitarian intervention" and "preventive diplomacy"; and there are tensions, as well, between the national sovereignty of states, and the principle that people should have the right to remain in their own countries and to be protected by their governments. Given that these are fundamental principles, it is not surprising that there is a new concern for the ethical dimensions of population movements.

Developed and developing countries, or rich and poor countries in developing areas, may have deeply divergent interests regarding how these tensions are resolved. Indeed, the issue of human population movements has the potential to divide countries along these cleavages, as the issue of fertility did at the World Conference on Population in Bucharest in 1974 or as the issue of the environment did at the Rio Earth Summit in 1992. Nonetheless, the sensitivities surrounding the issue of migration cannot justify a failure to face the challenges it poses.

To deal constructively with migration issues, policymakers must first recognize that there is a limit to what policies explicitly directed toward population movements can do. For example, those affecting internal movements have proven notoriously ineffective and can be costly. Policies to control immigration from other countries *can* have substantial impacts upon such flows if sufficient resources are provided, but they may also have unintended effects, such as increased clandestine migration.

Second, policymakers must recognize that a wide range of *other* policy measures can and do affect population movements; the migration consequences of these "collateral policies" need to be spelled out. Within countries, for example, agricultural pricing policies, or economic policies that lead to depletion of natural resources, or decisions regarding the location of domestic and foreign investments all affect migration.

Between countries, trade policies may have the most profound effects on voluntary population movements in the long term by gradually reducing economic differentials. However, these may meet with strong opposition from domestic constituencies in the short term.

Third, countries that fund official development assistance (ODA) are beginning to recognize the importance of development in both stimulating and reducing international migration. At the same time, all parties need to be realistic about the time frames involved and the potential of ODA to influence migration. The volume of net remittances — money sent home by immigrants — from developed to developing countries was US\$ 37 billion in 1990. This amount was roughly 70 percent of the value of ODA, which was US\$ 54 billion in 1992. If we were to include *unofficial* remittances sent through informal channels, remittance figures would be considerably higher. Policymakers should be cautioned, however, that remittances do not automatically translate into development, and labor migration is not a long-term solution to problems of unemployment. As the experience of the Gulf War reminds us, dependence on migration and remittances can make countries vulnerable to sudden changes.

Fourth, policymakers in developing countries can recognize that there is more to be done to stimulate the return of remittances and to harness international migration to national development strategies. These include broad macroeconomic, political, and institutional measures, such as adopting sound exchange rate, monetary, and economic policies, facilitating the provision of commercial banking services that enable the safe and timely transfer of migrants' funds, and promoting the conditions necessary to increase domestic savings and channel them to productive investment.

Fifth and finally, policymakers must remember that migration is a complex and diverse process, and remain skeptical in the face of ideological conclusions and value judgments about migration processes. Because the causes and implications of population movements vary considerably, sweeping generalizations about migration are suspect, with one important exception: The volume of migration is likely to increase and continue to pose challenges.

FRANK SHARRY

Immigration and Population in the United States: Collision or Consensus?

....

"More and more countries, most of them poor and less developed, are reaching the point of excessive population, resource depletion, and economic stagnation. Their 'huddled masses' cast longing eyes on the apparent riches of the industrial west. The developed countries lie directly in the path of a great storm from the Third World."

— Dr. John Tanton, founder of the
Federation for American Immigration Reform

"[The] social qualities [of immigrants]— hard work, creative thinking, strong social bonds, and especially optimism in the future — are already saving the U.S. from decline. But as they do so, they will radically change the visual face of the U.S. And it is fear of this change that is often cloaked in images of the population bomb rather than expressed directly in cultural or ethnic terms....Like the West's own modernization, the rise of a world of color is bringing with it large-scale population growth. But the West, especially the U.S., can meet this challenge by finally conceding that it has to share power, wealth, and space with that new world."

— Franz Schurmann, professor of history and sociology,
University of California, Berkeley

"We can confidently predict that removing the excess fertility from a poor and overpopulated country will produce a rise in fertility. Accepting the

Frank Sharry is the executive director of the National Immigration Forum.

*"superfluous" emigrants is no way to help a poor country solve its popula-
tion problem! And what about us, the receiving nation? Will more millions
of immigrants put an end to our traffic jams? Increase the speed and safety
of our commuting?...Decrease the size of our ghettos? Decrease the crime that
comes with crowding? The answers are surely obvious....Our present population of a quarter of a billion is more than enough to exploit the resources with
which we have been blessed."*
    — Garret Hardin, Professor Emeritus of Human Ecology,
    University of California, Santa Barbara

*"Isn't it great that you can be moral and defend your advantages as an
affluent Westerner. I find it repugnant that anybody could have such utter
contempt for other human beings."*
    — Charles Keely, Herzberg Professor of International Migration,
    Georgetown University

In the United States, the collision of population and immigration issues
may be right around the corner. In preparation for the International
Conference on Population and Development (ICPD), to be held in
Cairo in September of 1994, major environmental organizations, family planning groups, foreign policy specialists, women's organizations, and
other interested parties are beginning to seriously consider how migration intersects with overpopulation, environmental strains, underdevelopment, and political stability.

These sectors are being courted by groups with environmentally
friendly names such as Population-Environment Balance, the Carrying
Capacity Network, and Negative Population Growth, all of which are
closely connected to the Federation for American Immigration Reform
(FAIR), the lead immigration restriction lobby in the United States.
These groups contend that, as long as the U.S. provides a "safety valve"
for overpopulated developing countries by admitting significant numbers
of refugees and immigrants, such nations will delay confronting their
population pressures. Moreover, they argue, admitting immigrants prevents the U.S. from achieving population stabilization, which undercuts
our ability to lead the world towards the same.

On the other side of the immigration debate, advocates for generous
refugee and immigration policies fear that it is only a matter of time
before some environmentalists and population advocates adopt the
rhetoric and positions of those who want to restrict immigration. Some
suggest that it is racist to blame low-income people of color for environmental problems, when the real problem is the wasteful consumption
practiced by those trying to close the doors. Sound bites are being

sharpened. Environmentalists have been accused of "brown-shirted scapegoating coated in green." Others might charge that, "The people who claim that admitting refugees and immigrants contributes to over-population and environmental problems usually own two houses, four cars, and produce more waste in a week than ten frugal immigrant fami-lies do in a month." Lines are being drawn, and the logic of confronta-tion is beginning to take hold. If it comes to pass, the battle will generate more heat than light.

It need not turn out this way. A polarized, reductive debate is not inevitable. There is ample room and opportunity for a broader under-standing, common ground, and even joint action.

I say this as one of the leaders of the pro-immigration movement in this country. Over the years, many of my colleagues and I — representing civil rights groups, ethnic associations, trade unions, religious organiza-tions, and grassroots service providers throughout the United States — have recognized the need to move beyond the rather simplistic pro/con debate about whether or not the U.S. should admit immigrants. We believe that the human rights of refugees must be protected, that close family members deserve the opportunity to be reunited, that newcomers contribute mightily to their new communities and country, and that the diversity that results from sensible immigration policies strengthens us as a nation. However, we do not believe in open borders, nor are we so nar-row in our thinking that we would argue the solution to political instabili-ty, underdevelopment, population growth, and environmental destruc-tion throughout the world is more migration to the United States.

What is needed is rational deliberation based on an examination of facts, figures, and findings; a deeper understanding of the root causes of population growth and international migration; and a commitment to forge a consensus to attack those causes. Let us try to light the road to Cairo with such a spirit.

THE FACTS ABOUT IMMIGRATION

Without a doubt, the nexus of population growth and international migration is attracting a great deal of attention. Migration has been iden-tified as one of the major issue areas for the Cairo Conference. In July 1993, the United Nations Population Fund released a report noting that migration might well be "the crisis of our age." As if to underline the con-cern, at the 1993 G-7 Summit in Tokyo, the leaders of the industrialized world suggested that the threat of uncontrolled migration may be poten-tially more destabilizing and threatening than the proliferation of nuclear weapons or the prospect of terrorism. In the U.S., polls show that 60-65 percent of the American people want a decrease in current immigration levels, and a number of politicians are staking their political futures on a "get-tough" approach to illegal immigration.

Let's address the key questions related to migration in an attempt to determine whether the highly visible and emotionally charged debate in Europe and North America is justified or exaggerated.

WHY ARE PEOPLE MIGRATING?

The root causes of migration are underdevelopment, political upheavals, oppression, population pressures, and environmental destruction. On a personal level, most people migrate by choice. They do so to provide for their families, to seek freedom and opportunity, to reunite with family members, and to give their children a brighter future. Others leave out of necessity. They leave to escape the knock on the door in the middle of the night, to flee the bombs and bullets of civil war, to get out from under the grinding boot of oppression and tyranny. Migration, particularly across national boundaries, is not generally for the tired and weak, but for the strong and courageous. Giving up the known for the unknown is an option chosen by risk-takers. Not surprisingly, most international migrants are young. Women make up nearly half of the migrant population, and trends indicate the percentage of women on the move will increase in the coming decade.

HOW MANY PEOPLE ARE ON THE MOVE ACROSS THE GLOBE AND WHERE DO THEY GO?

Most migration takes place within, rather than across, national borders: In fact, most migration is from rural to urban areas. According to the United Nations Population Fund (UNFPA), in 1950 over 80 percent of the developing world's population lived in rural areas. By the early in the next century, more than half of the world's people will live in cities.

Most international migrants stay within the less developed world: According to a 1992 World Bank estimate, there are 100 million international migrants of all kinds, which represents one out of 50 people on the globe. Of that total, approximately two thirds reside in less developed countries, with the rest settling or working in Western Europe, North America, and Australia. On an annual basis, the United States receives approximately 1.1. million newcomers: 700,000 legal immigrants, 125,000 refugees, and approximately 300,000 undocumented immigrants. Even when one adds the 450,000 workers, trainees, and their family members who are admitted for temporary employment each year, the U.S. permits roughly 1.5 percent of the world's international migrants to enter the country each year.

A note on the difference between those admitted with our consent and those who enter illegally: According to estimates made by the U.S.

Immigration and Naturalization Service (INS), 8 out of 11 newcomers enter the U.S. each year legally. Those residing here as undocumented immigrants represent only 1.25 percent of the U.S. population. In California, where the debate is especially heated, undocumented immigrants comprise four percent of the population.

Most people uprooted by war and persecution are displaced in their own countries; most refugees (those who cross international boundaries in search of safety) seek protection in developing countries: The United Nations High Commissioner for Refugees (UNHCR) estimates that 24 million people are internally displaced, and almost 20 million are counted as refugees. This means that one in every 130 people on earth has been forced to flee because of massive human rights violations, civil war, ethnic violence, or religious and political persecution. Of the total refugee population, approximately 90 percent live in developing countries. The United States resettles a total of 120,000 refugees a year (the refugees are screened and accepted from outside the U.S.) and currently receives approximately 100,000 applications for asylum each year (those who come to the U.S. and apply for refugee status once here). This means that while the U.S. has one of the world's most generous refugee resettlement programs, no more than one percent of the world's refugee population finds its way to the U.S. each year.

WHAT ARE THE IMPACTS OF MIGRATION FOR THE SENDING AND RECEIVING COUNTRIES?

Are newcomers taking jobs, depressing wages, draining budgets, and refusing to integrate? In the United States, an impressive body of knowledge and research indicates that newcomers make great contributions to the countries that receive them. This is because most are young, hard working, entrepreneurial, and reluctant to use government services. As a result, the overall impact on economic growth and government costs and revenues is positive. Furthermore, refugees and immigrants are integrating quickly into American life. One out of three immigrants marries outside of their ethnic group, and within 15 years after arrival, 9 out of 10 are using English on a regular basis.

For the sending countries, the impact is more mixed. International migrants send money home in staggering numbers. An estimated $67 billion in remittances — second only to oil in its value in international trade, and more than international development assistance ($46 billion) — keeps millions of families and numerous nations financially viable. Much of the money is used for family expenses that help create demand and jobs in sending countries. This "perking up" of spending does not have the spectacular results often associated with large-scale development projects, but since World War II has enabled two or three generations in many areas of the world to grow up healthier, better educated, and more able to contribute to their countries as adults. On the other hand, some remittances are squandered on useless imports. And the loss of skilled workers, technicians, and managers (the so-called "brain drain") hurts prospects for development.

WHAT ARE THE IMPLICATIONS OF THESE FACTS, FIGURES, AND FINDINGS?

It is clear that migration is expanding and accelerating; more people are on the move than ever before in the history of the world. Migration is a legitimate and important issue that must be understood and addressed by those who are concerned about the future. On the other hand, it is equally clear that most migrants — internal and international — remain in developing countries. Consequently, the argument that the most effective approach to tackling population growth and uncontrolled migration is to limit immigration into the United States and other wealthy industrialized countries is illogical at best, and mean-spirited at worst. If we could wave a magic wand and eliminate all international migration from the South to the North, the problems of population growth and uncontrolled migration would remain unaddressed, and perhaps be exacerbated.

This is not to suggest that the wealthy industrialized countries should ignore the problem of unauthorized migration. To the contrary, I believe that it is the right and duty of sovereign nations to regulate who enters their territory and uphold the rule of law. I also believe that we have an obligation to enforce our laws in a humane fashion consistent with democratic standards. For example, I support carefully conceived enforcement policies directed at deterring people at "points of entry" (border crossings and airports) as long as the enforcement officials are trained to respect human rights and are held accountable for the way they treat applicants for admission. On the other hand, I strongly disfavor enforcement policies which target undocumented immigrants already residing in-country. Attempts have been made to root people out of the workforce, schools, health clinics, and neighborhoods by introducing national identification cards, denying basic services, and carrying out raids. These measures do more to discriminate against legal residents who look or sound "foreign," terrorize undocumented immigrants, and create community tensions than they do to reduce illegal immigration.

The fact is that policies that treat international migration as a domestic enforcement issue have little or no impact on population growth and uncontrolled migration. Clearly, the key to reducing migration and population pressures is to confront their root causes. Grappling with causes rather than symptoms is neither easy nor politically popular. Nevertheless, it is where the possibility of progress lies.

The questions we should focus on, then, include: How can migration pressures be reduced by dealing with the "push" factors of poverty and political oppression? What is the relationship between population growth, environmental degradation, underdevelopment, political instability, civil strife, and migration? What policies and initiatives — such as trade, aid, debt relief, conflict prevention and resolution, voluntary family planning, comprehensive community development, and democratic institution-building — might be carefully integrated and targeted as a means to give people hope, security, and opportunity at home, thereby removing the need to migrate? Would such an approach also serve to reduce population and environmental pressures? Is it possible for those concerned with population, the environment, development, trade, and foreign policy to establish a common language, vision, and agenda?

SEARCHING FOR COMMON GROUND

I believe that the answers to these questions would enable an unprecedented range of groups to forge a consensus on the most pragmatic way to reduce both population and migration pressures. In particular, the upcoming Cairo Conference and discussions leading up to it provide a unique opportunity for sectors that do not know each other well to develop relationships, find common ground, and develop a concrete plan of action.

For example, I can envision a broad consensus which would support sustainable development, respect for human rights, regulated migration flows, and efforts to mitigate the root causes of population growth and uncontrolled migration. Common goals might include:

- Reduced population growth through humane and effective efforts;
- Development that arrests resource depletion;
- Increased attention and support to those traditionally marginalized because of gender, race, or ethnicity;
- Protection of the fundamental human rights of all migrants, refugees, and the internally displaced, regardless of status;
- Prevention of political violence that causes significant refugee movements;
- Respect for cultural pluralism, coupled with promotion of integration and citizenship rights;
- Continuation of regulated admissions which reunite close family members and attract small numbers of specially skilled workers;
- Enforcement of restrictions on unauthorized migration, without "police-state" tactics;
- Support for repatriation programs for refugees, return-of-talent programs for international migrants, and innovative programs to channel remittances for productive investment in development schemes; and

- International mechanisms to address issues of migration within a global context.

Is this asking too much? I think not. The purpose of conferences such as the International Conference on Population and Development is to break out of our parochial thinking and challenge ourselves to be responsible stewards of our future. Let us meet the challenge by thinking big, in order to forge a common vision which can produce lasting results.

Population and National Security

....

THOMAS F. HOMER-DIXON, JEFFREY H. BOUTWELL, AND GEORGE W. RATHJENS

Environmental Change and Violent Conflict

....

Within the next 50 years, the human population is likely to exceed nine billion, and global economic output may quintuple. Largely as a result of these two trends, scarcities of renewable resources may increase sharply. The total area of highly productive agricultural land will drop, as will the extent of forests and the number of species they sustain. Future generations will also experience the ongoing depletion and degradation of aquifers, rivers and other bodies of water, the decline of fisheries, further stratospheric ozone loss and, perhaps, significant climatic change.

As such environmental problems become more severe, they may precipitate civil or international strife. Some concerned scientists have warned of this prospect for several decades, but the debate has been constrained by lack of carefully compiled evidence. To address this shortfall of data, we assembled a team of 30 researchers to examine a set of specific cases. In studies commissioned by the University of Toronto and the American Academy of Arts and Sciences, these experts reported their initial findings.

The evidence that they gathered points to a disturbing conclusion: scarcities of renewable resources are already contributing to violent conflicts in many parts of the developing world. These conflicts may foreshadow a surge of similar violence in coming decades, particularly

Thomas F. Homer-Dixon, Jeffrey H. Boutwell, and George W. Rathjens are co-directors of the Project on Environmental Change and Acute Conflict. The Project is jointly sponsored by the American Academy of Arts and Sciences and the University of Toronto. This chapter is reprinted from Scientific American, *February 1993.*

in poor countries where shortages of water, forests and, especially, fertile land, coupled with rapidly expanding populations, already cause great hardship.

Before we discuss the findings, it is important to note that the environment is but one variable in a series of political, economic and social factors that can bring about turmoil. Indeed, some skeptics claim that scarcities of renewable resources are merely a minor variable that sometimes links existing political and economic factors to subsequent social conflict.

The evidence we have assembled supports a different view. Such scarcity can be an important force behind changes in the politics and economics governing resource use. It can cause powerful actors to strengthen, in their favor, an inequitable distribution of resources. In addition, ecosystem vulnerability often contributes significantly to shortages of renewable resources. This vulnerability is, in part, a physical given: the depth of upland soils in the tropics, for example, is not a function of human social institutions or behavior. And finally, in many parts of the world, environmental degradation seems to have passed a threshold of irreversibility. In these situations, even if enlightened social change removes the original political, economic and cultural causes of the degradation, it may continue to contribute to social disruption. In other words, once irreversible, environmental degradation becomes an independent variable.

Skeptics often use a different argument. They state that conflict arising from resource scarcity is not particularly interesting, because it has been common throughout human history. We maintain, though, that renewable-resource scarcities of the next 50 years will probably occur with a speed, complexity and magnitude unprecedented in history. Entire countries can now be deforested in a few decades, most of a region's topsoil can disappear in a generation, and acute ozone depletion may take place in as few as 20 years.

Unlike nonrenewable resources—including fossil fuels and iron ore—renewable resources are linked in highly complex, interdependent systems with many nonlinear and feedback relations. The overextraction of one resource can lead to multiple, unanticipated environmental problems and sudden scarcities when the system passes critical thresholds.

Our research suggests that the social and political turbulence set in motion by changing environmental conditions will not follow the commonly perceived pattern of scarcity conflicts. There are many examples in the past of one group or nation trying to seize the resources of another. For instance, during World War II, Japan sought to secure oil, minerals and other resources in China and Southeast Asia.

Currently, however, many threatened renewable resources are held in common—including the atmosphere and the oceans—which makes them unlikely to be the object of straightforward clashes. In addition, we have come to understand that scarcities of renewable resources often produce insidious and cumulative social effects, such as population displacement and economic disruption. These events can, in turn, lead to clashes between ethnic groups as well as to civil strife and insurgency. Although such conflicts may not be as conspicuous or dramatic as wars over scarce resources, they may have serious repercussions for the security interests of the developed and the developing worlds.

Human actions bring about scarcities of renewable resources in three principal ways. First, people can reduce the quantity or degrade the quality of these resources faster than they are renewed. This phenomenon is often referred to as the consumption of the resource's "capital": the capital generates "income" that can be tapped for human consumption. A sustainable economy can therefore be defined as one that leaves the capital intact and undamaged so that future generations can enjoy undiminished income. Thus, if topsoil creation in a region of farmland is 0.25 millimeter per year, then average soil loss should not exceed that amount.

The second source of scarcity is population growth. Over time, for instance, a given flow of water might have to be divided among a greater number of people. The final cause is change in the distribution of a resource within a society. Such a shift can concentrate supply in the hands of a few, subjecting the rest to extreme scarcity.

These three origins of scarcity can operate singly or in combination. In some cases, population growth by itself will set in motion social stress. Bangladesh, for example, does not suffer from debilitating soil degradation or from the erosion of agricultural land: the annual flooding of the Ganges and Brahmaputra rivers deposits a layer of silt that helps to maintain the fertility of the country's vast floodplains.

But the United Nations predicts that Bangladesh's current population of 120 million will reach 235 million by the year 2025. At about 0.08 hectare per capita, cropland is already desperately scarce. Population density is 785 people per square kilometer (in comparison, population density in the adjacent Indian state of Assam is 284 people per square kilometer). Because all the country's good agricultural land has been exploited, population growth will cut in half the amount of cropland available per capita by 2025. Flooding and inadequate national and community institutions for water control exacerbate the lack of land and the brutal poverty and turmoil it engenders.

Over the past 40 years, millions of people have migrated form Bangladesh to neighboring areas of India, where the standard of living is often better. Detailed data on the movements are few: The Bangladeshi government is reluctant to admit there is significant migration because the issue has become a major source of friction with India. Nevertheless, one of our researchers, Sanjoy Hazarika, an investigative journalist and reporter at the *New York Times* in New Delhi, pieced together demographic information and experts' estimates. He concludes that Bangladeshi migrants and their descendants have expanded the population of neighboring areas of India by 15 million. (Only one to two million of those people can be attributed to migrations during the 1971 war between India and Pakistan that resulted in the creation of Bangladesh.)

This enormous flux has produced pervasive social changes in the receiving Indian states. Conflict has been triggered by altered land distribution as well as by shifts in the balance of political and economic power between religious and ethnic groups. For instance, members of the Lalung tribe in Assam have long resented Bengali Muslim migrants: They accuse them of stealing the area's richest farmland. In early 1983, during a bitterly contested election for federal offices in the state, violence finally erupted. In the village of Nellie, Lalung tribespeople massacred nearly 1,700 Bengalis in one five-hour rampage.

In the state of Tripura the original Buddhist and Christian inhabitants now make up less than 30 percent of the population. The remaining percentage consists of Hindu migrants from either East Pakistan or Bangladesh. This shift in the ethnic balance precipitated a violent insurgency between 1980 and 1988 that was called off only after the government agreed to return land to dispossessed Tripuris and to stop the influx of Bangladeshis. As the migration has continued, however, this agreement is in jeopardy.

Population movements in this part of South Asia are, of course, hardly new. During the colonial period, the British imported Hindus from Calcutta to administer Assam, and Bengali was made the official language. As a result, the Assamese are particularly sensitive to the loss of political and cultural control in the state. And Indian politicians have often encouraged immigration in order to garner votes. Yet today changes in population density in Bangladesh are clearly contributing to the exodus. Although the contextual factors of religion and politics are important, they do not obscure the fact that a dearth of land in Bangladesh has been a force behind conflict.

In other parts of the world the three sources of scarcity interact to produce discord. Population growth and reductions in the quality and quantity of renewable resources can lead to large-scale development projects that can alter access to resources. Such a shift may lead to decreased supplies for poorer groups whose claims are violently opposed by powerful elites. A dispute that began in 1989 between Mauritanians and Senegalese in the Senegal River valley, which demarcated the common border between these countries, provides an example of such causality.

Senegal has fairly abundant agricultural land, but much of it suffers from severe wind erosion, loss of nutrients, salinization because of overirrigaton and soil compaction caused by the intensification of agriculture. The country has an overall population density of 38 people per square kilometer and a population growth rate of 2.7 percent; in 25 years the population may double. In contrast, except for the Senegal River valley along its southern border and a few oases, Mauritania is for the most part arid desert and semiarid grassland. Its population density is very low, about two people per square kilometer, and the growth rate is 2.8 percent a year. The U.N. Food and Agriculture Organization has included both Mauritania and Senegal in its list of countries whose croplands cannot support current or projected populations without a large increase in agricultural inputs, such as fertilizer and irrigation.

Normally, the broad floodplains fringing the Senegal River support productive farming, herding and fishing based on the river's annual floods. During the 1970s, however, the prospect of chronic food shortages and a serious drought encouraged the region's governments to seek international financing for the Manantali Dam on the Bafing River tributary in Mali and for the Diama salt-intrusion barrage near the mouth of the Senegal River between Senegal and Mauritania. The dams were designed to regulate the river's flow for hydropower, to expand irrigated agriculture and to raise water levels in the dry season, permitting year-round barge transport from the Atlantic Ocean to land-locked Mali, which lies to the east of Senegal and Mauritania.

But the plan had unfortunate and unforeseen consequences. As anthropologist Michael M. Horowitz of the State University of New York at Binghamton has shown, anticipation of the new dams raised land values along the river in areas where high-intensity agriculture was to become feasible. The elite in Mauritania, which consists primarily of white Moors, then rewrote legislation governing land ownership, effectively abrogating the rights of black Africans to continue farming, herding and fishing along the Mauritanian riverbank.

There has been a long history of racism by white Moors in Mauritania toward their non-Arab, black compatriots. In the spring of 1989 the killing of Senegalese farmers by Mauritanians in the river basin triggered explosions of ethnic violence in the two countries. In Senegal almost all of the 17,000 shops owned by Moors were destroyed, and their owners were deported to Mauritania. In both countries several hundred people were killed, and the two nations nearly came to war. The Mauritanian regime used this occasion to activate the new land legislation, declaring the black Mauritanians who lived alongside the river to be "Senegalese," thereby stripping them of their citizenship; their property was seized. Some 70,000 of the black Mauritanians were forcibly expelled to Senegal, from where some launched raids to retrieve expropriated cattle. Diplomatic relations between the two countries have now been restored, but neither has agreed to allow the expelled population to return or to compensate them for their losses.

We see a somewhat different causal process in many parts of the world: unequal access to resources combines with population growth to produce environmental damage. This phenomenon can contribute to economic deprivation that spurs insurgency and rebellion. In the Philippines, Spanish and American colonial policies left beind a grossly inequitable distribution of land. Since the 1960s, the introduction of green revolution technologies has permitted a dramatic increase in lowland production of grain for domestic consumption and of cash crops that has helped pay the country's massive external debt.

This modernization has raised demand for agricultural labor. Unfortunately, though, the gain has been overwhelmed by a population growth rate of 2.5 to 3.0 percent. Combined with the maldistribution of good cropland and an economic crisis in the first half of the 1980s, this growth produced a surge in agricultural unemployment.

With insufficient rural or urban industrialization to absorb excess labor, there has been unrelenting downward pressure on wages. Economically desperate, millions of poor agricultural laborers and landless peasants have migrated to shantytowns in already overburdened cities, such as Manila; millions of others have moved to the least productive—and often most ecologically vulnerable—territories, such as steep hillsides.

In these uplands, settlers use fire to clear forested or previously logged land. They bring with them little ability to protect the fragile ecosystem. Their small-scale logging, charcoal production and slash-and-burn farming often cause erosion, landslides and changes in hydrologic patterns. This behavior has initiated a cycle of falling food production, the clearing of new plots and further land degradation. Even marginally fertile land is becoming hard to find in many places, and economic conditions are critical for peasants.

The country has suffered from serious internal strife for many decades. But two researchers, Celso R. Roque, the former undersecretary of the environment of the Philippines, and his colleague Maria I. Garcia, conclude that resource scarcity appears to be an increasingly powerful force behind the current communist-led insurgency. The upland struggle—including guerrilla attacks and assaults on military stations—is motivated by the economic deprivation of the landless agricultural laborers and poor farmers displaced into the hills, areas that are largely beyond the control of the central government. During the 1970s and 1980s, the New People's Army and the National Democratic Front found upland peasants receptive to revolutionary ideology, especially where coercive landlords and local governments left them little choice but to rebel or starve. The revolutionaries have built on indigenous beliefs and social structures to help the peasants focus their discontent.

Causal processes similar to those in the Philippines can be seen in many other regions around the planet, including the Himalayas, the Sahel, Indonesia, Brazil and Costa Rica. Population growth and unequal access to good land force huge numbers of people into cities or onto marginal lands. In the latter case, they cause environmental damage and become chronically poor. Eventually these people may be the source of persistent upheaval, or they may migrate yet again, stimulating ethnic conflicts or urban unrest elsewhere.

The short but devastating "Soccer War" in 1969 between El Salvador and Honduras involved just such a combination of factors. As William H. Durham of Stanford University has shown, changes in agriculture and land distribution beginning in the mid-19th century concentrated poor farmers in El Salvador's uplands. Although these peasants developed some understanding of land conservation, their growing numbers on very steep hillsides caused deforestation and erosion. A natural population growth rate of 3.5 percent further reduced land availability, and as a result many people moved to neighboring Honduras. Their eventual expulsion from Honduras precipitated a war in which several thousand people were killed in a few days. Durham notes that the competition for land in El Salvador leading to this conflict was not addressed in the war's aftermath and that it powerfully contributed to the country's subsequent, decade-long civil war.

In South Africa the white regime's past apartheid policies concentrated millions of blacks in the country's least productive and most ecologically sensitive territories. High natural birth rates exacerbated population densities. In 1980 rural areas of the Ciskei homeland supported 82 persons per square kilometer, whereas the surrounding Cape Province had a rural density of two. Homeland residents had, and have, little capital and few skills to manage resources. They remain the victims of corrupt and abusive local governments.

Sustainable development in such a situation is impossible. Wide areas have been completely stripped of trees for fuelwood, grazed down to bare dirt and eroded of topsoil. A 1980 report concluded that nearly 50 percent of Ciskei's land was moderately or severely eroded; close to 40 percent of its pasture was overgrazed. This loss of resources, combined with the lack of alternative employment and the social trauma caused by apartheid, has created a subsistence crisis in the homelands. Thousands of people have migrated to South African cities. The result is the rapid growth of squatter settlements and illegal townships that are rife with discord and that threaten the country's move toward democratic stability.

Dwindling natural resources can weaken the administrative capacity and authority of government, which may create opportunities for violent challenges to the state by political and military opponents. By contributing to rural poverty and rural-urban migration, scarcity of renewable resources expands the number of people needing assistance from the government. In response to growing city populations, states often introduce subsidies that distort prices and cause misallocations of capital, hindering economic productivity.

Simultaneously, the loss of renewable resources can reduce the production of wealth, thereby constraining tax revenues. For some countries, this widening gap between demands on the state and its capabilities may aggravate popular grievances, erode the state's legitimacy and escalate competition between elite factions as they struggle to protect their prerogatives.

Logging for export markets, as in Southeast Asia and West Africa, produces short-term economic gain for parts of the elite and may alleviate external debt. But it also jeopardizes long-term productivity. Forest removal decreases the land's ability to retain water during rainy periods. Flash floods then damage roads, bridges, irrigation systems and other valuable infrastructure. Erosion of hillsides silts up rivers, reducing their navigability and their capacity to generate hydroelectric power. Deforestation can also hinder crop production by altering regional hydrologic cycles and by plugging reservoirs and irrigation channels with silt.

In looking at China, Vaclav Smil of the University of Manitoba has estimated the combined effect of environmental problems on productivity. The main economic burdens he identifies are reduced crop yields caused by water, soil and air pollution; higher human morbidity resulting from air pollution; farmland loss because of construction and erosion; nutrient loss and flooding caused by erosion and deforestation; and timber loss arising from poor harvesting practices. Smil calculates the current annual cost to be at least 15 percent of China's gross domestic product; he is convinced the toll will rise steeply in the coming decades. Smil also estimates that tens of millions of Chinese will try to leave the country's impoverished interior and northern regions—where water and fuelwood are desperately scarce and the land often badly damaged—for the booming coastal cities. He anticipates bitter disputes among these regions over water sharing and migration. Taken together, these economic and political stresses may greatly weaken the Chinese state.

Water shortages in the Middle East will become worse in the future and may also contribute to political discord. Although figures vary, Miriam R. Lowi of Princeton University estimates that the average amount of renewable fresh water available annually to Israel is about 1,950 million cubic meters (mcm). Sixty percent comes from groundwater, the rest from river flow, floodwater and wastewater recycling. Current Israeli demand—including that of settlements in the occupied territories and the Golan Heights—is about 2,200 mcm. The annual deficit of about 200 mcm is met by overpumping aquifers.

As a result, the water table in some parts of Israel and the West Bank has been dropping significantly. This depletion can cause the salinization of wells and the infiltration of seawater from the Mediterranean. At the same time, Israel's population is expected to increase from the present 4.6 million to 6.5 million people in the year 2020, an estimate that does not include immigration from the former Soviet Union. Based on this projected expansion, the country's water demand could exceed 2,600 mcm by 2020.

Two of the three main aquifers on which Israel depends lie for the most part under the West Bank, although their waters drain into Israel. Thus, nearly 40 percent of the groundwater Israel uses originates in occupied territory. To protect this important source, the Israeli government has strictly limited water use on the West Bank. Of the 650 mcm of all forms of water annually available there, Arabs are allowed to use only 125 mcm. Israel restricts the number of wells Arabs can drill in the territory, the amount of water Arabs are allowed to pump and the times at which they can draw irrigation water.

The differential in water access on the West Bank is marked: on a per capita basis, Jewish settlers consume about four times as much water as Arabs. Arabs are not permitted to drill new wells for agricultural purposes, although Mekorot (the Israeli water company) has drilled more than 30 for settlers. Arab agriculture in the region has suffered because some Arab wells have become saline as a result of deeper Israeli wells drilled nearby. The Israeli water policy, combined with the confiscation of agricultural land for settlers as well as other Israeli restrictions on Palestinian agriculture, has encouraged many West Bank Arabs to abandon farming. Those who have done so have become either unemployed or day laborers within Israel.

The entire Middle East faces increasingly grave and tangled problems of water scarcity, and many experts believe these will affect the region's stability. Concerns over water access contributed to tensions preceding the 1967 Arab-Israeli War; the war gave Israel control over most of the Jordan Basin's water resources. The current Middle East peace talks include multilateral meetings on water rights, motivated by concerns about impending scarcities.

Although "water wars" are possible in the future, they seem unlikely given the preponderance of Israeli military power. More probably, in the context of historical ethnic and political disputes, water shortages will aggravate tensions and unrest within societies in the Jordan River basin.

In recent U.S. congressional testimony, Thomas Naff of the University of Pennsylvania noted that "rather than warfare among riparians in the immediate future...what is more likely to ensue from water-related crises in this decade is internal civil disorder, changes in regimes, political radicalization and instability."

Scarcities of renewable resources clearly can contribute to conflict, and the frequency of such unrest will probably grow in the future. Yet some analysts maintain that scarcities are not important in and of themselves. What is important, they contend, is whether people are harmed by them. Human suffering might be avoided if political and economic systems provide the incentives and wherewithal that enable people to alleviate the harmful effects of environmental problems.

Our research has not produced firm evidence for or against this argument. We need to know more about the variables that affect the supply of human ingenuity in response to environmental change. Technical ingenuity is needed for the development of, for example, new agricultural and forestry technologies that compensate for environmental deterioration. Social ingenuity is needed for the creation of institutions that buffer people from the effects of degradation and provide the right incentives for technological innovation.

The role of social ingenuity as a precursor to technical ingenuity is often overlooked. An intricate and stable system of markets, legal regimes, financial agencies and educational and research institutions is a prerequisite for the development and distribution of many technologies—including new grains adapted to dry climates and eroded soils, alternative cooking technologies that compensate for the loss of firewood and water-conservation technologies. Not only are poor countries ill endowed with these social resources, but their ability to create and maintain them will be weakened by the very environmental woes such nations hope to address.

The evidence we have presented here suggests there are significant causal links between scarcities of renewable resources and violence. To prevent such turmoil, nations should put greater emphasis on reducing such scarcities. This means that rich and poor countries alike must cooperate to restrain population growth, to implement a more equitable distribution of wealth within and among their societies, and to provide for sustainable development.

EMMA ROTHSCHILD

Population and Common Security

······················

C ommon security is a way of living in society. The idea that people
form societies in order to ensure their security is as old as the
political philosophies of Aristotle and Kautilya. "The state or
republic," wrote the great German philosopher Leibniz in 1705,
"is a great society of which the object is common security."[1]

In the early 1980s, at one of the most tense periods of the Cold War,
the Palme Commission on Disarmament and Security Issues applied
these old ideas of security to international relations, in particular to
the prevention of nuclear war. In its report, *Common Security*, the
Commission observed that because no country could win a nuclear
war, and since nuclear coercion would risk catastrophe, "States can no
longer seek security at each other's expense; it can be attained only
through cooperative undertakings."[2]

The Palme Commission also pointed towards two more comprehensive conceptions of common security. The first is that security must be
thought of in terms of economic and political, as well as military objectives; that military security is a means to an end, while the economic
security of individuals, or the social security of citizens "to chart futures
in a manner of their own choosing," or the political security which follows when "the international system [is] capable of peaceful and orderly change," are ends in themselves.

---

*Emma Rothschild is director of the Centre for History and Economics, King's
College, Cambridge, England, and Chair of the Research Council of the Common
Security Forum.*

The second conception is that lasting security must be founded on an effective system of international order. As Cyrus Vance wrote in his introduction to the U.S. edition of Common Security, "the problems of nuclear and conventional arms are reflections of weaknesses in the international system. It is a weak system because it lacks a significant structure of laws and norms of behavior which are accepted and observed by all states."

In the post-Cold War world, it is these more extensive ideas of common security which have assumed particular political importance. The threat of worldwide nuclear war is less in the 1990s than it has been since the mid-1950s. But it is not clear that societies or groups or individuals actually feel — or have reason to feel — more secure. There are new threats to individual and national security, or at least old threats which have become more frightening. Even the most immediate threats to individual security, such as unemployment, drugs, and infectious diseases, have international as well as domestic causes. The agenda of common security — a recognition that security is economic, social, and environmental as well as military, and that international cooperation is a necessary condition for security — is at the heart of foreign policy in the 1990s.

POPULATION GROWTH AND INTERNATIONAL SOCIETY

The subject to be addressed at the Cairo Conference — population and development — has recently been seen as a prime illustration of the new, non-military threats to security. The suggestion is that periods of sustained increase in population densities, and continuing change in the composition of population by age, region, and culture, are likely to be periods of social insecurity. People will be more violent, it is suggested, and more favorable to sudden political change; they may also be more likely to become migrants, both within and between countries.

This argument is hardly new in the late 20th century. John Maynard Keynes speculated in 1920 that "the disruptive powers of excessive national fecundity" may have helped to bring about the Russian Revolution. "That vast upheaval of Society," he wrote, "may owe more to the deep influences of expanding numbers than to Lenin or Nicholas."[3] What is novel, now, is that the consequences of population growth are expected to be international; to influence the security not only of groups within a society but of other, distant societies as well. A world in which population is growing very fast, it is suggested, will be one in which many people are poor, deprived, and living in conditions of social disruption. This is bad in itself. It is also bad because such a world, or such an international society, is likely to be insecure. The problem with this general account is that it takes as a given the very entity that is in question: the "international society."

There is oddly little historical evidence of the ways in which societies change in periods of very rapid population growth. The great historical cause of international migration, for example, is not population growth but war; as Malthus wrote in 1798, "few persons will leave their families, connections, friends, and native land" without very strong reasons.[4] Political tension, too, can result from the perception that different groups within a society are growing (or contracting) at different rates. All nations are shared by different groups, and an increase in the relative population of one group can be seen as a threat to the security — or the "territory" — of other groups. Such concerns were at the heart of the eugenicists' efforts in the 1930s to improve the "quality of the stock." The rhetoric of eugenic improvement may even be returning, in the 1990s, to the domestic politics of several countries.

The statement on demographic dynamics produced at the 1992 Earth Summit repeatedly uses the phrase "freedom, dignity and personally held values."[5] The official prose, here, points to the difficulty of interpreting population growth as a threat to security. For example, the individual security of a young Chinese woman might well be threatened by the social and environmental disruption associated with increases in the population of her "own" or contiguous groups. But it would be threatened in a more immediate respect by coercive policies to prevent such increases: the physical coercion of enforced contraception and abortion; or the moral coercion of enforced changes in "personally held values."

The substantial consensus in current population studies is that improvements in the lives of poor women — in their education, their social dignity, the survival of their children, and in their opportunities for employment — are the most important factors in preventing the undesirable consequences of rapid population growth. But even these improvements can be a source of conflict. Present institutions, in which women's opportunities are restricted, may reflect "personally held" values. Fertility decisions are not taken by unitary "households," but are the outcome of conflicting or convergent interests within families. The interests of husbands and wives, or of sisters and brothers, do not dissolve into collective preferences. Even within individual households, the "international community" is made up of different, dissonant voices.

THE MILITARY METAPHOR

The perception of population growth as a threat to international security has led, in the period since the end of the Cold War, to a renewed interest in quasi-military policies to defeat the threat. One metaphor is of defense against a turbulent outside world. The countries of the secure, low-fertility "core" should simply defend their borders, it is suggested, against the insecure, high-fertility "periphery." The problem with such a strategy is that the Maginot Lines of the 1990s would turn the societies inside the fortifications into unrecognizable fortress cities. The defense against migration is hardly beyond the technological capabilities of most developed countries. The defense against threats to human health presents more serious difficulties. The defense against ideas — including the idea of one's country as part of an international society — would itself be a threat to the foundations of economic and social development.

A second military metaphor is of offense against the supposed sources of turbulence. The simpler notions of direct intervention of the 1960s (the rapid deployment of civilian divisions with contraceptive devices) have been discredited, in part by evidence of the extent to which the contraceptives deployed remained unused. Interventions which took the form of exerting pressure on governments to deploy their own coercive population policies have also had limited policy success. The country that has been most violently coercive in its population policies, China, has done so with little influence (or help) from outside).

The coercive population policies of the 1990s will be those that reduce the extent of individual choice or consent in contraceptive decisionmaking. People will make fertility choices once every several years (or have choices made for them), as with the NORPLANT contraceptive implant. But coercion is not a sufficient condition for reducing fertility, as the experience of Indira Gandhi's forced sterilization policies suggests. Nor is it necessary; the Indian state of Kerala has achieved levels of fertility as low as in China, with lower per capita income, with no coercive policies at all.[6]

There is a third, more benign military metaphor, of massive technical and scientific mobilization to slow population growth. But the precedent of the space program, or of the Manhattan Project, which led to the development of the atomic bomb, is less compelling in relation to population problems than the precedent of the "War on Cancer" of the 1970s. The cumulative effect of research supported under that initiative is of great importance. But it did not amount to a war, or victory. To understand the social foundations of technical change requires "mobilization," no doubt, but of a different and more open sort than the mobilization for military victory.

The final and most repugnant military metaphor is of triage. The self-perception of rich societies as the equivalent of battlefield surgeons, choosing who should live, who should be cared for, and who should be left to die, is being evoked once more in the 1990s. It is argued that rich "donor" countries should withhold foreign aid, including humanitarian aid and aid to reduce child mortality, in an attempt to interrupt the slow process of demographic transition. On the one hand, donors would induce the governments of poor "recipient" countries to adopt coercive policies to reduce fertility; on the other hand, they would contribute directly to reducing population growth, not by reducing the rate of births, but by increasing the rate of deaths.

It is demeaning, after 200 years of the systematic study of populations and their transformation, to marshal responses to arguments of this sort. Malthus himself observed what he called "the dependence of the births on the deaths," and he quoted a Swiss study of 1766 which asked, "Whence comes it, that the country where children escape the best from the dangers of infancy...should be precisely that in which the fecundity is the smallest?"[7] Declines in death rates precede declines in birth rates, as the experience of both Kerala and China shows. Increases in death rates, by contrast, have often preceded increases in birth rates. A recent study of the Occupied Territories in the West Bank and Gaza describes "a rising birth rate since the outbreak of Intifada."[8] The outcome of "battlefield non-intervention" would be a world that was even poorer, more insecure, and more populous.

COMMON SECURITY

The best prospect for the coming half century of transition lies in preventing insecurity, and not in trying to defend against it. The environmental, social and political conflicts which may follow from rapid population growth are, in at least some respects, of common concern to the different constituents of the international community. They are more likely to be prevented by cooperation (or by different communities acting together) than by persuasion (or by one community acting upon another).

The difficulty, in the 1990s, is to invent the procedures for such cooperation. The conventional idea of cooperation to insure international security is of meetings of senior diplomatic representatives, in public or secret sessions — something like the Versailles Peace Treaty or the Israeli-PLO talks. Cooperation to increase social and environmental security, and to reduce population growth rates, will look very different. The eventual decision-makers, after all, are not "senior," or "representatives"; they are the many hundreds of millions of individuals and households on whose choices the course of the transition will depend.

Populations change very slowly, and the demographic history of the next half century is in this sense already visible. The "momentum" of population growth is such that projections of growth for the coming two or three decades are the subject of considerable consensus. There are two distinct, related objectives for cooperative policies. The first is to reduce the insecurity that may follow from the population growth which is now predicted. The second is to influence population growth in subsequent periods. The first objective is an end in itself; it is to increase the quality of people's lives. But it is also likely to be the best instrument for achieving the second objective.

Policies for social and individual security — within countries, within communities, and even within families — are essential to the effort to reduce international insecurity. Some of those policies are beyond the influence of the international community. Others, including the expenditures of international organizations and efforts to promote education and reduce discrimination, are already the subject of international discussion. The 1995 U.N. World Summit for Social Development, which will address education, discrimination, and unemployment, may have a more lasting effect on population growth than the deliberations of the U.N. Conference on Population and Development.

The study of population has always been a subject that makes people uncomfortable. One fundamental reason for this is that it treats people as numbers, or as a "stock," and not as individuals. Both population policies and the scientific study of population were associated for much of the later 19th and the earlier 20th centuries with eugenicist and racist ideas. But even in their most enlightened forms, they have tended, as perhaps they must, to disregard the subtlety and poignancy of individual choices. The great merit of the new population policies, centered on rights and security, is that they see people as individuals and not as numbers. This is a necessary, if not sufficient, condition for an international population policy that is truly the policy of an international society, or a policy for common security.

Endnotes

....

Laurie Ann Mazur, "Beyond the Numbers: An Introduction and Overview"

1   Population Reference Bureau, *1993 World Population Data Sheet* (Washington, D.C.: 1993).

2   United Nations Population Fund, *Population and the Environment: The Challenges Ahead* (New York: United Nations Population Fund, 1991) p. 14.

3   William Murdoch, *The Poverty of Nations: The Political Economy of Hunger and Population* (Baltimore: The Johns Hopkins University Press, 1980) pp. 284-289.

4   Robert Repetto, "Population, Resource Pressures, and Poverty," in *The Global Possible: Resources, Development, and the New Century*, Robert Repetto, ed. (New Haven: Yale University Press, 1985).

5   United Nations Population Fund, 1991, *op. cit.* p. 14.

6   R. Paul Shaw, "The Impact of Population Growth on the Environment: The Debate Heats Up," *Environmental Impact Assessment Review*, 12 (1/2) March-June 1992, p.17.

7   Donella H. Meadows, Dennis L. Meadows, Jorgen Randers, *Beyond the Limits: Confronting Global Collapse, Envisioning a Sustainable Future* (Vermont: Chelsea Green Publishing Co., 1992) p. 75.

8   The Pew Global Stewardship Initiative Program Paper, October 1993.

9   Lester Thurow, *Technology Review* (August/September 1986).

10   Julian Simon, *The Ultimate Resource* (Princeton: Princeton University Press, 1981).

11   Lester R. Brown, "Facing Food Insecurity," in *State of the World 1994* (New York: W.W. Norton, 1994) p.177.

12   Paul Harrison, "Sex and the Single Planet," *Amicus Journal*, Winter 1994, p. 23.

13   Harrison, ibid.

14   Environmentalists, of course, are not a monolith. Many grassroots and environmental justice groups were founded out of concern for human health, rather than from concern about nature as distinct from human communities.

15   Nathan Keyfitz, "The Growing Human Population," *Scientific American*, September 1989, p. 122.

16   United Nations Population Fund 1991, op. cit., p.20.

17   Quoted in ibid., p. 21.

18   National Research Council, *Population Growth and Economic Development: Policy Questions* (Washington, D.C.: National Academy Press, 1986) p. 90.

19   James P. Grant, *The State of the World's Children 1994* (Oxford: Oxford University Press, 1994) p. 29.

20   Susan George and Nigel Paige, *Food for Beginners* (London: Writers and Readers Publishing Cooperative Society Ltd., 1982) p. 143.

21   The Development Group for Alternative Policies, "The World Bank and Hunger: Facing the Facts," fact sheet, November, 1993.

22   Quoted by Douglas Hellinger, Managing Director of the Development Group for Alternative Policies, in testimony before the Senate Committee on Foreign Relations, Subcommittee on International Economic Policy, Trade, Oceans and the Environment, May 27, 1993.

23   Jodi Jacobson, *Gender Bias: Roadblock to Sustainable Development* (Washington, D.C.: Worldwatch Institute, 1992) p. 9.

24   Cited in ibid., p.13.

25   Ibid., p.7.

26   Ibid., p.13.

27   Peter J. Donaldson, *Nature Against Us: The United States and the World Population Crisis*, 1960-1980 (Chapel Hill, NC: University of North Carolina Press, 1990).

28   Harrison, op. cit., p. 26.

29   William K. Stevens, "Poor Lands' Success In Cutting Birth Rate Upsets Old Theories," *The New York Times*, January 2, 1994, p. A1.

30   Population Reference Bureau, "Bangladesh: The Matlab Maternal and Child Health/Family Planning Project," in *Family Planning Programs: Diverse Solutions for a Global Challenge* (Washington, D.C.: Population Reference Bureau, 1993).

31   Ibid.

32   John Bongaarts, W. Parker Mauldin, and James Phillips, "The Demographic Impact of Family Planning Programs," *Studies in Family Planning*, Vol. 21, No. 6, Nov. Dec. 1990, pp. 302-303.

33   Personal communication, March 4, 1994.

34   "Population Council Researcher Urges Three Policy Options To Reduce Birth Rates in Developing Countries," Press release issued by the Population Council, February 11, 1994.

35   Margaret Catley-Carlson, "Explosions, Eclipses and Escapes: Charting a Course on Global Population Issues," The 1993 Paul Hoffman Lecture, United Nations Development Program, New York, June 7, 1993.

36   Grant, op. cit.

37   Margaret Catley-Carlson, op. cit.

38   Ibid.

39   Testimony of Judith Bruce before the Senate Foreign Relations Committee, February 22, 1994.

40   Catley-Carlson, op. cit.

41   Timothy E. Wirth, address to the Council on Foreign Relations, New York, October 21, 1993.

42   Grant, op. cit., p.11.

43   Lester R. Brown, Hal Kane, Ed Ayres, *Vital Signs 1993* (New York: W.W. Norton Co., 1993) p. 80.

44   Grant, op. cit., p.54.

45   Al Gore, *Earth in the Balance* (New York: Plume Books, 1993) pp.297-307.

46   Pew Global Stewardship Initiative Program Paper, op. cit., p.10.

SECTION I: POPULATION, CONSUMPTION, DEVELOPMENT AND THE ENVIRONMENT
Mark Sagoff, "Population, Nature and the Environment"

REFERENCES:

L. H. Bailey, *The Holy Earth* (Scribner's Sons, 1915)

Lester Brown, "Vital Signs 1993: The Trends That Are Shaping Our Future," *The Washington Post* (July 18, 1993)

Paul Ehrlich, *The Population Bomb* (Ballantine, 1968); Barry Commoner, *Making Peace with the Planet* (Pantheon, 1990)

Paul Ehrlich and Anne Ehrlich, *The Population Explosion* (Simon and Schuster, 1990)

J. P. Holdren and Paul Ehrlich, "Human Population and the Global Environment," *American Scientist*, vol. 62 (1974)

John P. Holdren, "Population and the Energy Problem," *Population and Environment*, vol. 12, no. 3 (1991)

Daniel Janzen, letter to the editor, *Science* (June 5, 1987)

Jessica Tuchman Mathews, "Redefining Security," *Foreign Affairs* (Spring 1989)

Noel Perrin, "Forever Virgin: The American View of America," in *On Nature: Nature, Landscape, and Natural History*, Daniel Halpern, ed. (North Point Press, 1987)

R. Paul Shaw, "The Impact of Population Growth on Environment: The Debate Heats Up," *Environmental Impact Assessment Review*," vol. 12 (1992)

United Nations Fund for Population Activities, *Population and the Environment* (1992)

John E. Young, *Mining the Earth*, Worldwatch Paper 109 (July 1992)

Alan Thein Durning, "The Conundrum of Consumption"

1   Sidney Quarrier, geologist, Connecticut Geological & Natural History Survey, Hartford, Conn., private communication, February 25, 1992.

2   Ibid.

3   Ibid.

4   Ibid.

5   Lebow in *Journal of Retailing*, quoted in Vance Packard, *The Waste Makers* (New York: David Mckay, 1960).

6   Billionaires from Jennifer Reese, "The Billionaires: More Than Ever in 1991," *Fortune*, September 9, 1991; millionaires estimated from Kevin R. Phillips, "Reagan's America: A Capital Offense," *New York Times Magazine*, June 18, 1990; homelessness from U.N. Centre for Human Settlements, New York, private communication, November 1, 1989; luxury goods from "The Lapse of Luxury," *Economist*, January 5, 1991; gross national product from *United Nations Development Programme, Human Development Report 1991* (New York: Oxford University Press, 1991); member countries in United Nations from U.N. Information

Center, Washington, D.C., private communication, January 14,1992; world average income from 1987, in 1987 U.S. dollars adjusted for international variations in purchasing power, from Ronald V.A. Sprout and James H. Weaver, "International Distribution of Income: 1960-1987," Working Paper No. 159, Department of Economics, American University, Washington, D.C., May 1991; U.S. 1987 poverty line for an individual from U.S. Bureau of the Census, Statistical Abstract of the United States: 1990 (Washington, D.C.: U.S. Government Printing Office, 1990).

7   Estimated annual earnings per family member, in 1988 U.S. dollars of gross domestic product (GDP) per capita adjusted for international variations in purchasing power, and share of world income from Ronald V.A. Sprout and James H. Weaver, "1988 International Distribution of Income" (unpublished data) provided by Ronald V.A. Sprout, U.N. Economic Commission for Latin America and the Caribbean, Washington Office, Washington, D.C., private communication, January 2, 1992. Sprout and Weaver combined income distribution data and purchasing-power adjusted GDP per capita data to disaggregate 127 countries into five classes each and reaggregate these segments into five global classes; see Ronald V.A. Sprout and James H. Weaver, "International Distribution of Income: 1960-1987," Working Paper No. 159, Department of Economics, American University, Washington, D.C., May 1991. Number in each class adjusted to mid-year 1992 population from Machiko Yanagishita, demographer, Population Reference Bureau, Washington, D.C., private communication, February 26, 1992.

8   Income range and share of world income estimated from Sprout and Weaver, "1988 International Distribution of Income"; Chinese appliances from "TV Now in 50% of Homes," China Daily, February 15, 1988.

9   Income range and share of world income from Sprout and Weaver, "1988 International Distribution of Income"; comparison to U.S. poverty line from U.S. Bureau of the Census, Statistical Abstract of the United States: 1990 (Washington, D.C.: U.S. Government Printing Office, 1990). As used in this chapter, "consumers," "consumer class," and "global consumer society" are synonymous and refer to the richest fifth of humanity as measured by per capita income or life-style. The global consumer society, of course, does not share the institutions that a national society does, but it does share a way of life and many values.

10 Sprout and Weaver, "1988 International Distribution of Income." See also Nathan Keyfitz, "Consumerism and the New Poor," Society, January/February 1992.

11 Carbon emissions exclude the 7-33 percent that originate from forest clearing. Although this somewhat biases the figures against the consumer class — forest clearing emissions are concentrated in rural areas of developing countries, where many of the poor live — emissions of other greenhouse gases, such as chloro-fluorocarbons, are more concentrated in the consumer society than fossil-derived carbon dioxide. Thus, fos-sil-fuel carbon emissions are a relatively good overall indicator of responsibility for global warming. The esti-mates of emissions by class assume — plausibly — that carbon emissions and world income distribution coincide, and were calculated by combining income distribution data from World Bank, World Development Report 1991 (New York: Oxford University Press, 1991), with carbon emissions data from Gregg Marland et al., Estimates of CO2 Emissions from Fossil Fuel Burning and Cement Manufacturing, Based on the United Nations Energy Statistics and the U.S. Bureau of Mines Cement Manufacturing Data (Oak Ridge, Tenn.: Oak Ridge National Laboratory, 1989), and from Thomas Boden et al., Trends '91 (Oak Ridge, Tenn.: Oak Ridge National Laboratory, in press), and comparing them with Ronald V.A. Sprout and James H. Weaver, "1988 International Distribution of Income" (unpublished data) provided by Ronald V.A. Sprout, U.N. Economic Commission for Latin America and the Caribbean, Washington Office, Washington, D.C., private communication, January 2, 1992.

12 Acid rain, hazardous chemicals, and chlorofluorocarbons are Worldwatch Institute estimates based on World Resources Institute, World Resources 1990-91; nuclear warheads from Swedish International Peace Research Institute, SIPRI Yearbook 1990: World Armaments and Disarmament (Oxford: Oxford University Press, 1990); radioactive waste is Worldwatch Institute estimate based on cumulative nuclear-power electricity production from International Atomic Energy Agency, Nuclear Power Reactors in the World (Vienna: 1991).

13 Energy intensity and toxics emissions from Michael Renner, Jobs in a Sustainable Economy, Worldwatch Paper 104 (Washington, D.C.: Worldwatch Institute, September 1991); air pollution from U.S. Environmental Protection Agency, Office of Air Quality Planning and Standards, National Air Pollution Estimates 1940-89 (Washington, D.C.: 1991).

14 Herman Daly, "Environmental Impact Identity — Orders of Magnitude" (draft), World Bank, Washington, D.C., 1991; also see Ekins, "The Sustainable Consumer Society: A Contradiction in Terms?"

15 Worldwatch Institute estimate of consumption since 1950 based on gross world product data from Angus Maddison, The World Economy in the 20th Century (Paris: Organisation for Economic Co-operation and Development, 1989); minerals from Ralph C. Kirby and Andrew S. Prokopovitsh, "Technological Insurance Against Shortages in Minerals and Metals," Science, February 20, 1976; opinion surveys from Michael Worley, National Opinion Research Center, University of Chicago, Chicago, Ill., private communication, September 19, 1990; gross national product per capita and personal consumption expenditures are adjust-

ed for inflation from U.S. Bureau of the Census, *Statistical Abstract of the United States: 1991* (Washington, D.C.: U.S. Government Printing Office, 1991).

16 Four-and-a-half times richer from Angus Maddison, *The World Economy in the 20th Century* (Paris: Organisation for Economic Co-operation and Development, 1989).

17 Alan Durning, *Poverty and the Environment: Reversing the Downward Spiral*, Worldwatch Paper 92 (Washington, D.C.: Worldwatch Institute, November 1989).

18 Henry David Thoreau, *Walden* (1854; reprint, Boston: Houghton Mifflin, 1957).

Sandra Postel "Carrying Capacity: The Earth's Bottom Line"

1 David R. Klein, "The Introduction, Increase, and Crash of Reindeer on St. Matthew Island," *Journal of Wildlife Management*, April 1968.

2 The resource projections for 2010 are not predictions but extrapolations based largely on recent trends — primarily those observed from 1980 to 1990 — and current knowledge of the resource base, as discussed and documented later in this chapter.

3 Thomas F. Homer-Dixon et al., "Environmental Change and Violent Conflict," *Scientific American*, February 1993; Norman Myers, *Ultimate Security: The Environmental Basis of Political Security* (New York: W.W. Norton & Company, 1993).

4 These proportions are based on per capita gross national product data that are unadjusted for purchasing power parity (PPP), which measures the relative domestic purchasing powers of national currencies and therefore may convey a more realistic comparison of actual living standards.

5 Alan Thein Durning, *How Much is Enough? The Consumer Society and the Future of the Earth* (New York: W.W. Norton & Company, 1992); Alan Durning, *Poverty and the Environment: Reversing the Downward Spiral*, Worldwatch Paper 92 (Washington, D.C.: Worldwatch Institute, November 1989).

6 Sandra Postel, *Last Oasis: Facing Water Scarcity* (New York: W.W. Norton & Company, 1992); 1.2 billion figure from Joseph Christmas and Carel de Rooy, "The Decade and Beyond: At a Glance," *Water International*, September 1991.

7 Number of people without enough food and percentage of necessary calories consumed by the average African from Kevin Cleaver and Gotz Schreiber, *The Population, Agriculture, and Environment Nexus in Sub-Saharan Africa* (Washington, D.C.: World Bank, 1992); fat-laden diets in rich countries from Alan Durning and Holly Brough, *Taking Stock: Animal Farming and the Environment*, Worldwatch Paper 103 (Washington, D.C.: Worldwatch Institute, July 1991); increase in global harvest is a Worldwatch Institute estimate, based on U.S. Department of Agriculture (USDA), *World Grain Database* (unpublished printout) (Washington, D.C.: 1992), and on Population Reference Bureau (PRB), *1990 World Population Data Sheet* (Washington, D.C.: 1990).

8 Gross world product in 1950 from Herbert R. Block, *The Planetary Product in 1980: A Creative Pause?* (Washington, D.C.: U.S. Department of State, 1981); gross world product in 1990 from International Monetary Fund (IMF), *World Economic Outlook: Interim Assessment* (Washington, D.C.: 1993); increase in value of internationally traded goods from $308 million in 1950 to $3.58 trillion in 1992 (in 1990 dollars) is a Worldwatch Institute estimate, based on IMF, Washington, D.C., unpublished data base; World Bank, Washington, D.C., unpublished data base.

9 Industrial roundwood from United Nations, *Statistical Yearbook, 1953* (New York: 1954), and from U.N. Food and Agriculture Organization (FAO), *1991 Forest Products Yearbook* (Rome: 1993); water from Postel, op. cit. note 6; oil from American Petroleum Institute, *Basic Petroleum Data Book* (Washington, D.C.: 1992).

10 U.S. Bureau of the Census, Department of Commerce, International Data Base (unpublished printout) (Washington, D.C., November 2, 1993); population of Mexico from PRB, *1992 World Population Data Sheet* (Washington, D.C.: 1992).

11 Shiro Horiuchi, "Stagnation in the Decline of the World Population Growth Rate During the 1980s," *Science*, August 7, 1992.

12 Figure for 2030 from Larry Heligman, Estimates and Projections Section, Population Division, United Nations, New York, private communication, November 3, 1993; United Nations, Department of Economic and Social Affairs, *Long-Range World Population Projections: Two Centuries of Population Growth 1950-2150* (New York: 1992).

13 IUCN—The World Conservation Union, *The IUCN Sahel Studies 1991* (Gland, Switzerland: 1992).

14 Ibid.

15 Maria Concepcion Cruz et al., *Population Growth, Poverty, and Environmental Stress: Frontier Migration in the Philippines and Costa Rica* (Washington, D.C.: World Resources Institute, 1992).

16 Peter M. Vitousek et al., "Human Appropriation of the Products of Photosynthesis," *BioScience*, June 1986.

17   Ibid.

18   Cropland expansion from FAO, *Production Yearbook* 1991 (Rome: 1992); figure of 76 million from "Crops from Pasture Land," *International Agricultural Development*, March/April 1993, and from Richard J. Thomas, Centro Internacional de Agricultura Tropical, Cali, Colombia, private communication, July 21, 1993.

19   Vaclav Smil, "China's Environment in the 1980s: Some Critical Changes," *Ambio*, September 1992; cropland areas of European countries from FAO, op. cit. note 18; number of Chinese people that 35 million hectares could support is Worldwatch Institute estimate, based on USDA, op. cit. note 7.

20   L.R. Oldeman et al., "The Extent of Human-Induced Soil Degradation," in L.R. Oldeman et al., *World Map of the Status of Human-Induced Soil Degradation* (Wageningen, Netherlands: United Nations Environment Programme and International Soil Reference and Information Centre, 1991).

21   FAO, op. cit. note 18.

22   Oldeman et al., op. cit. note 20.

23   Fish catch from FAO, Fisheries Department, "Global Fish and Shellfish Production in 1991," COFI Support Document: Fishery Statistics, Rome, March 1993; percentage of human protein consumption from FAO, *Food Balance Sheets* (Rome: 1991); contribution to coastal diet from FAO, "Marine Fisheries and the Law of the Sea: A Decade of Change," Fisheries Circular No. 853, Rome, 1993.

24   Fivefold increase in world fish catch from FAO, *Yearbook of Fishery Statistics: Catches and Landings* (Rome: various years); other figures from M. Perotti, chief, Statistics Branch, Fisheries Department FAO, Rome, private communication, November 3, 1993.

25   Advent of fisheries technologies and assessment of all fishing areas from FAO, "Marine Fisheries and the Law of the Sea," op. cit. note 23; FAO, "World Review of High Seas and Highly Migratory Fish Species and Straddling Stocks," FAO Fisheries Circular No. 858 (preliminary version), Rome, 1993.

26   Estimate of potential catch from FAO-sponsored book by J.A. Gulland ed., *The Fish Resources of the Ocean* (West Byfleet, Surrey, U.K.: Fishing News (Books) Ltd.: 1971); 2010 projection is Worldwatch Institute estimate, based on data from FAO, "Marine Fisheries and the Law of the Sea," op. cit. note 23.

27   Postel, op. cit. note 6.

28   Ibid.

29   Annual loss of irrigated land to salinization from Dina L. Umali, *Irrigation-Induced Salinity*, World Bank Technical Paper No. 215 (Washington, D.C.: World Bank, 1993).

30   1700 figure does not include woodlands or shrublands, from R.A. Houghton et al., "Changes in the Carbon Content of Terrestrial Biota and Soils Between 1860 and 1980: A Net Release of CO_2 to the Atmosphere," *Ecological Monographs*, September 1983. FAO and the U.N. Economic Commission for Europe/FAO (UNECE/FAO) used somewhat different definitions in their 1990 forest assessments, which precludes a strict comparison of the data between tropical and temperate zones when calculating the change in world forest cover between 1980 and 1990. For both regions, "other wooded areas" have been excluded from all calculations. Because of gaps and discrepancies in the data, FAO does not plan to make any estimate of global deforestation at the conclusion of the 1990 assessment (in progress as of October 1993), according to Klaus Janz, senior forestry officer, Resources Appraisal and Monitoring, Forestry Department, FAO, Rome, private communication, October 29, 1993. The estimated net loss of forests refers to the conversion of forests to an alternative land use, minus the net addition of plantations in tropical regions. As defined by FAO, loss of forests does not include forest that was logged and left to regrow, even if it was clear-cut (unless the forest cover is permanently reduced to less than 10 percent). Thus, the statistics fail to reflect the fragmentation or degradation of forests.

     Given these limitations, the 1990 figure is a preliminary and rough Worldwatch Institute estimate, based on: FAO, *1961-1991...2010: Forestry Statistics Today for Tomorrow* (Rome: 1993); FAO, "Areas of Woody Vegetation at End 1980 for Developing and Developed Countries and Territories by Region," Table 1, in *An Interim Report on The State of Forest Resources in the Developing Countries* (Rome: 1988); all tropical countries from FAO, *Forest Resources Assessment 1990—Tropical Countries*, Forestry Paper No. 112 (Rome: 1993); Australia, Europe, Japan, and New Zealand from UN-ECE/FAO, *The Forest Resources of the Temperate Zones: the UN-ECE/FAO 1990 Forest Resource Assessment, Vol. 1: General Forest Resource Information* (New York: United Nations, 1992); Canada from Canadian Council of Forest Ministers, *Compendium of Canadian Forestry Statistics 1992*, National Forestry Database, (Ottawa: 1993); Joe Lowe, Forest Inventory and Analysis Project, Petawawa National Forestry Institute, Canadian Forest Service, Chalk River, Ontario, unpublished printout and private communication, November 4, 1993; United States from USDA, Forest Service, *An Analysis of the Timber Situation in the United States: 1952-2030*, Forest Resource Report No. 23 (Washington, D.C.: 1982); Karen Waddell, Daniel Oswald, and Douglas Powell, Pacific Northwest Experiment Station, Forest Service, *Forest Statistics of the United States 1987*, Research Bulletin 168 (Portland, OR: USDA, 1989); James Bones, branch chief, Forest Inventory Research and Analysis, Forest

Service, USDA, Washington, D.C., private communication, October 22, 1993; former Soviet Union from Anatoly Shvidenko, International Institute for Applied Systems Analysis (IIASA), Laxenburg, Austria, unpublished printout and private communication, July 18, 1993; China from Smil, op. cit. note 19; Argentina from The Republic of Argentina, National Commission on the Environment, National Report to the United Nations Conference on Environment and Development, July 1991; rough estimates for the other temperate developing countries based on trends and projections in FAO, Forest Resources Division, *An Interim Report on the State of Forest Resources in the Developing Countries* (Rome: 1988), and on J.P. Lanly, *1980 Forest Resources Assessment* (Rome: FAO, 1983); area of tropical plantations from FAO, *Forest Resources Assessment 1990—Tropical Countries*, op. cit. in this note. The estimated loss of 148 million hectares of natural forest was offset by the 18-million-hectare gain in tropical plantations to arrive at the figure of 130-million-hectare net loss between circa 1980 and circa 1990.

31 Tropical natural forest loss and plantation gain from FAO, *Forest Resources Assessment 1990 — Tropical Countries*, op. cit. note 30; China from Smil, op. cit. note 19. The former Soviet Union reported a net gain of 22.7 million hectares between 1978 and 1988, from Shvidenko, op. cit. note 30. This figure is the only one available, but it is problematic. It can be assumed that the reported increase has been in low-productive sites with limited growing stocks, while excessive harvesting has taken place in the mature and old-growth forests, according to Sten Nilsson, Forestry and Climate Change Project, IIASA, Laxenburg, Austria, private communication, September 30, 1993. 2010 projection is Worldwatch Institute estimate, based on two decades of loss in forests calculated for 1980-90; 1990 global population from PRB, op. cit. note 7; 2010 population from Bureau of the Census, op. cit. note 10.

32 USDA, op. cit. note 7.

33 1992 grain production figure from Francis Urban, section leader, Markets and Competition, Economic Research Service, USDA, Washington, D.C., private communication, October 20, 1993; 1984 grain figure from USDA, op. cit. note 7; population figures from U.S. Bureau of the Census, in Francis Urban and Ray Nightingale, *World Population by Country and Region, 1950-1990 and Projections to 2050* (Washington, D.C.: USDA, Economic Research Service, 1993).

34 Genetic yield potential from Lloyd T. Evans, *Crop Evolution, Adaptation and Yield* (Cambridge: Cambridge University Press, 1993); new crop strains from Donald O. Mitchell and Merlinda D. Ingco, International Economics Department, *The World Food Outlook: Malthus Must Wait* (Washington, D.C.: World Bank, July 1993 (draft)).

35 Engineering of maize from Gabrielle J. Persley, *Beyond Mendel's Garden: Biotechnology in the Service of World Agriculture* (Wallingford, U.K.: CAB International, 1990); Gabrielle J. Persley, World Bank, Washington, D.C., private communications, July 1993.

36 R.P.S. Malik and Paul Faeth, "Rice-Wheat Production in Northwest India," in Paul Faeth, ed., *Agricultural Policy and Sustainability: Case Studies from India, Chile, the Philippines, and the United States* (Washington, D.C.: World Resources Institute, 1993).

37 Willem Van Tuijl, *Improving Water Use in Agriculture: Experiences in the Middle East and North Africa* (Washington, D.C.: World Bank, 1993).

38 Postel, op. cit. note 6.

39 For an overview of some traditional methods and their use, see Chris Reij, *Indigenous Soil and Water Conservation in Africa* (London: International Institute for Environment and Development, 1991); Will Critchley, *Looking After Our Land: Soil and Water Conservation in Dryland Africa* (Oxford: Oxfam, 1991).

40 Sandra Postel and John C. Ryan, "Reforming Forestry," in Lester R. Brown et al., *State of the World 1991* (New York: W.W. Norton & Company, 1991).

41 Ed Ayres, "Making Paper Without Trees," *World Watch*, September/October 1993; 9 percent figure from FAO, *Pulp and Paper Capacities 1992-1997* (Rome: 1993).

42 Quotation from Donella H. Meadows, Dennis L. Meadows, and Jorgen Randers, *Beyond the Limits* (Post Mills, VT: Chelsea Green Publishing Company, 1992).

43 Population and densities from PRB, *1993 World Population Data Sheet* (Washington, D.C.: 1993); area of cropland from FAO, op. cit. note 18; grain imports are Worldwatch Institute estimates, based on USDA, op. cit. note 7; wood imports from FAO, op. cit. note 9.

44 Real GDP per capita figures are adjusted for purchasing power parity, from U.N. Development Programme, *Human Development Report 1993* (New York: Oxford University Press, 1993). See note 4 for description of PPP.

45 Figure of 24 million hectares from Mathis Wackernagel et al., "How Big is Our Ecological Footprint?: A Handbook for Estimating A Community's Appropriated Carrying Capacity," discussion draft, Task Force on Planning Healthy and Sustainable Communities, University of British Columbia, Vancouver, B.C., July 1993; area of Dutch croplands, pastures, and forests from FAO, op. cit. note 18.

46 Herman Daly, *Steady-State Economics* (Washington, D.C.: Island Press, 2nd ed., 1991).

47 Worldwatch Institute estimates, based on FAO, AGROSTAT-PC 1993, Forest Products Electronic Data Series (Rome: 1993); Taiwan from Philip Wardle, senior forestry economist, FAO, Rome, unpublished printout from FAO Forestry Products Database, September 21, 1993; broad conversion factors for converting non-round-wood products into green solid wood equivalent, which differs from roundwood raw material equivalent, were supplied by Philip Wardle, unpublished printout and private communication, July 1, 1993.

48 He Bochuan, *China on the Edge: The Crisis of Ecology and Development* (San Francisco: China Books and Periodicals, Inc., 1991); rise in Chinese demand is Worldwatch Institute estimate for 1991, based on sources and methodology cited in footnote 47.

49 Gretchen C. Daily and Paul R. Ehrlich, "Population, Sustainability, and Earth's Carrying Capacity," *BioScience*, November 1992.

50 Postel and Ryan, op. cit. note 40.

51 Net trade in forest products is a Worldwatch Institute estimate, based on sources and methodology from FAO, op. cit. note 47; estimates of sustained yield vary, depending on assumptions of how fast cutover forests grow wood and how much forest is to be opened to logging; timber production in Sarawak, Malaysia's primary timber-exporting region, exceeds sustainable yield by four times, from Kieran Cooke, "Warfare Escalates Over Use of the Tropical Forests," *Financial Times*, August 31, 1993; timber production exceeds sustained yield in Malaysia by 71-228 percent, according to François Nectoux and Yoichi Kuroda, *Timber from the South Seas: An Analysis of Japan's Timber Trade and Its Environmental Impact* (Gland, Switzerland: World Wide Fund for Nature International, 1989).

52 Hilary F. French, *Costly Tradeoffs: Reconciling Trade and the Environment*, Worldwatch Paper 113 (Washington, D.C.: Worldwatch Institute, March 1993).

53 This analogy is borrowed from Herman Daly, senior economist, World Bank; Herman E. Daly, "Allocation, Distribution, and Scale: Towards an Economics That is Efficient, Just, and Sustainable," *Ecological Economics*, December 1992.

54 Union of Concerned Scientists, "World's Leading Scientists Issue Urgent Warning to Humanity," Washington, D.C., press release, November 18, 1992.

55 External debt in developing countries from IMF, *Annual Report of the Executive Board for the Fiscal Year Ended April 30, 1993* (Washington, D.C.: 1993).

56 Wangari Maathai, "The Green Belt Movement for Environment & Development," presented at the International Conference on Environmentally Sustainable Development, World Bank, Washington, D.C., October 1, 1993.

57 United Nations, Statistical Division, *Integrated Environment and Economic Accounting: Handbook of National Accounting,* Studies in Methods (New York: forthcoming); Peter Bartelmus, Officer in Charge of the Environment and Energy Statistics Branch, Statistical Division of the United Nations, New York, private communication, October 21, 1993.

58 Unmet family planning needs from John Bongaarts, "The KAP-Gap and the Unmet Need for Contraception," *Population and Development Review, June 1991*; efforts needed for women from Jodi L. Jacobson, *Gender Bias: Roadblock to Sustainable Development*, Worldwatch Paper 110 (Washington, D.C.: Worldwatch Institute, September 1992); Jodi L. Jacobson, *Women's Reproductive Health: The Silent Emergency*, Worldwatch Paper 102 (Washington, D.C.: Worldwatch Institute, June 1991).

59 Restoration of international family planning funding from Susan Cohen, Senior Policy Associate, Alan Guttmacher Institute, Washington, D.C., private communication, October 21, 1993; Timothy Wirth, Counselor, U.S. Department of State, Speech at Second Preparatory Committee for the International Conference on Population and Development, New York, May 11, 1993.

F. Landis MacKellar and David E. Horlacher, "Population, Living Standards ,and Sustainability: An Economic View"

1 Is global ecological sustainability, in fact, at risk? We do not know, and this appears to us to be a matter better addressed by natural and environmental scientists than by economists. The more mundane question of whether economic scarcity of natural resources is increasing or decreasing was an active area of research a few years ago, with most economists reaching conclusions in the moderately optimistic to moderately pessimistic range (MacKellar and Vining 1987 and 1988). The economic literature on global environmental trends is spotty because of data deficiencies.

2 Kelley, A. 1988, "Economic consequences of population change in the third world." *Journal of Economic Literature 26*(4): 685-728; Horlacher, D. and L. MacKellar 1988, "Population growth versus economic growth (?)" In D. Salvatore, ed., *World Population Trends and their Impact on Economic Development*, pp. 25-44. New York: Greenwood Press; United States National Research Council *1986, Rapid population*

growth and economic development: policy questions, Washington, D.C.: National Academy Press

3   Teitenberg, T. 1992, *Environmental and Natural Resource Economics.* New York: Harper Collins; World Bank 1992; *World Development Report.* New York: Oxford University Press.

4   If it did, it seems fair to speculate, the combined effect of increases in income per capita and population since the seventeenth century would have led to ecological collapse several times over.

5   All data used in this section are from the statistical appendices of the 1992 and 1993 World Bank World Development Reports.

6   The same argument may be made for food, in which case, after the diet attains a sufficiently high level of calories and meat protein content, further increments to income will be spent on non-food items. The observation that the share of food in the consumer's budget declines as income rises is called Engels' Law after the nineteenth-century German statistician who discovered it.

7   Meadows, D., D. Meadows, J. Randers and W. Behrens, *The Limits to Growth,* New York: Universe Books, 1972.

8   How would the price system function in the case of the environment where, as we have pointed out, it tends to operate poorly in the absence of some public intervention? Presumably, as the deteriorating environment eroded economic productivity, including the health of workers, and raised production costs, producers, consumers and public authorities would take notice and initiate appropriate policies.

9   As the Report points out, it may be unfair to make conclusions about CO_2 emissions based on data from the 1980s, as global warming was only an emerging concern at the time.

10 While the composition of output differs drastically at the national level, roughly the same mix of activities are pursued in commercial city centers, so differences in air quality reflect mostly differences in income; i.e., demand for air quality. What is entirely uncontrolled for in Figure 4 is, of course, variation due to topographical and climatic factors.

11 OECD, *The State of the Environment,* Paris: OECD, 1991.

12 Cleaver, K. and G. Schreiber, "The population, agriculture and environment nexus in sub-Saharan Africa," Africa Region Technical Department *Agriculture and Rural Development Series,* No. 9. Washington, D.C.: World Bank, 1993.

13 Simon, Julian, *The Ultimate Resource,* Princeton: Princeton University Press, 1981.

14 Lee, R. , "Evaluating externalities to childbearing in developing countries: the case of India." In *Consequences of Rapid Population Growth in Developing Countries: Proceedings of the United Nations / Institut national d'études démographiques Expert Group Meeting,* New York, 23-26 August, 1988, pp. 297-342. New York: Taylor and Francis, 1991.

15 The sad story of environmental policy in the formerly centrally planned economies, where the institutions necessary for eliminating externalities were suppressed, is well known.

16 Rodgers, G., *Population and Poverty.* Geneva: International Labour Office, 1984.

17 United Nations, "Report of the Expert Group Meeting on Population, Environment and Development, U.N. Headquarters, 20-24 January, 1992," *Population Bulletin of the United Nations* 34/35: 19-34, 1993.

18 United Nations Secretariat, "Relationships between population and environment in rural areas of developing countries," *Population Bulletin of the United Nations* 31/32: 52-69, 1991.

19 Ehrlich, P., A. Ehrlich and G. Daily 1993, "Food security, population and environment," *Population and Development Review* 19(1): 1-32.

20 MacKellar, L. and D. Vining, "Natural resource scarcity: a global survey," In G. Johnson, G. and R. Lee, eds., *Population Growth and Economic Development: Issues and Evidence,* pp. 259-327. Madison: University of Wisconsin Press, 1987.

21 United Nations Secretariat, 1991, op. cit.

22 Bilsborrow, R., "Population, development and deforestation: some recent evidence," *Carolina Population Center Papers,* No. 92-04. Chapel Hill, North Carolina: Unversity of North Carolina, 1992.

23 Vincent J. 1992, "The tropical timber trade and sustaianble development," *Science* 256: 1951-55.

24 World Resources Institute data cited in Falkenmark, M. and C. Widstrand, "Population and water resources: a delicate balance." *Population Bulletin* 47(3) Washington, D.C.: Population Reference Bureau, 1992, p. 14.

25 Unrealistic pricing policies are not limited to water intended for irrigation. Failure to invest in adequate urban water and sanitation systems in LDCs can be partly explained by the fact that, based on a survey of World Bank-funded water projects, urban users only pay about a third of the supply costs. The remainder must be borne by persons who reap no benefit from the investment.

26 Bongaarts, J. , "Global warming," *Population and Development Review* 18(2): 299-319, 1992; Kolsrud, G. and B. Torrey, "The importance of population growth in future commercial energy consumption," In *Global Climate Change,* ed. J. White. New York: Plenum Press, 1992.

27 United Nations Secretariat 1993, op. cit.

28 Kolsrud and Torrey, op. cit.

29 Forests are a sink for carbon and thus play a significant, though secondary, part in the global CO2 equation.

30 Nordhaus, W. 1992, "An optimal path for controlling greenhouse gases, *Science* 258: 1315-19; William Cline, 1992, *Global Warming: the Economic Stakes*, Washington D.C.: Institute for International Economics.

SECTION II: POPULATION GROWTH AND STRUCTURE
Carl Haub and Martha Farnsworth Riche, "Population by the Numbers: Trends in Population Growth and Structure"

REFERENCES:

Jean-Claude Chenais, *La Démographie* (Paris: Presses Universitaires de France, 1990)

1993 World Population Data Sheet (Washington, D.C.: Population Reference Bureau, 1993)

Population Projections of the United States, by Age, Sex, Race and Hispanic Origin: 1993 to 2050 (Washington, D.C.: U.S. Bureau of the Census, forthcoming)

World Population Prospects: 1992 Revisions (New York: United Nations, 1993)

ENDNOTES:

1   Michael S. Teitelbaum, "The Population Threat," *Foreign Affairs*, Winter, 1992/93, pp. 63-78.

2   Goubert, *Beauvais et le Beauvaisis,II.* Cited in Wrigley, E.A., Population and History, McGraw Hill, New York, !969.

3   Wrigley, E.A., op. cit.

4   The total fertility rate Is the average number of children a woman would have in her lifetime based on the pace of childbearing in a given year.

5   In general, contraceptive use levels of about 70 percent or a little higher represent the upper limit of women likely to use family planning at any one time.

6   For an important discussion of the effect of population trends on the global labor force, see "Labor and the Emerging World Economy" by David E. Bloom and Adi Brender, *Population Bulletin*, Vol. 48, No.2 (Washington, D.C.: Population Reference Bureau, Inc., October 1993)

SECTION III: HISTORY AND ANALYSIS OF POPULATION AND FAMILY PLANNING PROGRAMS
Peter J. Donaldson and Amy Ong Tsui,"The International Family Planning Movement"

1   Fryer, Peter. *The Birth Controllers* (New York: Stein and Day, 1965).

2   Gillespie, Duff G. and Judith R. Seltzer. "The Population Assistance Program of the U.S. Agency for International Development," in Helen M. Wallace and Kanti Giri (eds.), *Health Care of Women and Children in Developing Countries* (Oakland, California: Third Party Publishing, 1990) p. 562.

3   For example see Bogue, Donald (ed.). *Sociological Contributions to Family Planning Research* (Chicago: Community and Family Study Center, University of Chicago, 1967).

4   Crane, Barbara and Jason Finkle. "Ideology and Politics at Mexico City: The United States at the 1984 International Conference on Population," *Population and Development Review*, Vol. 11 (March 1985) pp.145-151.

5   Bongaarts, John. "A Framework for Analyzing the Proximate Determinants of Fertility," *Population and Development Review*, Vol. 4 (March 1978) pp. 105-132.

6   David, Henry P. and Robert J. McIntyre. *Reproductive Behavior: Central and Eastern European Experience* (New York: Springer, 1981) pp. 190-191.

7   Lapham, Robert and W. Parker Mauldin. "The Effects of Family Planning on Fertility: Research Findings," in Robert J. Lapham and George B. Simmons (eds.), *Organizing Effective Family Planning Programs* (Washington, D.C.: National Academy Press, 1987).

8   Mauldin, Parker and Bernard Berelson. "Conditions of Fertility Decline in Developing Countries," *Studies in Family Planning*, Vol. 9, No. 5 (May 1978) pp. 90-147.

9   Cutright, Phillips and William R. Kelly. "The Role of Family Planning Programs in Fertility Declines in Less Developed Countries, 1958-1977," *International Family Planning Perspectives*, Vol. 7 (December 1981) pp. 145-151; and John Bongaarts, W. Parker Mauldin, and James Phillips, "The Demographic Impact of Family Planning Programs," paper prepared for the OECD-Sponsored Meeting on Population and Development, Paris, April 1990.

10 Entwisle, Barbara. "Measuring Components of Family Planning Program Effort," *Demography*, Vol. 26

(February 1989) pp. 53-76; and Albert I Hermalin, "Issues in the Comparative Analysis of Techniques for Evaluating Family Planning Programmes," in United Nations, *Evaluation of the Impact of Family Planning Programmes on Fertility: Sources of Variance.* Population Studies No. 76 ST/ESA/SER. A/76. (New York: United Nations, 1982) pp. 29-40.

11   Phillips, James F., Ruth Simmons, Michael Koenig, and J. Chakraborty. "Determinants of Reproductive Change in a Traditional Society: Evidence from Matlab, Bangladesh," *Studies in Family Planning,* Vol. 19, No. 6 (November/December 1988) pp. 313-334.

12   Lapham, Robert J. and George Simmons (eds.), *Organizing for Effective Family Planning Programs* (Washington, D.C.: National Academy Press, 1987); and George Simmons, "Family Planning Programs," in Jane Menken (ed.), *World Population and U.S. Policy: The Choices Ahead.*

Sharon L. Camp, "The Politics of U.S. Population Assistance"

1   Michele McKeegan, *Abortion Politics — Mutiny on the Ranks of the Right* (New York: Free Press, 1992), pp. 77-94.

2   See for example, Betsy Hartmann, *Reproductive Rights and Wrongs — The Global Politics of Population Control and Contraceptive Choice* (New York: Harper and Row, 1987); and Jacqueline Kasun, *The War Against Population — The Economics and Ideology of World Population Control* (San Francisco: Ignatius Press, 1988).

3   United Nations, Department of International Economic and Social Affairs, *World Population Monitoring 1991* (Population Studies No. 126), 1992, pp. 94-100.

4   See International Planned Parenthood Federation, *Family Planning in Five Continents* (London: IPPF, 1991).

5   Estimate based on United Nations Population Fund (UNFPA), *Global Population Assistance Report 1982-1990* (April 1992).

6   U.S., Department of State, "U.S. National Report on Population" (Report in preparation for the 1994 International Conference on Population prepared by the Population Reference Bureau), October 1993, p. 65.

7   Shanti R. Conly and J. Joseph Speidel, *Global Population Assistance — A Report Card on the Major Donor Countries* (Washington, D.C.: Population Action International, 1993), p. 29.

8   UNFPA, *Global Population Assistance Report 1982-1990,* pp. 26.

9   UNFPA, *Annual Reports,* 1981 and 1991.

10   Gordon S. Black Corporation, "U.S. Attitudes on Population" (Public opinion survey conducted for the Population Crisis Committee), November 1991.

11   For a history of U.S. population assistance policy see Phyllis Tilson Piotrow, *World Population Crisis — The United States Response* (New York: Praeger, 1974); and Peter J. Donaldson, *Nature Against Us — The United States and the World Population Crisis 1965-1980* (Chapel Hill: University of North Carolina Press, 1990).

12   Barbara Crane and Jason Finkle, "The Conservative Transformation of Population Policy," *Governance* (Winter/Spring 1987): 9-14.

13   U.S., Department of State, "U.S. National Report on Population," p. 65.

14   Carroll J. Doherty, "Aid Bill Moves Smoothly Into Law Despite Crisis in Russia," *Congressional Quarterly 51* (October 2, 1993), p. 2658.

15   Planned Parenthood Federation of America, "The Far Right — What They Say About Your Rights!" (pamphlet, 1985).

16   Rev. Paul Marx, "From Contraception to Abortion" (pamphlet published by Human Life International, undated).

17   Robert A. Hatcher et al., *Contraceptive Technology 1990-1992* (New York: Irvington Publishers, 1990), pp. 228-229, 356-357; and *Contraceptive Technology: International Edition* (Atlanta: Printed Matter, Inc., 1989), pp. 238-239.

18   PAI, "Expanding Access to Safe Abortion: Key Policy Issues," *Population Policy Information Kit 8* (September 1993).

19   Alan Guttmacher Institute, "Domestic and International Family Planning Under Fire," *Issues in Brief* (September 1987).

20   McKeegan, *Abortion Politics — Mutiny on the Ranks of the Right,* pp. 77-94.

21   Sharon L. Camp and Craig R. Lasher, "International Family Planning Policy — A Chronicle of the Reagan Years" (unpublished manuscript), February 1989.

22   U.S., *Policy Statement of the United States of America at the International Conference on Population* (Second Session), Mexico, D.F., August 6-13, 1984.

23 U.S., Agency for International Development, "AID Statement on International Planned Parenthood Federation" (Press release), 12 December 1984.

24 Camp and Lasher, "The Reagan Years," pp. 67-71.

25 "Limits on Abortion in Foreign Aid Are Upheld," *New York Times*, 9 March 1990.

26 Camp, "The Impact of the Mexico City Policy on Women and Health Care in Developing Countries," *New York University Journal of International Law and Politics* 20 (Fall 1987): 35-52.

27 *Congressional Quarterly, Almanac — 102nd Congress, 1st Session*, Volume XLVII (Washington, D.C.: CQ, 1992), pp. 470-478.

28 *Rust v. Sullivan*, 111 S. Ct. 1759 (1991).

29 "Clinton Orders Reversal of Abortion Restrictions Left by Reagan and Bush," *New York Times*, 23 January 1993, p. A1, A10.

30 U.S., AID, "U.S. Withholds $10 Million from United Nations Fund for Population Activities" (Press release), 30 March 1985.

31 *Congressional Quarterly, Almanac — 101st Congress, 1st Session*, Volume XLV (Washington, D.C.: CQ, 1990), pp. 799-800.

32 John Felton, "Congress Facing Bitter Battle on Population Control Abroad," *Congressional Quarterly* 42 (September 1, 1984), p. 2145.

33 Conly, Speidel, and Camp, *U.S. Population Assistance: Issues for the 1990s* (Washington, D.C.: Population Crisis Committee, 1991), p. 15.

34 UNFPA, *Global Population Assistance Report 1982-1990*, p. 5.

35 "Pledge Fall Short of UNFPA Needs," *Population,* December 1991, p. 1.

36 Steven W. Sinding and Sheldon J. Segal, "Birth Rate News," *New York Times*, 19 December 1991.

37 U.S., Department of State, Bureau of Public Affairs, "FY 1994 International Affairs Budget" (Press background memorandum), 8 April 1993, p. 10.

38 H. Patricia Hynes, "The Pocketbook and the Pill: Reflections on Green Consumerism and Population Control," *Issues in Reproductive and Genetic Engineering* 4 (1991): 47-51.

39 For examples of viewpoints on population on opposite ends of the political spectrum, see Garrett Hardin, "The Toughlove Solution," *Newsweek*, 26 October 1981, 45; and Committee on *Women, Population and the Environment, Women, Population and the Environment: A Call for a New Approach* (August 1992).

40 Marguerite Holloway, "Population Pressures — The Road From Rio is Paved with Factions," *Scientific American* (September 1992): 36, 38.

41 John Bongaarts, "The Fertility Impact of Family Planning Programs," paper presented at the International Planned Parenthood Federation Family Planning Congress, New Delhi, October 1992.

42 *Institute for Resource Development, Kenya Demographic and Health Survey 1989* (October 1989), pp. 50-51; and *World Fertility Survey, Kenya Fertility Survey 1977-1978*, vol. 1 (February 1980), pp. 114-115.

43 Susan H. Cochrane, "The Effects of Education, Health and Social Security on Fertility in Developing Countries," *World Bank Population, Health and Nutrition Working Paper No. 93* (September 1988).

44 Thomas W. Merrick, "Meeting the Challenge of Unmet Need for Family Planning," paper presented at the International Planned Parenthood Federation Family Planning Congress, New Delhi, October 1992.

45 PCC, *Access to Affordable Contraception* (1991), based on survey research conducted by the Population Council in 1989.

46 Judith Bruce, "Fundamental Elements of the Quality of Care: A Simple Framework," *Population Council Staff Working Paper*, No. 1 (May 1989), pp. 1-7.

47 Lawrence H. Summers, "Investing in All the People: Educating Women in Developing Countries," presentation for the Development Economics Seminar at the 1992 World Bank Annual Meetings, September 1992, p. 6.

48 PAI, *Closing the Gender Gap: Educating Girls* (forthcoming 1994).

49 United Nations Children's Fund, *The State of the World's Children 1991* (New York: Oxford University Press, 1991), p. 15.

50 Nafis Sadik, *Investing in Women: The Focus of the '90s* (New York, UNFPA, 1990).

51 United Nations, *The World's Women — Trends and Statistics 1970-1990* (Social Statistics and Indicators, Series K, No. 8), p. 18.

52 Perdita Huston, *Third World Women Speak Out* (New York: Praeger, 1979), pp. 63-84.

SECTION IV: POPULATION POLICY, REPRODUCTIVE HEALTH AND REPRODUCTIVE RIGHTS
Mahmoud F. Fathalla, "From Family Planning to Reproductive Health"

1   Short, R.V. 1976. *The Evolution of Human Reproduction.* Proc. R. Soc. Lond. B. 195:3-25

2   Veil, S. 1978. "Human Rights, Ideologies and Population Policies" *Population and Development Review* 4,313-321

3   Fathalla, M.F. 1993. "Family Planning and Reproductive Health — A Global Overview." Paper presented to the Population Summit of the World's Scientific Academies. New Delhi 24 to 27 October, 1993.

4   Hord, C., David, H.P., Donnay, F. & Wolf, M. 1991. "Reproductive Health in Romania: Reversing the Ceausescu Legacy," *Studies in Family Planning* 22, 231-240.

5   Fathalla, M.F. 1989. "New Contraceptive Methods and Reproductive Health." In *Demographic and Programmatic Consequences of Contraceptive Innovations* (eds. S.J. Segal, A.O. Tsui, & S. M. Rogers), pp 153-176.

6   World Bank (1993). *World Development Report 1993. Investing in Health.* Oxford University Press. pp 10 & 215-225.

7   World Health Organization. 1978. *Primary Health Care. Report of the International Conference on Primary Health Care,* Alma Ata, USSR, 6-12 September 1978. World Health Organization, 1987. "Health for All" Series, No. 1.

8   World Health Organization. 1992. *Global health situation and projections. Division of Epidemiological Surveillance and Health Situation and Trend Assessment.* pp 41-44. WHO/HST/92.1. WHO, Geneva.

9   World Health Organization. 1991. *Maternal mortality ratios and rates. A tabulation of available information. Third edition.* WHO/MCH/MSM/91.6. Geneva.

10 Fathalla, M.F. (1993) "The Tragedy of Maternal Mortality in Developing Countries: A Health Problem or a Human Rights Issue? *News on Health Care in the Developing Countries,* 7:4-6.

Judith Bruce, "Population Policy Must Encompass More than Family Planning Services"

1   Bongaarts, John. "Population Policy Options in the Developing World," *Science,* vol. 263, 11 February, 1994, pp. 771-776.

2   Morris, Leo. 1993. "Determining Male Fertility Through Surveys: Young Adult Reproductive Health Surveys in Latin America," Presented as Side-Meeting of the IUSSP Committee on Anthropology and Demography. General Conference of the International Union for the Scientific Study of Population. Montreal, Canada, 24 August - 1 September.

3   Buvinic, Mayra and J.P. Valenzuela. 1992. "The Fortunes of Adolescent Mothers and Their Children: A Case Study on the Transmission of Poverty in Santiago, Chile," from the joint Population Council/International Center for Research on Women working paper series, "Family Structure, Female Headship and Maintenance of Families and Poverty."

4   Russell-Brown, Pauline, Patrice Engle, and John Townsend. 1992. "The Effects of Early Childbearing on Women's Status in Barbados," from the joint Population Council/International Center for Research on Women working paper series, "Family Structure, Female Headship and Maintenance of Families and Poverty."

5   Ashurst, Balkaran and Casterline. 1987. "Fertility Behaviour in the Context of Development: Evidence from the World Fertility Survey," (New York: United Nations).

6   Bruce, Judith and Cynthia B. Lloyd. 1992. "Finding the Ties That Bind: Beyond Headship and Household," from the joint Population Council/International Center for Research on Women working paper series on "Family Structure, Female Headship and Maintenance of Families and Poverty."

7   Blanc, Ann. (1993). "Determining Male Fertility Through Surveys: The DHS Experience." Presented at the General Conference of the International Union for the Scientific Study of Population. Montreal, Canada. August 24 - September 1.

8   Lloyd, Cynthia B. and Sonalde Desai. (1992). "Children's Living Arrangements in Developing Countries," Joint Population Council/International Center for Research on Women Working Paper.

Steven W. Sinding, John Ross and Allan Rosenfield, "Seeking Common Ground: Unmet Need and Demographic Goals"

1   Davis, K. (1967). Population Policy: Will Current Programs Succeed? *Science* 158:730-92

2   Conly, Shanti R. and J. Joseph Speidel (1993). *Global Population Assistance: A Report Card on the Major Donor Countries.* (Washington, D.C.: Population Action International)

3   Westoff, Charles (1990). "Reproductive intentions and fertility rates," *International Family Planning Perspectives,* 16, pp. 84-89

4   Bongaarts, J., Mauldin, W.P., and Phillips, J.F. (1990). The Demographic Impact of Family Planning Programs. *Studies in Family Planning* 21, 6:299-310

5 Cleland, J. and Wilson, C. (1987). Demand theories of the fertility transition: An iconoclastic view. *Population Studies* 41:5-30

6 Mazur, L. (1992). *Population and the Environment: A Grantmakers Guide.* (New York: Environmental Grantmakers Association)

7 See, for example, Westoff, Charles and Ann Pebley ... and Westoff and Luis Ochoa (1991). "Unmet need and the demand for family planning," (Columbia, Md., Institute for Resource Development, Inc., *Demographic ands Health Surveys*, Comparative Studies, No. 5.

8 Dixon-Mueller, R., and Germain, A. (1992). "Stalking the Elusive 'Unmet Need' for Family Planning." *Studies in Family Planning* 23, 5:330-335

9 Bongaarts, J. (1991). "The KAP-Gap and Unmet Need for Contraception," *Population and Development Review* 17, 2:293-313

10 Amsterdam Declaration (1989). "A Better Life for Future Generations." Adopted by the International Forum on Population in the Twenty-First Century. Amsterdam, The Netherlands, 6-9 November, 1989

11 UNICEF. (1991). *The State of the World's Children.* See Table 1, p. 102-103.

12 Royston, E. and S. Armstrong. 1989. *Preventing Maternal Deaths.* (Geneva: World Health Organization);16 Royston, E. and A. Lopez. 1987. "On the Assessment of Maternal Mortality." *World Health Statistics Quarterly* 40:211-213

13 Maine ref - 1/2 of maternal deaths preventable from contraception.

Jodi L. Jacobson, "Abortion and the Global Crisis in Women's Health"

1 According to calculations by the World Health Organization, supported by work done by Stanley Henshaw of the Alan Guttmacher Institute, the annual number of abortions worldwide ranges from 36 to 51 million. Henshaw's most recent calculations suggest that about 21 million are illegal procedures. Stanley Henshaw, personal communication, January 4, 1994.

2 Henshaw, personal communication, January 4, 1994.

3 The U.N. General Assembly declared in 1968 that it is the basic right of all couples to "decide freely and responsibly on the number and spacing of their children." This right as recognized was extended to include "individuals" as well as couples at the International Conference on Population in Mexico City, June 1984.

4 John Paxman, "Abortion in Latin America," paper prepared for Population Council, meeting on abortion trends, Bogota, Columbia, October 1988.

5 Wanda Nowicka, "The New Abortion Law in Poland," *Planned Parenthood in Europe*, Vol. 22, No.3, October 1993; Evert Ketting, "Abortion in Europe: Current Status and Major Issues," *Planned Parenthood in Europe*, Vol. 22, No.3, October 1993.

6 Tomas Frejka, "Induced Abortion and Fertility," *International Family Planning Perspectives*, December 1985.

7 See Jodi L. Jacobson, *The Global Politics of Abortion*, Worldwatch Paper 97, July 1990.

8 Abortion-related mortality rates in the U.S. from W. Cates et al., "Legalized Abortion: Effect on National Trends of Maternal and Abortion-Related Mortality," *American Journal of Obstetricians and Gynecologists*, Volume 132, 1978.

9 Stanley K. Henshaw, "Induced Abortion: A World Review, 1990," *Family Planning Perspectives*, March/April 1990.

10 Jacobson, *The Global Politics of Abortion.*

11 IMAP Statement on Abortion, IPPF Medical Bulletin, Volume 27, No. 4, August 1993.

12 Henshaw, "Induced Abortion."

13 Nowicka, "The New Abortion Law in Poland."

14 Henshaw, "Induced Abortion."

15 Julie DeClerque, "Unsafe Abortion Practices in sub-Saharan Africa and Latin America: A Call to Policymakers," presentation at the Panel on Culture, Public Policy, and Reproductive Health, Association for Women in Development Conference, Washington, D.C. November 17-19, 1989; Royston and Armstrong, *Preventing Maternal Deaths*; Henshaw, "Induced Abortion."

16 Jacobson, *The Global Politics of Abortion*; Henshaw, personal communication, January 4, 1994.

17 Adrienne Germain, *Reproductive Health and Dignity: Choices by Third World Women* (New York: International Women's Health Coalition).

18 John M. Paxman, Alberto Rizo, Laura Brown, and Janie Benson, "The Clandestine Epidemic: The Practice of Unsafe Abortion in Latin America," *Studies in Family Planning*, Volume 24, No. 4, July/August 1993.

19 Fred T. Sai and Janet Nassim, "The Need for a Reproductive Health Approach," *International Journal of*

Gynecology and Obstetrics, Supplement 3, 1989; Erica Royston and Sue Armstrong, eds., "Preventing Maternal Deaths" (Geneva: World Health Organization (WHO), 1989); Frejka and Atkin, "The Role of Induced Abortion in the Fertility Transition of Latin America."

20 Khama Rogo, "Induced Abortion in Africa" (unpublished draft), prepared for the Population Association of America Annual Meeting, Toronto, Canada, May 2-3, 1990.

21 Irene Figa-Talamanca et al., "Illegal Abortion: An Attempt to Assess its Cost to the Health Services and Its Incidence in the Community," *International Journal of Health Services*, Volume 16, No. 3, 1986.

22 Rogo, "Induced Abortion in Africa"; Royston and Armstrong, *Preventing Maternal Deaths*.

23 Determinants and Consequences of Pregnancy Wastage in Zaire: A Study of Patients with Complications Requiring Hospital Treatment in Kinshasa, Matadi and Bukavu" (draft), prepared by Family Health International, Research Triangle Park, NC, and Comite National des Naissances Desirables, Kinshasa, Zaire.

24 Judith A. Fortney, "The Use of Hospital Resources to Treat Incomplete Abortions: Examples from Latin America," *Public Health Reports*, November/December 1981.

25 As cited in Paxman et al., "The Clandestine Epidemic."

26 Tongplaew Narkavonnakit and Tony Bennett, "Health Consequences of Induced Abortion in Rural Northeast Thailand," *Studies in Family Planning*, February 1981.

27 Frejka and Atkin, "The Role of Induced Abortion in the Fertility Transition in Latin America."

28 IPPF cited in Tomas Frejka et al., Program Document: Research Program for the Prevention of Unsafe Induced Abortion and Its Adverse Consequences in Latin America and the Caribbean, Center for Policy Studies, Working Paper No. 23 (Mexico City: The Population Council, 1989).

29 Frejka and Atkin, "The Role of Induced Abortion in the Fertility Transition of Latin America".

30 Paxman et al., "The Clandestine Epidemic."

31 Frejka and Atkin, "The Role of Induced Abortion in the Fertility Transition of Latin America."

32 Margoth Mora and Jorge Villarreal, "Unwanted Pregnancy and Abortion: Bogota, Columbia," *Reproductive Health Matters*, No 2, November 1993.

33 N.N. Mashalaba, "Commentary on the Causes and Consequences of Unwanted Pregnancy from an African Perspective," *International Journal of Gynecology and Obstetrics,* Supplement 3, 1989.

Christopher J. Elias, "AIDS: An Agenda for Population Policy"

1   World Health Organization. 1993. December 1. Press Release.

2   Alexander, V. 1991. "Black women and HIV/AIDS." *SIECUS Report* 19, 2: 8-10.

3   Center for International Research. 1992. "Recent seroprevalence by country." *Research Note No. 6.* Washington, DC: U.S. Bureau of the Census.

4   World Development Report. 1993. *Investing in Health: World Development Indicators.* Washington DC: The World Bank.

5   World Health Organization. 1990. December 20. Press Release.

6   Dixon-Mueller, R. and J. Wasserheit. 1991. *The Culture of Silence: Reproductive Tract Infections Among Women in the Third World.* New York: International Women's Health Coalition.

7   Younis, N. et. al.1993. "A community study of gynecological and related morbidities in rural Egypt." *Studies in Family Planning,* Vol. 24, No. 3, pp. 175-186.

8   Elias, C.J. and L. Heise. 1993. *The Development of Microbicides: A New Method of HIV Prevention for Women.* Programs Division Working Paper No. 6. New York: The Population Council.

9   Elias, C.J. and L. Heise. 1994. "Challenges for the development of female-controlled vaginal microbicides." *AIDS* 8: 1-9.

Ruth Macklin, "Ethical Issues in Population and Reproductive Health"

1   At a meeting of Health Sector Personnel, Mexico City, February 1993, part of the Ford Foundation project.

2   T.K. Sundari Ravidran, "Women and the Politics of Population and Development in India," *Reproductive Health Matters,* Number 1 (May 1993), p. 29.

3   Ibid., pp. 29-30.

4   Allan Rosenfield, "RU 486," *American Journal of Public Health,* Vol. 82, No. 10, October 1992, p. 1325.

5   Lynn P. Freedman and Stephen L. Isaacs, "Human Rights and Reproductive Choice," *Studies in Family Planning,* Vol. 24 (January/February 1993).

Lynn P. Freedman, "Law and Reproductive Health"

1   For an excellent summary of the literature on these connections, see Ruth Dixon-Mueller, *Population Policy and Women's Rights*. Westport, Conn: Praeger. 1993.

2   John Ratcliffe, "Social Justice and the demographic transition: lessons from India's Kerala State," in Morley, Rohde, and Williams, eds. *Practising Health for All*. New York: Oxford University Press. 1983; John C. Caldwell, "Routes to Low Mortality in Poor Countries," *Population and Development Review*. 12: 171-220. (1986); Lynn Freedman and Deborah Maine, "Women's Mortality: A Legacy of Neglect," in Koblinsky, Timyan, and Gay, eds. *The Health of Women: A Global Perspective*. Boulder, Colorado: Westview Press, 1993. See generally, Cornia G., Jolly R. and Stewart F., eds. *Adjustment with a Human Face: Protecting the Vulnerable and Promoting Growth*. Oxford: Clarendon Press. 1987; UNICEF. *The Invisible Adjustment: Poor Women and the Economic Crisis*. Santiago, Chile. 1989.

3   Stephen L. Isaacs and Andrea Irvin. *Population Policy: A Manual for Policymakers and Planners*. Washington, DC: The Futures Group. 1991.

4   "China Weighs Using Sterilization and Abortions to Stop 'Abnormal' Births," *The New York Times*. December 22, 1993. For detailed information on the history of China's population policy see H. Yuan Tien, *China's Strategic Demographic Initiative*. New York: Praeger (1991).

5   The most dramatic proof of this fact is found in the history, in industrialized countries, of changes in infant versus maternal mortality rates. While infant mortality dropped dramatically in the first decades of the century with improvements in overall living conditions, such as better nutrition and sanitation, maternal mortality did not decrease in any significant amount until the 1940s when, for the first time, there was widespread access to specific medical interventions for treating obstetric complications, such as antibiotics, blood transfusions, and caesarean sections. Irvine Loudin, "On Maternal and Infant Mortality 1900-1960," *Social History of Medicine* 4:29-73 (April 1991).

6   Lynn P. Freedman, *Women and the Law in Asia and the Near East*. GENESYS Special Studies No. 1. Washington, DC: The Futures Group. 1991; Lawrence M. Friedman, "Legal Culture and Social Development," *Law and Society Review*. 4:29-44 (1969).

7   Sandra Gifford, "The Meaning of Lumps: A Case Study of the Ambiguities of Risk " in James, C.R., Stall, R. and Gifford, S.M., eds. *Anthropology and Epidemiology: Interdisciplinary Approaches to the Study of Health and Disease*. Dordrecht, Holland: D. Reidel, 1986.

8   Deborah Maine, Lynn Freedman, Farida Shaheed and Schuyler Frautschi, "Risk, Reproduction, and Rights: The Uses of Reproductive Health Data," in Robert Cassen, ed. *Population and Development: Old Debates, New Conclusions*. Overseas Development Council (forthcoming 1994).

9   Johnson Controls' policy was challenged under the federal laws prohibiting sex discrimination in employment. The choice between keeping one's childbearing capacity and keeping one's job was one that only women, and not men, were forced to make. The Supreme Court therefore ruled, unanimously, that such fetal protection policies violate equal employment laws and so, in the United States, are now illegal. See Joan E. Bertin, "A Womb of One's Own" in J. Callahan, ed. *Reproduction, Ethics and the Law*. Indiana University Press (forthcoming) for an analysis of the legal and scientific evidence considered in the *Johnson Controls* case.

10 See Rebecca J. Cook. *Human Rights in Relation to Women's Health: The Promotion and Protection of Women's Health through International Human Rights Law*. Geneva: World Health Organization. 1993.

SECTION V: POPULATION, GENDER AND THE FAMILY
Ruth Dixon-Mueller, "Women's Rights and Reproductive Choice: Rethinking the Connections"

1   Huston, Perdita. 1978. *Message from the Village*, (New York: Epoch B Foundation).

2   United Nations, Department of International Economic and Social Affairs. 1989a. *Adolescent Reproductive Behavior: Evidence from Developing Countries, vol.* 2 (New York: United Nations), p.34.

3   Blake, Judith. 1961. *Family Structure in Jamaica: The Social Context of Reproduction*, (New York: Free Press).

4   Cain, Mead. 1986. "The Consequences of Reproductive Failure: Dependence, Mobility, and Mortality Among the Elderly of Rural South Asia," *Population Studies*, vol. 40, no. 3 (Nov.), pp. 375-388.

5   Nichols, Douglas, O. A. Ladipo, John M. Paxman, and E. O. Otolorin. 1986. "Sexual Behavior, Contraceptive Practice, and Reproductive Health Among Nigerian Adolescents," *Studies in Family Planning*, vol. 17, no. 2 (March/April), pp. 100-106.

6   Caldwell, John C., P. H. Reddy, and Pat Caldwell. 1983. "The Cause of Marriage Change in South India," *Population Studies* vol. 37, no. 3 (Nov.), pp. 343-362.

7   Salaff, Janet. 1981. *Working Daughters of Hong Kong: Filial Piety or Power in the Family?*, (Cambridge: Cambridge University Press).

8   Greenhalgh, Susan. 1985. "Sexual Stratification: The Other Side of 'Growth with Equity' in East Asia,'' *Population and Development Review,* vol. 11, no. 2 (June), pp. 265-314.

9   Birdsall, Nancy, and Lauren A. Chester. 1987. "Contraception and the Status of Women: What is the Link?" *Family Planning Perspectives,* vol. 19, no. I (Jan./Feb.), pp. 14-18.

10   Anker, Richard, and Catherine Hein, eds. 1986. *Sex Inequalities in Urban Employment in the Third World,* (London: Macmillan).

11   Ware, Helen. 1977. "Women's Work and Fertility in Africa," in Stanley Kupinsky, ed., *The Fertiity of Working Women,* (New York: Praeger), pp. 1-34.

12   Oppong, Christine. 1983. "Women's Roles, Opportunity Costs, and Fertility," in Rodolfo A. Bulatao and Ronald D. Lee, eds., *Determinants of Fertility In Developing Countries,* vol. I (New York: Academic Press), pp. 553-554.

13   Knodel, John, and Malinee Wongsith. 1991. "Family Size and Children's Education in Thailand," *Demography,* vol. 28, no. I (Feb.), pp. II9-131.

14   Cochrane, Susan H. 1979. *Fertility and Education: What Do We Really Know?,* (Baltimore: Johns Hopkins University Press for The World Bank).

15   Ibid.

16   Cleland, John, and German Rodriguez. 1988. "The Effect of Parental Education on Marital Fertility in Developing Countries," *Population Studies,* vol. 42, no. 3 (Nov.), pp. 419-442.

17   United Nations, Department of International Economic and Social Affairs. 1985. *Socio-Economic Differentials in Child Mortality in Developing Countries,* (New York: United Nations), pp.286-287.

18   Caldwell, John C. 1986. "Routes to Low Mortality in Poor Countries," *Population and Development Review,* vol. 12, no. 2 (June), pp. 171-220.

19   United Nations, Department of International Economic and Social Affairs. 1987. *Fertility Behaviour in the Context of Development: Evidence from the World Fertility Survey,* (New York: United Nations), pp. 214-254.

20   Ibid.: 225.

21   Ibid.

22   Stycos, Joseph M. 1968. *Human Fertility in Latin America,* (Ithaca: Cornell University Press), p. 269.

23   United Nations, Department of International Economic and Social Affairs. 1987. *Fertility Behaviour in the Context of Development: Evidencefrom the World Fertility Survey,* (New York: United Nations), pp. 260-262.

24   Safilios-Rothschild, Constantina. 1969. "Sociopsychological Factors Affecting Fertility in Urban Greece,'' *Journal of Marriage and the Family,* vol. 31, no. 3 (Aug.), pp. 595-598.

25   Stycos, Joseph M., and Robert N. Weller. 1967. "Female Working Roles and Fertility," *Demography,* vol. 4, no. 1, pp. 210-217.

26   Piepmeier, K. B., and T. S. Adkins. 1973. "The Status of Women and Fertility,'' *Journal of Biosocial Science,* vol. 5, no. 4 (Oct.), pp. 507-520.

27   McCabe, James L., and Mark R. Rosenzweig. 1976. "Female Employment Creation and Family Size," in Ronald Ridker, ed., *Population and Development: The Search for Selective Interventions,* (Baltimore: Johns Hopkins University Press for Resources for the Future), pp. 322-355.

28   Dixon, Ruth B. 1978. *Rural Women at Work: Strategies for Development in South Asia,* (Baltimore: The Johns Hopkins University Press for Resources for the Future).

29   Safilios-Rothschild, Constantina.1982. "Female Power, Autonomy and Demographic Change in the Third World," in Richard Anker, Mayra Buvinic, and Nadia H. Youssef, eds., *Women's Roles and Population Trends in the Third World,* (London: Croom Helm), pp. 117-132.

30   Oppong, Christine. 1983. "Women's Roles, Opportunity Costs, and Fertility," in Rodolfo A. Bulatao and Ronald D. Lee, eds., *Determinants of Fertility in Developing Countries,* vol. I (New York: Academic Press), pp. 547-589.

31   Standing, Guy. 1983. "Women's Work Activity and Fertility," in Rodolfo A. Bulatao and Ronald D. Lee, eds., *Determinants of Fertility in Developing Countries,* vol. I (New York: Academic Press), pp. 517-546.

32   Ware, Helen.1981. *Women, Demography, and Development,* (Canberra: Australian National University), p. 101.

33   Dixon, Ruth B. 1978. *Rural Women at Work: Strategies for Development in South Asia,* (Baltimore: The Johns Hopkins University Press for Resources for the Future).

34   Mies, Maria. 1982. *The Lacemakers of Narsapur: Indian Housewives Produce for the Worldmarket,* (London: Zed Books).

35   Greenhalgh, Susan. 1985. "Sexual Stratification: The Other Side of 'Growth with Equity' in East Asia,'' *Population and Development Review,* vol. 11, no. 2 (June), pp. 265-314.

36 Dwyer, Daisy, and Judith Bruce, eds. 1988. *A Home Divided: Women and Income in the Third World,* (Stanford: Stanford University Press).

37 United Nations, Department of International Economic and Social Affairs. 1991. *The World's Women: Trends and Statistics 1970-1990,* (New York: United Nations), pp. 108-111.

38 Schultz, T. Paul. 1990. "Women's Changing Participation in the Labor Force: A World Perspective," *Economic Development and Cultural Change,* vol. 38, no. 3 (April), pp. 457-488.

39 Henry, Alice, and Phyllis T. Piotrow. 1979. "Age at Marriage and Fertility," *Population Reports, Series M,* no. 4 (Nov.), pp. 106-159.

40 United Nations, Department of International Economic and Social Affairs. 1989a. *Adolescent Reproductive Behavior: Evidence from Developing Countries,* vol. 2 (New York: United Nations); 1990a. *Patterns of First Marriage: Timing and Prevalence,* (New York: United Nations); 1991. *The World's Women: Trends and Statistics 1970-1990,* (New York: United Nations).

41 United Nations, Department of International Economic and Social Affairs. 1987. *Fertility Behaviour in the Context of Development: Evidence from the World Fertility Survey,* (New York: United Nations), p. 96.

42 Zimicki, Susan. 1989. "The Relation between Fertility and Maternal Mortality," in Allan M. Parnell, ed., *Contraceptive Use and Controlled Fertility: Health Issues for Women and Children,* (Washington, D.C.: National Academy Press), pp. 1-47.

43 Duncan, M. Elizabeth, Gerard Tibaux, Andre Pelzer, Karin Reimann, John F. Peutherer, Peter Simmonds, Hugh Young, Ysamin Jamil, and Sohrab Daroughar. 1990. "First Coitus Before Menarche and Risk of Sexually Transmitted Disease," *The Lancet,* vol. 335 (Feb. 10), pp. 339.

44 Cain, Mead. 1984. *Women's Status and Fertility in Developing Countries: Son Preference and Economic Security,* World Bank Staff Working Paper no. 682 (Washington, D.C.: The World Bank), p. 41.

45 Beckman, Linda J. 1983. "Communication, Power, and the Influence of Social Networks in Couple Decisions on Fertility," in Rodolfo A. Bulatao and Ronald D. Lee, eds., *Determinants of Fertility in Developing Countries,* vol. 2 (New York: Academic Press), pp. 415-443.

46 Hollerbach, Paula E. 1983. "Fertility Decision-Making Processes: A Critical Essay," in Rodolfo A. Bulatao and Ronald D. Lee, eds., *Determinants of Fertility in Developing Countries,* vol. 2 (New York: Academic Press), pp. 340-380.

47 Hull, Terence H. 1983. "Cultural Influences on Fertility Decision Styles," in Rodolfo A. Bulatao and Ronald D. Lee, eds., *Determinants of Fertility in Developing Countries,* vol. 2 (New York: Academic Press), pp. 381-414.

48 Weller, Robert H. 1968. "The Employment of Wives, Role Incompatibility and Fertility: A Study Among Lower and Middle Class Residents of San Juan, Puerto Rico," *Milbank Memorial Fund Quarterly,* vol. 46, no. 4 (Oct.), pp. 507-526.

49 Safilios-Rothschild, Constantina. 1982. "Female Power, Autonomy and Demographic Change in the Third World," in Richard Anker, Mayra Buvinic, and Nadia H. Youssef, eds., *Women's Roles and Population Trends in the Third World,* (London: Croom Helm), pp. 117-132.

50 Beckman, Linda J. 1983. "Communication, Power, and the Influence of Social Networks in Couple Decisions on Fertility," in Rodolfo A. Bulatao and Ronald D. Lee, eds., *Determinants of Fertility in Developing Countries,* vol. 2 (New York: Academic Press), pp. 417-418.

51 Pariani, Siti, David M. Heer, and Maurice D. Van Arsdol, Jr. 1991. "Does Choice Make a Difference to Contraceptive Use? Evidence from East Java," *Studies in Family Planning,* vol. 22, no. 6 (Nov./Dec.), pp. 384-390.

52 Rainwater, Lee. 1965. *Family Design: Marital Sexuality, Family Size, and Contraception,* (Chicago: Aldine).

53 Hollerbach, Paula E. 1980. "Power in Families, Communication, and Fertility Decision Making," *Population and Environment,* vol. 3, pp. 146-173.

54 Beckman, Linda J. 1983. "Communication, Power, and the Influence of Social Networks in Couple Decisions on Fertility," in Rodolfo A. Bulatao and Ronald D. Lee, eds., *Determinants of Fertility in Developing Countries,* vol. 2 (New York: Academic Press), pp. 415-443.

55 United Nations, Department of International Economic and Social Affairs. 1991. *The World's Women: Trends and Statistics 1970-1990,* (New York: United Nations), pp. 26-29.

56 Cain, Mead. 1978. "The Household Life Cycle and Economic Mobility in Rural Bangladesh," *Population and Development Review,* vol. 4, no. 3 (Sept.), pp. 421-438.

57 Population Crisis Committee. 1988. "Country Rankings of the Status of Women: Poor, Powerless and Pregnant," Population Briefing Paper no. 20 (June), p. 5

58 Buvinic, Mayra, and Nadia H. Youssef. 1978. *Women-Headed Households: The Ignored Factor in Development Planning,* (Washington, D.C.: International Center for Research on Women).

59 United Nations, Department of International Economic and Social Affairs. 1991. *The World's Women: Trends and Statistics 1970-1990*, (New York: United Nations), pp. 226-229.

60 Cain, Mead. 1978. "The Household Life Cycle and Economic Mobility in Rural Bangladesh," *Population and Development Review*, vol. 4, no. 3 (Sept.), pp. 421-438; 1981. "Risk and Insurance: Perspectives on Fertility and Agrarian Change in India and Bangladesh," *Population and Development Review*, vol. 7, no. 3 (Sept.), pp. 435-474; 1984. *Women's Status and Fertility in Developing Countries: Son Preference and Economic Security*, World Bank Staff Working Paper no. 682 (Washington, D.C.: The World Bank).

61 Ahmad, Alia. 1991. *Women and Fertility in Bangladesh*, (New Delhi, Sage).

62 Mauldin, W. Parker, Nazli Choucri, Frank W. Notestein, and Michael Teitelbaum. 1974. "A Report on Bucharest: The World Population Conference and the Population Tribune, August 1974," *Studies in Family Planning*, vol. 5, no. 12 (Dec.), p. 386.

63 Isaacs, Stephen L., and Rebecca J. Cook. 1984. "Laws and Policies Affecting Fertility: A Decade of Change," *Population Reports, Series E*, no. 7 (Nov.), p. 139.

64 Mauldin, W. Parker, Nazli Choucri, Frank W. Notestein, and Michael Teitelbaum. 1974. "A Report on Bucharest: The World Population Conference and the Population Tribune, August 1974," *Studies in Family Planning*, vol. 5, no. 12 (Dec.), p.387.

65 Isaacs, Stephen L., and Rebecca J. Cook. 1984. "Laws and Policies Affecting Fertility: A Decade of Change," *Population Reports, Series E*, no. 7 (Nov.), p. 139.

Cynthia B. Lloyd, "Family and Gender Issues for Population Policy"

1   United Nations. (1975). *Report of the United Nations World Population Conference*, 1974, New York; United Nations. (1984). *Report of the International Conference on Population*, 1984, New York.

2   Findley, Sally E. and Lindy Williams. (1991). "Women Who Go and Women Who Stay: Reflections of Family Migration Processes in a Changing World," Population and Labor Policies Programme, ILO, Working Paper, No. 176.

3   Safilios-Rothschild, Constantina. (1985). "The Persistence of Women's Invisibility in Agriculture: Theoretical and Policy Lessons from Lesotho and Sierra Leone," *Economic Development and Cultural Change*, Vol. 33(2), pp. 299-317.

4   Macunovich, Diane J. and Richard A. Easterlin. (1990). "How Parents Have Coped: The Effect of Life Cycle Demographic Decisions on the Economic Status of Pre-School Age Children, 1964-87," *Population and Development Review*, Vol. 16(2), pp. 301-325; Garfinkel, Irwin and Sara S. McLanahan. (1986). *Single Mothers and Their Children: A New American Dilemma*. Washington, D.C.: The Urban Institute Press; Duncan, Greg J. and Willard L. Rodgers. (1988). "Longitudinal Aspects of Childhood Poverty," *Journal of Marriage and the Family*, Vol. 50, pp. 1007-1021; Weiss, Robert S. (1984). "The Impact of Marital Dissolution on Income and Consumption in Single-Parent Households," *Journal of Marriage and the Family*, Vol. 46(1), pp. 115-127.

5   Peterson, James L. and Christina Winquist Nord. (1990). "Steps to the Regular Receipt of Child Support," *Journal of Marriage and the Family*, Vol. 52, pp. 539-552.

6   Lloyd, B. and Sonalde Desai. (1992). "Children's Living Arrangements in Developing Countries," *Population and Development Review*, Vol. 11, pp. 193-216

7   Lloyd, Cynthia B. and Anastasia J. Gage-Brandon. (1992). "Does Sibsize Matter?: The Implications of Family Size for Children's Education in Ghana," Population Council Research Division Working Paper No. 45.

8   Ibid.

9   Ezeh, Alex C. (1991). "Gender Differences in Reproductive Orientation in Ghana: A New Approach to Understanding Fertility and Family Planning Issues in Sub-Saharan Africa," paper presented at the Demographic and Health Surveys World Conference, August 1991, Washington, D.C.

10 Ross, John A., W. Parker Mauldin, Stephen R. Green, and E. Romana Cooke. (1992). *Family Planning and Child Survival Programs as Assessed in 1991*, Population Council, New York; United Nations. (1984). *Recent Levels and Trends of Contraceptive Use as Assessed in 1983*, New York; United Nations. (1989). *Levels and Trends in Contraceptive Use as Assessed in 1988*, New York.

11 Bongaarts, John. (1991). "The KAP-gap and the Unmet Need for Contraception," *Population and Development Review*, Vol. 17(2), pp. 293-313; Westoff, Charles F. (1988). "Is the KAP-gap Real?" *Population and Development Review*, Vol. 14(2), pp. 225-232.

12 Mauldin, W. Parker and John A. Ross. (1992). "Contraceptive Use and Commodity Costs in Developing Countries, 1990-2000," *International Family Planning Perspectives*, Vol. 18(1), pp. 4-9.

13 United Nations. (1992). *World Contraceptive Use Data Diskettes*, 1991, New York.

14 Mason, Karen Oppenheim and Anju Malhotra Taj. (1987). "Differences Between Women's and Men's Reproductive Goals in Developing Countries," *Population and Development Review*, Vol. 13(4), pp. 611-638.

15 Ezeh (1991) op. cit.

16 Mott and Mott. op. cit.

17 Institute of Population Studies, Chulalongkorn University, and Population Survey Division, National Statistical Office. (1977). *The Survey of Fertility in Thailand: Country Report, Bangkok.*

18 Central Agency for Public Mobilisation and Statistics. (1983). *The Egyptian Fertility Survey, 1980, Volume III, Socio-Economic Differentials and Comparative Data from Husbands and Wives*, World Fertility Survey/International Statistical Institute.

19 Republique du Mali. (1989). *Enquete Demographique et de Sante au Mali*, 1987, Centre d'Etudes et de Recherches sur la Population pour le Developpement, Institut du Sahel, Bamako, Mali and Institute for Resource Development/Westinghouse, Columbia, Maryland.

20 Mbizvo, Michael T. and Donald J. Adamchak. (1991). "Family Planning Knowledge, Attitudes, and Practices of Men in Zimbabwe," *Studies in Family Planning*, Vol. 22(1), pp. 31-38.

21 Khalifa, Mona A. (1988). "Attitudes of Urban Sudanese Men Toward Family Planning," *Studies in Family Planning*, Vol. 19(4), pp. 236-243.

22 Ezeh, Alex C. (1992). "Contraceptive Practice in Ghana: Does Partner's Attitude Matter?" paper presented at the Annual Meeting of the Population Association of America, Denver.

23 Singh, Susheela. (1987). "Additions to the Core Questionnaires," in John Cleland and Chris Scott (eds.), *The World Fertility Survey: An Assessment*. New York: Oxford University Press.

24 Joesoef, Mohamad R., Andrew L. Baughman, and Budi Utomo. (1988). "Husband's Approval of Contraceptive Use in Metropolitan Indonesia: Program Implications," *Studies in Family Planning*, Vol. 19(3), pp. 162-168.

25 Ezeh (1991) op. cit.

26 Johansson, S. Ryan. (1991). "'Implicit' Policy and Fertility During Development," *Population and Development Review*, Vol. 17(3), pp. 377-414.

27 United Nations (1984a) op. cit.

28 Lloyd, Cynthia B. (1991). "The Contribution of the World Fertility Surveys to an Understanding of the Relationship Between Women's Work and Fertility," *Studies in Family Planning*, Vol. 22(3), pp. 144-161; Mason, Karen Oppenheim. (1985). *The Status of Women: A Review of Its Relationships to Fertility and Mortality*, The Rockefeller Foundation, New York.

29 Lloyd (1991) op. cit.

30 UNFPA. (1990). *Investing in Women: The Focus of the 90s*, New York; Weeden, Donald et al. (1986). "Community Development and Fertility Management in Rural Thailand," *International Family Planning Perspectives*, Vol. 12(1), pp. 11-16.

31 Greenhalgh, Susan. (1991). "Women in the Informal Enterprise: Empowerment or Exploitation?" *Population Council Research Division Working Paper No. 33.*

32 Greenhalgh (1991) op. cit.

33 Boserup, Ester. (1970). *Woman's Role in Economic Development*, London: George Allen and Unwin; Oppong, Christine (ed.). (1983). *Female and Male in West Africa.* London: George Allen and Unwin; Joekes, S.P. (1987). *Women in the World Economy: An INSTRAW Study.* New York: Oxford University Press.

34 Desai, Sonalde. (1992). "Children at Risk: The Role of Family Structure in Latin America and West Africa," *Population and Development Review*, Vol. 18(4), pp. 689-717.

35 Fapohunda, Eleanor R. and Michael P. Todaro. (1988). "Family Structure, Implicit Contracts, and the Demand for Children in Southern Nigeria," *Population and Development Review*, Vol. 14(4), pp. 571-594.

36 Bledsoe, Caroline. (1988). "The Politics of Polygamy in Mende Education and Child Fosterage Transactions," in Barbara D. Miller (ed.), *Gender Hierarchies.* Chicago: University of Chicago Press.

37 Desai (1992) op. cit.

38 United Nations. (1991). *The World's Women 1970-1990: Trends and Statistics*, New York.

39 Bruce, Judith and Cynthia B. Lloyd. (1992). "Finding the Ties that Bind: Beyond Headship and Household," Population Council Research Division Working Paper No. 41.

40 E.g., Knodel, John, Napaporn Havanon, and Werasit Sittitrai. (1990). "Family Size and the Education of Children in the Context of Rapid Fertility Decline," *Population and Development Review*, Vol. 16(1), pp. 31-62.

41 D'Souza, Stan and Lincoln C. Chen. (1980). "Sex Differentials in Mortality in Rural Bangladesh," *Population*

and Development Review, Vol. 6(2), pp. 257-270; Das Gupta, Monica. (1987). "Selective Discrimination Against Female Children in Rural Punjab, India," *Population and Development Review*, Vol. 13 (1), pp. 77-100.

42 Chen, Lincoln C., Emdadul Huq, and Stan D'Souza. (1981). "Sex Bias in the Family Allocation of Food and Health Care in Rural Bangladesh," *Population and Development Review*, Vol. 7(1), pp. 55-70.

43 Levine, Nancy E. (1987). "Differential Child Care in Three Tibetan Communities: Beyond Son Preference," *Population and Development Review*, Vol. 13(2), pp. 281-304.

44 Jejeebhoy, Shireen. (1992). "Family Size, Outcomes for Children, and Gender Disparities: The Case of Rural Maharashtra," paper prepared for the Population Council seminar on Fertility, Family Size, and Structure: Consequences for Families and Children, New York, 9-10 June 1992.

45 Sathar, Zeba. (1992). "Micro-Consequences of High Fertility: The Case of Child Schooling in Rural Pakistan," paper prepared for the Population Council seminar on Fertility, Family Size, and Structure: Consequences for Families and Children, New York, 9-10 June 1992; Lloyd and Gage-Brandon (1992) op. cit.; Basu, Alaka Malwade. (1992). "Family Size and Child Welfare in an Urban Slum: Some Disadvantages of Being Poor but 'Modern'," paper prepared for the Population Council seminar on Fertility, Family Size, and Structure: Consequences for Families and Children, New York, 9-10 June 1992.

46 Levine (1987) op. cit.

47 Parish, William L. and Robert J. Willis. (1992). "Daughters, Education and Family Budgets: Taiwan Experiences," Discussion Paper Series, Economic Research Center, NORC, University of Chicago.

48 Lloyd, Cynthia B. and Anastasia Gage-Brandon. (1993). "High Fertility and the Intergenerational Transmission of Gender Inequality: Children's Transition to Adulthood in Ghana," IUSSP Seminar on Women and Population in Sub-Saharan Africa, Dakar, March 3-6.

M. Teresita De Barbieri, "Gender and Population Policy"

1 This section summarizes a previous work (Teresita De Barbieri, "Sobre la Categoría Género: Una Introducción Teórico-Metodológica," Sao Paulo: The Program of Research on Reproductive Rights in Latin America [PRODIR], 1991), which synthesized ideas from existing literature, from a continuing dialogue with colleagues — especially Mary Goldsmith, María Antonieta Torres Arias, and Nelson Minello — and from my own sociological research. For an anthropological analysis and a review of the concept of gender, see Marta Lamas, "La Antropología Feminista y la Categoría 'Género'," *Nueva Antropología*, No. 30, Nov.-Dec. 1986.

2 Gayle Rubin, "El Tráfico de Mujeres: Notas para una 'Economía Política' del Sexo," *Nueva Antropología*, No. 30, Nov.-Dec. 1986. This work first appeared in 1975 in English.

3 See Michel Foucault, "El Sexo Verdadero," in Michel Foucault and Antonio Serrano, *Herculine Barbin Llamada Alexina B.*, Editorial Revolución, Madrid, 1985; and Elisabeth Badinter, *L'un est l'autre*, Paris: Odile Jacob, 1986.

4 There is a heated debate in Latin American between those who would radically redistribute income before taking steps to decrease birth rates and those who believe that reducing family size could alleviate the need for such policies. See Angel Fucaraccio, *La Resurrección del Control Natal: Discusión Crítica de Argumentos Científicos*, CELADE-PISPAL, Documento de Trabajo, No. 18, 1977; and Angel Fucaraccio and Fernando González, *Notas para una Discusión acerca de la Ley de Población en Marx*, Consejo Latino-Americano de Ciencias Sociales (CLACSO), Comisión de Población y Desarrollo-Problemas de Población Relevantes para Políticas de Población en América Latina (PISPAL). Documento de Trabajo No. 11, Santiago, Chile, 1975.

   For a study of the failure of population policies to solve poverty problems, see Raúl Benítez Zenteno, "Políticas de Población para el Siglo XXI," *La Jornada Semanal*, Feb. 1991; and Demos 1990: "Editorial. Hacia el Siglo XXI," *Demos* 1990, No. 2, México.

5 Since the late 1980s, population agencies have realized that it is in their interest to hire, in positions of responsibility, Third World feminists with international experience of grassroots women's issues and with critical perspectives on population policies and birth control programs. Is this a cooptation of rebellion and independent thought? It is too early to tell. (Thank you to Brígida García.)

6 *Informe de la Encuesta sobre Conocimiento, Actitud y Práctica en el Uso de Métodos Anticonceptivos de la Población Masculina Obrera del Area Metropolitana de la Ciudad de México* (ENCAPO), Subsecretaría de Salud, Dirección de Planificación Familiar, Sept. 1990.

7 Carmen Barroso, "As Mulheres e as Naçoes Unidas: As Linhagens do Plano Mundial de Populaçao," *Tempo Social*, Vol. 1, No. 1, 1989; Carmen Barroso, "Fecundidade e Políticas Públicas," *Sao Paulo em Perspectiva*, Vol. 3, No. 3, July-Sept. 1989 (Barroso 1989b); and Teresita De Barbieri, "Las Políticas de Población y el Derecho al Propio Cuerpo," paper presented at the Fourth National Conference of Demographic Research in Mexico, Mexico City, April 23-27, 1990.

8 Secretaría de Salud, *Encuesta Nacional sobre Fecundidad y Salud 1987* (ENFES 1987), Memoria de la Reunión Celebrada el 30 de Sept. de 1988.

9 Fay O. Ridwine, "Embarazo en Mujeres Mayores de 35 Años," *Mundo Médico*, Vol. 17, No. 118, Dec. 1989.

10 Comisión Mexicana de Derechos Humanos and *Debate Feminista*, "Razones de Salud o Razones de Estado?" *Debate Feminista*, Vol. 1, No. 2, Sept. 1990.

11 Alicia Pérez Duarte, *Viejos y Nuevos Problemas de la Reproducción desde el Punto de Vista Jurídico*, paper presented at the Fourth National Conference of Demographic Research in Mexico, Mexico City, April 23-27, 1990.

12 Carmen Barroso and Sonia Correa, "Servidores Públicos versus Profesionales Liberales," *Estudios Sociológicos*, Vol. 9, No. 25, Jan.-March 1991; and Patricia Camacho, "Nuestro Cuerpo, Base de la Meta Poblacional," *Doblejornada*, Vol. 3, No. 40, May 7, 1990.

13 Perla Oropeza, "Pocos Recursos y Mala Atención en Clínicas y Hospitales," *Doblejornada*, Vol. 3, No. 38, March 5, 1990.

14 De Barbieri, 1990, op. cit. [7].

15 Personal conversations with female university students.

16 Norma Mogrovejo, *Feminismo Popular en México*, thesis in social science, FLASCO, sede México, 1990.

17 I am indebted to María Antonieta Torres Arias for helping me understand these issues through numerous personal conversations and through her writings: "Nueva Identidad Femenina: El Dilema de las Diferencias," in Jennifer Cooper, et al., *Fuerza de Trabajo Femenina en México*, vol. 2, Coordinación de Humanidades UNAM-Miguel Angel Porrúa, México, 1989; and "La Procreación: Una Cuestión Más Allá de los Números," paper presented at the Fourth National Conference of Demographic Research in Mexico, Mexico City, April 23-27, 1990. See also Teresita De Barbieri, "El Filicidio, Tema que Horroriza," *Revista Interamericana de Sociología*, Vol. 1, No. 3, Sept.-Dec. 1987; interview with María Antonieta Torres Arias and Elsa Malvido.

18 Barroso, 1989b, op. cit. [7]. In the 1987 ENFES survey in Mexico, 24.7 percent of women who had been sterilized said that they had some medical complications, 4.7 percent had resulting personal, family, or other problems, and 9.2 percent had husbands who were unhappy that they had been sterilized.

19 UNICEF, El Ajuste Invisible: Los Efectos de la Crisis Económica en las Mujeres Pobres, UNICEF, Programa Regional para América Latina y el Caribe — Programa Regional Participación de la Mujer en el Desarrollo, Bogotá, 1989.

20 Ximena Bedregal, "Los Desafíos del MUP," *Doblejornada*, Vol. 1, No. 1, March 1987; and Mogrovejo, 1990, op. cit.

21 Torres Arias, in De Barbieri, 1987; and Francoise Dolto, *Sexualidad Femenina*, Paidós, 1986. It is an indictment of our societies that there are so many street children, aggressive youth gangs, and even children who are sold. While these problems are not inevitably the fault of the parents, they merit investigation as horrific examples of what can come of incompetent, abusive, or self-centered parenting.

Judith Lichtenberg, "Population Policy and the Clash of Cultures"

The author is indebted to anthropologists Alaka Wali and Lynne E. Greabell for their contributions to the section on cultural integrity and cultural change, and for providing many useful examples.

Other sources:

Betsy Hartmann, *Reproductive Rights and Wrongs: The Global Politics of Population Control and Contraceptive Choice* (Harper and Row, 1987)

Nicholas D. Kristof, "Peasants of China Discover New Way to Weed Out Girls," *The New York Times* (July 21, 1993)

Kerry Lauerman, "Still Ticking," *Mother Jones* (March/April 1993)

Barbara Miller, "Female Infanticide and Child Neglect in Rural North India," in *Gender in Cross-Cultural Perspective*, Caroline B. Brettell and Carolyn F. Sargent, eds. (Prentice-Hall, 1993)

Martha Nussbaum, "Human Functioning and Social Justice: In Defense of Aristotelian Essentialism," *Political Theory*, vol. 20 (1992)

Donald Warwick, *Bitter Pills: Population Policies and Their Implementation in Eight Developing Countries* (Cambridge University Press, 1982)

Denese Shervington,,"Reflections on African-American Resistance to Population Policies and Birth Control"

1 "We Remember," brochure, National Black Women's Health Project.

2 Angela Davis, *Women, Race and Class* (New York: Random House, 1981) .

3 Quoted in Barbara Kantrowitz and Pat Wingert, "The Norplant Debate," *Newsweek*, February 15, 1993, p.40.

4 Cited in Mireya Navarro, "A Condom for Women is Winning Favor," *The New York Times*, December 15, 1993, p.B1.

5   Kantrowitz and Wingert, p.37.

6   The 1991-1992 Women of Color Reproductive Health Poll, Communications Consortium, Washington, D.C.

Mencer Donahue Edwards, "People of Color and the Discussion of Population"

1   The 1991-1992 Women of Color Reproductive Health Poll, Communications Consortium, Washington, D.C.

2   W. E. B. DuBois, "Black Folk and Birth Control," *Birth Control Review*, Vol. 17, 8, 90 (May, 1938).

Russel Lawrence Barsh, "Indigenous Peoples, Population, and Sustainability"

REFERENCES

R.B. Alley et al. 1993. *Abrupt Increase in Greenland Snow Accumulation at the End of the Younger Dryas Event.* Nature 362(April 8):527-529.

Jeanne E. Arnold. 1992. Complex Hunter-Gatherer-Fishers of Prehistoric California: Chiefs, Specialists, and Maritime Adaptations of the Channel Islands. *American Antiquity* 57(1):6-84.

Arsenio M. Balisacan. 1993. Agricultural Growth, Landlessness, Off-Farm Employment, and Rural Poverty in the Philippines. *Economic Development & Cultural Change* 41(3):533-562.

Russel L. Barsh. 1990a. Ecocide, Nutrition, and the "Vanishing Indian." Pp. 221-252 in Pierre van den Berghe, ed., *State Violence and Ethnicity* (Niwot, CO: University Press of Colorado).

Russel L. Barsh. 1990b. The Substitution of Cattle for Bison on the Great Plains. Pp. 103-126 in Paul A. Olson, ed., *The Struggle for the Land* (Lincoln, NE: University of Nebraska Press).

Russel L. Barsh. 1991. Backfire from Boldt: The Judicial Transformation of Coast Salish Proprietary Fisheries into a Commons. *Western Legal History* 4(1):85-102.

Russel L. Barsh. 1992. Indigenous Peoples' Role in Achieving Sustainability. Pp. 25-34 in *Green Globe Yearbook 1992* (Oslo: The Fridtjof Nansen Institute).

Clifford Behrens. 1992. Labor Specialization and the Formation of Markets for Food in a Shibipo Subsistence Economy. *Human Ecology* 20(4):435-462.

Lewis R. Binford and W.J. Chasko, Jr. 1976. Nunamiut Demographic History: A Provocative Case. Pp. 63-143 in E.B.W. Zurbow, ed., *Demographic Anthropology: Quantitative Approaches* (Albuquerque: University of New Mexico Press).

June L. Collins. 1986. Smallholder Settlement of Tropical South America: The Social Causes of Ecological Destruction. *Human Organization* 45(1):1-10.

D.L. Coppock, D.M. Swift, and J.E. Ellis. 1986. Livestock Feeding Ecology and Resource Utilization in a Nomadic Pastoral Ecosystem. *Journal of Applied Ecology* 23(4):573-583.

Bruce A. Cox. 1985. Prospects for the Northern Canadian Native Economy. *Polar Record* 22(139):393-400.

Peter D. Dwyer and Monica Minnegal. 1992. Ecology and Community Dynamics of Kubo People in the Tropical Lowlands of Papua New Guinea. *Human Ecology* 20(1):21-55.

Marcel Fafchamps. 1992. Solidarity Networks in Preindustrial Societies: Rational Peasants with a Moral Economy. *Economic Development & Cultural Change* 41(1):147-174.

James Ferguson. 1992. The Cultural Topography of Wealth: Commodity Paths and the Structure of Property in Rural Lesotho. *American Anthropologist* 94(1):55-72.

R. Brian Ferguson. 1990. Blood of the Leviathan: Western Contact and Warfare in Amazonia. *American Ethnologist* 17(2):237-257.

Samuel Fortrey. 1663. *England's Interest and Improvement* (London: John Field).

Dov Friedlander. 1992. Occupational Structure, Wages, and Migration in Late Nineteenth-Century England and Wales. *Economic Development & Cultural Change* 40(2):295-318.

Dov Friedlander, Jona Schellekens, and Eliahu Ben-Moshe. 1991. The Transition from High to Low Marital Fertility: Cultural or Socioeconomic Determinants? *Economic Development & Cultural Change* 39(2):331-351.

Herve Fritz and Patrick Duncan. 1993. Large Herbivores in Rangelands. *Nature* 364 (July 22):292-293.

Warren M. Hern. 1992. Shibipo Polygyny and Patrilocality. *American Ethnologist* 19(3):501-522.

John Hoddinot. 1993. Rotten Kids or Manipulative Parents: Are Children Old Age Security in Western Kenya? *Economic Development & Cultural Change* 40(3):545-565.

Patricia L. Johnson. 1990. Changing Household Composition, Labor Patterns, and Fertility in a Highland New Guinea Population. *Human Ecology* 18(4):403-416.

James C. Knox. 1993. Large Increases in Flood Magnitude in Response to Modest Changes in Climate. *Nature* 361(February 4):430-432.

Shepard Krech III, ed. 1981. *Indians, Animals, and the Fur Trade* (Athens, GA: University of Georgia Press).

J. Stephen Lansing. 1991. *Priests and Programmers: Technologies of Power in the Engineered Landscape of Bali* (Princeton, NJ: Princeton University Press).

Daniel O. Larsen and Joel Michaelson. 1990. Impacts of Climatic Variability and Population Growth on Virgin Branch Anasazi Cultural Developments. *American Antiquity* 55(2):227-249.

Henry T. Lewis. 1982. Fire Technology and Resource Management in Aboriginal North America and Australia. Pp. 45-67 in Nancy M. Williams and Eugene S. Hunn, eds., *Resource Managers: North American and Australian Hunter-Gatherers* (Boulder, CO: Westview Press).

C.P. Lo. 1990. People and Environment in the Zhu Jiang Delta of South China. *National Geographic Research* 6(4):400-417.

B.R. Mitchell. 1962. *Abstract of British Historical Statistics* (Cambridge: Cambridge University Press).

Arthur E. Monroe. 1927. *Early Economic Thought* (Cambridge, MA: Harvard University Press).

G.E.B. Morren, Jr. 1986. *The Miyanmin: Human Ecology of a Papua New Guinea Society* (Ann Arbor, MI: UMI Research Press).

Monique Borgerhoff Mulder. 1992. Demography of Pastoralists: Preliminary Data on the Datoga of Tanzania. *Human Ecology* 20(4):383-405.

Thomas Mun. 1664. *England's Treasure by Forraign Trade* (1895 repr. New York: MacMillan & Co.).

A. Endre Nyerges. 1992. The Ecology of Wealth-in-People: Agriculture, Settlement, and Society on the Perpetual Frontier. *American Anthropologist* 94(4):860-881.

Sarah L. O'Hara, F. Alayne Street-Perrott and Timothy P. Burt. 1993. Accelerated Soil Erosion Around a Mexican Highland Lake Caused by Prehispanic Agriculture. *Nature* 362(March 4):48-51.

Eugene Parker. 1992. Forest Islands and Kayapo Resource Management in Amazonia: A Reappraisal of the Apete. *American Anthropologist* 94(3): 406-440.

Darrell A. Posey. 1992. Reply to Parker. *American Anthropologist* 94(3):441-443.

Doulas Southgate and Morris Whitaker. 1992. Promoting Resource Degradation in Latin America: Tropical Deforestation, Soil Erosion, and Coastal Ecosystem Distrubance in Ecuador. *Economic Development & Cultural Change* 40(4):787-807.

Daniel J. Stanley and Andrew G. Warne. 1993. Sea Level and Initiation of Predynastic Culture in the Nile Delta. *Nature* 363(June 3):435-438.

Glenn Davis Stone, Robert McC. Netting, and M. Priscilla Stone. 1990. Seasonality, Labor Scheduling, and Agricultural Intensification in the Nigerian Savanna. *American Anthropologist* 92(1):7-23,

K.C. Taylor et al. 1993. The "Flickering Switch" of Late Pleistocene Climate Change. *Nature* 361(February 4):432-436.

Russell Thornton, Tim Miller, and Jonathan Warren. 1991. American Indian Population Recovery Following Smallpox Epidemics. *American Anthropologist* 93(1):28-45.

Steve A. Tomka. 1992. Vicuñas and Llamas: Parallels in Behavioral Ecology and Implications for the Domestication of Andean Camelids. *Human Ecology* 20(4):407-433.

U.S. Bureau of the Census. 1960. *Historical Statistics of the United States; Colonial Times to 1957* (Washington, DC: Government Printing Office).

John Wacher. 1978. *Roman Britain* (London: J.M. Dent & Sons).

George W. Wenzel. 1983. The Integration of "Remote" Site Labor into the Inuit Economy of Clyde River, N.W.T. *Arctic Anthropology* 20(2): 79-91.

Paul N. Wilson and Gary D. Thompson. 1993. Common Property and Uncertainty: Compensating Coalitions by Mexico's Pastoral Ejidatarios. *Economic Development & Cultural Change* 41(2):299-318.

ENDNOTES

1 *Koniagmiut* is a generic *Alutiq* ("Aleut") term for people indigenous to Kodiak Island.

2 Archaeological exploration of Karluk was conducted in the 1980s by the late Richard H. Jordan of Bryn Mawr University in cooperation with the Kodiak Area Native Association (KANA). I was in Karluk in 1984-85 as an ecology consultant for KANA.

3 Except trauma, self-inflicted or accidental, which tends to be high among indigenous peoples who are suffering from continued intervention and lack of control over their own lives.

4 Compare the assumptions used by Thornton, Miller, and Wilder (1991), with studies of the dramatic increase of the Canadian Inuit population after centralization in the 1960s (Condon 1991:300; Binford and Chasko 1976). The Inuit cases suggest that breakdown of traditional controls on fertility offset higher mortality from contagious diseases and had a larger impact on fertility than access to Western medical care.

5   Wenzel (1983) and Cox (1985) note that seasonal "remote site labor" has enabled many Inuit communities to survive *in situ*. Work away from the village presumably results in *some* emigration, however, if workers develop tastes for the material benefits of full-time employment.

6   In the terminology of economic theory, the impact of each family's size on other households is not an *externality*. Families are directly concerned with one another's childbearing choices, through kinship and reciprocity.

7   It should be noted that coal was mined domestically, and used to produce manufactures for export beside heating British homes, while sugar and tea were imported entirely from abroad. There is no reliable data for coal before 1850.

8   As markets become globally integrated and technologies diffuse, the supplies of most goods increase and prices fall. Technological change also depresses prices by making most goods substitutable, or obsolete. In a growing and changing world economy, exporters of a particular raw material or product can only expect to enjoy a brief "flush."

9   These effects may be amplified in societies where cash-cropping and wage employment are dominated by men, displacing women from productive roles and increasing the importance placed on their function as child-bearers and child-rearers (*cf.* Johnson 1990; Friedlander et al. 1991).

SECTION VI: POPULATION AND RELIGION
L. Anathea Brooks and Teresa Chandler, "American Religious Groups and Population Policy"

Interview with Craig Lasher, Population Action International (Sept. 23, 1993)

"Religion in America," *The Gallup Report*, The Gallup Organization, in *Statistical Abstract of the United States 1991* (U.S. Department of Commerce, 1992)

From Belief to Commitment: The Activities and Finances of Religious Congregations in the United States, The Gallup Organization (1988)

Population and People of Faith: It's About Time, video produced by the Institute for Development Training (1991)

Garrett Hardin, "Living on a Lifeboat," in *Managing the Commons*, Garrett Hardin and John Baden, eds. (W.H. Freeman & Co., 1977)

The Egg: An Eco-Justice Quarterly, published by the Eco-Justice Project of the Center for Religion, Ethics, and Social Policy at Cornell University, vol. 12, no. 2 (Spring 1992)

Caring for Creation: Vision, Hope, and Justice, Evangelical Lutheran Church of America (1992)

Elizabeth Breuilly and Martin Palmer, eds., *Christianity and Ecology* (Cassell, 1992)

Fazlun M. Khalid and Joanne O. Brien, eds., *Islam and Ecology* (Cassell, 1992)

Richard Cizik, "Concern and Caution," and Billy A. Melvin, "One Perspective on the Environment," in *A Call to United Evangelical Action*, vol. 49, no. 3 (May/June 1990)

Marshall Massey, "Where are Our Churches Today? A Report on the Environmental Positions of the Thirty Largest Christian Denominations in the United States," *Firmament: The Quarterly of Christian Ecology*, vol. 2, no. 4 (Winter 1991)

Charles A. Donovan, "Population Panic and the Future of the Family," *Family Policy*, vol. 3, no. 3 (1990)

The Second Vatican Council, "Pastoral Constitution on the Church in the Modern World," in *1992 Catholic Almanac* (Our Sunday Visitor, 1992)

James A. Pike, "A Protestant's View," in *Our Crowded Planet: Essays on the Pressures of Population*, Fairfield Osborn, ed. (Doubleday, 1962)

Social Statements, Lutheran Church of America, Sixth Biennial Convention (1972)

Interview with Jaydee R. Hanson, Assistant General Secretary, Ministry of God's Creation, General Board of Church and Society, The United Methodist Church

"Population," in *The Book of Resolutions of the United Methodist Church* (United Methodist Publishing House, 1988)

Paul Boyer, *When Time Shall Be No More: Prophecy Belief in Modern American Culture* (Harvard University Press, 1992)

Pope Paul VI, "Humanae Vitae," in *1992 Catholic Almanac* (Our Sunday Visitor, 1992), and "On the Development of Peoples," *Populorum Progressio* (Edizioni Paoline, 1967)

"Renewing the Earth," statement of the U.S. Conference of Catholic Bishops, *Origins*, vol. 21, no. 27 (December 1991)

Catholic News Service, "Pope Urges Earth Summit Experts to Remember Ethical Issues," June 1, 1992

David M. Feldman, *Health and Medicine in the Jewish Tradition* (Crossroads, 1986)

Joseph Chamie, *Religion and Fertility: Arab Christian-Muslim Differentials* (Cambridge University Press, 1981)

Reverend James B. Martin-Schramm, "Population Policies and Christian Ethics"

1 For historical perspectives, see Peter J. Donaldson, *Nature Against Us: The United States and the World Population Crisis* (Chapel Hill: University of North Carolina Press, 1990); and Stanley P. Johnson, *World Population and the United Nations* (New York: Cambridge University Press, 1987). For an excellent feminist perspective, see Betsy Hartmann, *Reproductive Rights and Wrongs: The Global Politics of Population Control and Contraceptive Choice* (New York: Harper & Row, 1987).

2 For a broader discussion of the meaning of these key ethical terms, see Robert Veatch, ed., *Population Policy and Ethics: The American Experience* (New York: Irvington Publishers, 1977), 17-52; and Donald Warwick, "The Ethics of Population Control," *Population Policy: Contemporary Issues*, Godfrey Roberts, ed. (New York: Praeger Press, 1991), 22-23.

3 See Ronald Michael Green, *Population Growth and Justice: An Examination of Moral Issues Raised by Population Growth* (Missoula, Montana: Published by Scholars Press for Harvard Theological Review, 1976); Daniel Callahan, "Ethics and Population Limitation," *Ethics and Population,* Michael D. Bayles, ed. (Cambridge, MA: Schenkman Publishing Co., 1976); and Charles E. Curran, "Population Control: Methods and Morality," *Directions in Catholic Social Ethics* (Notre Dame: University of Notre Dame Press, 1985).

4 See Michael D. Bayles, *Morality and Population Policy* (Birmingham: University of Alabama Press, 1980); James M. Gustafson, "Population and Nutrition," *Ethics From a Theocentric Perspective,* vol. 2 (Chicago: University of Chicago Press, 1984); and John B. Cobb, Jr. and Herman E. Daly, "Population," *For the Common Good: Redirecting the Economy Toward Community, the Environment and a Sustainable Future* (Boston: Beacon Press, 1989).

5 Beverly Harrison makes this claim in her essay, "The Dream of a Common Language: Towards a Normative Theory of Justice," *Annual of the Society of Christian Ethics*, 1983 (Waterloo, Ontario: Council on the Study of Religion, 1983), 4.

6 Ibid.

7 Larry Rasmussen, "Creation, Church, and Christian Responsibility," *Tending the Garden: Essays on the Gospel and the Earth,* Wesley Grandberg-Michaelson, ed. (Grand Rapids: Eerdmans Publishing Co., 1987), 121

8 Stanley Hauerwas, *After Christendom?* (Nashville: Abingdon Press, 1991), 46.

9 Ibid., 68.

10 See Daniel Maguire, *A New American Justice* (San Francisco: Harper & Row, 1980), 58.

11 There has been some debate among Christian ethicists about whether one should speak of basic human rights and corresponding responsibilities, or whether it is more biblical to speak of basic Christian duties to meet corresponding human needs. See James A. Nash, *Loving Nature: Ecological Integrity and Christian Responsibility* (Nashville: Abingdon Press, 1991), 169-70; and Karen Lebacqz, Justice in an Unjust World (Minneapolis: Augsburg Press, 1987), 105-106.

12 See Karen Lebacqz, *Justice in an Unjust World*, 105-106. See Beverly Harrison on Christian solidarity in, *Our Right to Choose* (Boston: Beacon Press, 1983), 115.

13 Alan Durning, *Poverty and the Environment: Reversing the Downward Spiral* (Washington, DC: Worldwatch Institute, Worldwatch Paper #92, November 1989), 54.

14 See *Population Policies and Programmes: Lessons Learned From Two Decades of Experience*, Nafis Sadik, ed. (New York: UN Family Planning Association, New York University Press, 1991) 247; 267; 384.

15 Sandra Postel, "Denial in the Decisive Decade," *State of the World 1992* (New York: W.W. Norton & Co., 1992).

16 "Final Report of the Seventy-Seventh American Assembly," in Jessica Tuchman Mathews, *Preserving the Global Environment* (New York: W.W. Norton & Co., 1991), 327.

17 Third World debt figure from Jessica Tuchman Mathews, *Preserving the Global Environment,* 320. Global military expenditures figure from Ruth Leger Sivard, *World Military and Social Expenditures 1991* (Leesburg, VA: WMSE Publications, 1991), 11.

18 Jodi L. Jacobson, "Coming to Grips with Abortion," *State of the World 1991* (New York: W.W. Norton & Company, 1991), 114.

19 For a more specific discussion of the types of incentives and disincentives offered by nations, see *Population Policies and Programmes: Lessons Learned From Two Decades of Experience,* Nafis Sadik, ed., 120-123; see also Robert Veatch, "An Ethical Analysis of Population Policy Proposals," *Population Policy and Ethics: The American Experience,* 445-475.

20 See Herman Daly and John Cobb, *For the Common Good: Redirecting the Economy Toward Community, the Environment, and a Sustainable Future* (Boston: Beacon Press, 1989), 236-251.

21 Garrett Hardin, "There Is No Global Population Problem," *The Humanist*, vol. 49, July/August 1989, 11.

22 Beverly Harrison, "The Dream of a Common Language: *Towards a Normative Theory of Justice*," 14.

23 Lester Brown, "Launching the Environmental Revolution," *State of the World 1992*, 181.

Frances Kissling, "Theo-Politics: The Roman Catholic Church and Population Policy"

1 Thomas Aquinas, *The Commentary on Aristotle's Politics*, Book II, Chapter 6.

2 John T. Noonan, Jr., *Contraception: A History of Its Treatment by the Catholic Theologians and Canonists*, enlarged edition, (Cambridge, MA: Belknap Press/Harvard University Press) 1986, pp. 353.

3 Ibid., p. 414, quoting Cardinal Gaspar Mermillod, via R. Deppe, "Theologie pastorale," *Nouvelle revue theologique* 31 (1899), 455-456.

4 *Casti Connubii*, 13, December 31, 1930.

5 Noonan, *op. cit.*, p, 515, quoting John A. Ryan, "The Moral Aspects of Periodical Continence," *Ecclesiastical Review*, 89 (1933), 34.

6 *Origins* 21:333-34; *Detroit Free Press* and *St. Louis Post-Dispatch*, October 18, 1991.

7 Pope Pius XII, "Apostolate of the Midwife," address to the Italian Catholic Union of Midwives, October 29, 1951, printed in *The Catholic Mind*, January 1952, pp. 49-64, quotation on p. 57.

8 *Mater et Magistra*, 185, May 15, 1961.

9 *Populorum Progressio*, 37, March 26, 1967.

10 Statement of Archbishop Renato R. Martino, head of the Holy See Delegation to the United Nations Conference on Environment and Development, Rio de Janeiro, June 4, 1992.

11 For one of many examples, see the "Resolution on Abortion," approved November 7, 1989, by the National Conference of Catholic Bishops, printed in *Origins*, 19:24 (November 16, 1989) pp. 395-396.

12 Ari L. Goldman, "Focus of Earth Day Should Be on Man, Cardinal Cautions," *The New York Times*, April 22, 1990.

13 Augustine, *On Exodus*, 21:80.

14 John Connery, S.J., *Abortion: The Development of the Roman Catholic Perspective* (Chicago: Loyola University Press, 1977), pp. 107-110.

15 Alan Guttmacher Institute, "Facts in Brief: Abortion in the United States," fact sheet, February 11, 1991.

16 *The Guardian* (U.K.), May 29, 1992; and Catholic News Service, "Abortion, though illegal, believed chief birth control in Colombia," *Catholic Messenger*, June 4, 1992.

17 Stanley K. Henshaw, "Induced Abortion: A World Review, 1990," *Family Planning Perspectives*, vol. 22, no. 2 (March/April 1990) pp. 76-89; see p. 81. See also, World Health Organization/Division of Family Planning: Maternal and Child Health & Family Planning, *Abortion: A Tabulation of Available Data on the Frequency and Mortality of Unsafe Abortion* (WHO/MCH/90.14), Geneva: WHO, 1990.

18 For indications of the climate (commonly noted among theologians), see the extraordinary rebellion voiced in the January 1989 "Declaration of Cologne," in which 163 European Catholic theologians warned of Pope John Paul II's "creeping infallibilism" and of "a significant and dangerous intrusion into the freedom of research and teaching and into the dialogue-like structure of theological thinking"; from the English translation first published in the *London Tablet*, February 4, 1989.

19 I Timothy 2:11-15

20 Rosemary Radford Ruether, "Women, Sexuality, Ecology, and the Church," *Conscience*, July/Aug. 1990.

21 Beverly Wildung Harrison, *Our Right to Choose: Toward a New Ethic of Abortion*, (Boston: Beacon Press, 1983) p. 134.

22 Ranke-Heinemann, Uta, *Eunuchs for the Kingdom of Heaven*, (New York: Doubleday, 1990; trans. of German edition, Hamburg: Hoffmann und Campe Verlag, 1988) pp. 133-134.

23 Armstrong, Karen, *The Gospel According to Women: Christianity's Creation of the Sex War in the West*, (Garden City, NY: Anchor Press/Doubleday, 1987) p. 9-16.

24 Ibid., p. 21-26

25 Ranke- Heinemann, *op. cit.*, pp. 138-150.

26 "The U.S. and the Vatican on Birth Control," *Time*, February 24, 1992, p. 35. Also "Sex and Politics," *Mother Jones*, March/April 1993, pp. 20 ff.

27 "Vatican Seeks a Voice in Earth Summit Resolutions on Population," *The New York Times*, May 27, 1992; "Rome's population policy defies reality of Rio," *National Catholic Reporter*, June 19, 1992.

28 "Church and state go head-to-head over Philippine family planning," *Family Planning World*, March/April

1993; and Marilen J. Dañguilan, *Making Choices in Good Faith* (Quezon City: WomanHealth Philippines, 1993), chapters 6 and 7.

29 James F. Clarity, "Irish Begin to Liberalize Laws on Sex and Family," *The New York Times,* June 16, 1993.

30 In December 1990, the legislature of Chiapas decriminalized first-trimester abortions for "family planning and for unwed mothers." Within a month, after denunciations by Catholic leaders and the political party closest to the church, the legislature suspended the reforms and asked Mexico's National Commission on Human Rights to determine whether abortion violates human rights; shelved indefinitely, the legislative reforms in effect were killed ("Abortion controversy heats up in Mexico," *National Catholic Reporter,* Jan. 18, 1991; and Sara Lovera, "Doble moral de la Iglesia ante el aborto: F. Kissling," *La Jornada,* Feb. 6, 1991).

31 Tamar Lewin, "A.C.L.U. Lawyer Runs Afoul of Guam's New Abortion Act," *The New York Times,* March 21, 1990; and "Becoming a Radical: Puerto Rico's Luis Alberto Avilés," *Conscience,* Autumn 1992, pp. 36-37; Puerto Rican news clips.

32 Frances Kissling, "The Church's Heavy Hand in Poland," *Planned Parenthood in Europe,* May 1992; and Ann Snitow, "Poland's Abortion Law: The Church Wins, Women Lose," *The Nation,* April 26, 1993; and personal communications.

33 Denise Shannon, "Bishops on Birth Control: A Chronicle of Obstruction," *Conscience,* Winter 1992/93, pp. 29-37; Office of Government Liaison, USCC, "Re: Legislative Program, 102nd Congress," Feb. 1991; varied congressional testimony of the United States Catholic Conference/National Conference of Catholic Bishops.

34 Denise Shannon, op. cit.; Office of Government Liaison, USCC, "Re: Legislative Program, 102nd Congress," Feb. 1991; *Origins,* December 3, 1987 (Vol. 17:25), pp. 433-41; *Origins,* Sept. 3, 1987 (Vol. 17:12), pp. 187-90; "Mahony Rejects School Health Clinics Debate," *Los Angeles Times,* Nov. 11, 1986; "Matters of Urgency, Here and Abroad," *Catholic New York,* Sept. 19, 1991; "Law Firm Offers Free Aid to Parents Opposing School Condom Plan," *The Pilot,* Sept. 27, 1991. The bishops also exert pressure against condom advertising on television: "CBS Lifts Condom Ad Ban; Cardinal Opposed," *Chicago Tribune,* Feb. 20, 1987.

35 Denise Shannon, op. cit.; *London Independent,* January 30, 1990; *Washington Post,* November 16, 1989.

36 Vatican Congregation for the Doctrine of the Faith, March 10, 1987, *Donum Vitae,* p. 5.

Maura Anne Ryan, "Reflections on Population Policy from the Roman Catholic Tradition"

1 Lewis Carroll, "Alice's Adventures in Wonderland," in *The Annotated Alice* (New York: Bramhall House, 1960): 147-148. (First published, 1865.).

2 Bishop Jan Schotte, "Perspectives on Population Policy," complete text in *Origins* 14/13 (September 13, 1984): 205. (Statements delivered at subsequent conferences on population have had essentially the same content.).

3 Ibid., p. 206.

4 Ibid.

5 See for example, Marge Berer, "More Than Just Saying 'No': What Would A Feminist Population Policy Be Like?" and Rosemary Radford Ruether, "Women as Subjects, Not Objects," in *Conscience* XII/5 (September/October 1991): 1-5; 6-7. Also, Christine Overall, *Ethics and Human Reproduction: A Feminist Analysis* (Winchester, MA: Allen and Unwin, 1987): 166-196).

6 Pope John XXIII, *Mater et Magistra* (1961): #219.

7 For a comprehensive treatment of justice in Roman Catholic social teaching, see David Hollenbach, *Justice, Peace and Human Rights* (New York: The Crossroad Publishing Company, 1988): 16-33.

8 National Conference of Catholic Bishops, *Economic Justice for All: Pastoral Letter on Catholic Social Teaching and the U.S. Economy* (Washington, D.C.: United States Catholic Conference, 1986): # 7

9 Pope Paul VI, *Gaudium et Spes* (1965): 26.

10 NCCB, *Economic Justice for All,* #79.

11 Arguments for the protection of various human rights arise throughout the tradition. However, the 1963 encyclical *Pacem in Terris* is generally considered to contain the most comprehensive and representative table of human rights in the modern Catholic social tradition. See Pope John XXIII, *Pacem in Terris* (1963): #11-27.

12 *Pacem in Terris,* # 79.

13 Pope John Paul II, *Sollicitudo Rei Socialis* (December 30, 1987): #42. The concept of a "social mortgage" on the right to private property is also mentioned in the encyclicals *Gaudium et Spes* (#69) and *Populorum Progressio* (#22).

14 Pope John Paul II, *Centisimus Annus*, # 58.

15 Ibid., #58.

16 Ibid.

17 NCCB, *Economic Justice for All*, # 74.

18 Ibid., #86.

19 Pope Paul VI, *Octogesima Adveniens* (May 14, 1971), #23.

20 Shotte, p. 207.

21 Ibid.

22 See *Sollicitudo Rei Socilialis*, section 16.

23 For a helpful treatment of this issue, see Margaret A. Farley, "Ideology of the Family," in *The New Dictionary of Catholic Social Thought*, ed. by Judith Dwyer, (forthcoming).

24 See, for example, John Paul II, *Familiaris Consortio* (December 24, 1981).

25 An "overly physicalist account" would be one where the procreative meaning of the sexual act is given a disproportionate priority in analysis over all other meanings (i.e., where biology defines morality).

SECTION VII: POPULATION, DISTRIBUTION: URBANIZATION
AND INTERNATIONAL MIGRATION
Nancy Yu-Ping Chen and Hania Zlotnik, "Urbanization Prospects for the 21st Century"

1   Unless otherwise indicated, all the figures cited in the text are derived from the results of the most recent urban projections prepared by the United Nations. See United Nations, *World Urbanization Prospects: The 1992 Revision* (New York, 1993).

2   United Nations (1980). *Patterns of Urban and Rural Population Growth* (New York: United Nations).

3   Singelmann, Joachim (1991). "Global Assessment of Levels and Trends of Female Internal Migration, 1960-1980." Paper presented at the Expert Group Meeting on the Feminization of Internal Migration held in Aguascalientes, Mexico from 22 to 25 October 1991.

4   United Nations (1992). *World Population Monitoring 1991* (New York: United Nations).

5   Hardoy, Jorge E. and David Satterthwaite (1989). *Squatter Citizen: Life in the Urban Third World* (London: Earthscan Publications).

6   Economic Commission for Latin America and the Caribbean (1993). Population dynamics in the large cities of Latin America and the Caribbean. Paper presented at the Expert Group Meeting on Population Distribution and Migration held in Santa Cruz, Bolivia from 18 to 23 January 1993.

7   Economic Commission for Latin America and the Caribbean (1993). Op. cit. and Goldscheider, Calvin, ed. (1983). *Urban Migrants in Developing Nations: Patterns and Problems of Adjustment* (Westview Press: Boulder, Colorado).

8   Satterthwaite, David (1993). "The Social and Environmental Problems Associated with Rapid Urbanization." Paper presented at the Expert Group Meeting on Population Distribution and Migration held in Santa Cruz, Bolivia from 18 to 23 January 1993.

9   Economic Commission for Latin America and the Caribbean (1993). Op. cit.

10 Pernia (1993). Op. cit.

11 McGee, Terence G. and C. J. Griffiths (1993). "Global Urbanization: Towards the Twenty-first century." Paper presented at the Expert Group Meeting on Population Distribution and Migration held in Santa Cruz, Bolivia from 18 to 23 January 1993.

12 International Labour Organisation (1993). "Migration and Population Distribution in Developing Countries: Problems and Policies." Paper presented at the Expert Group Meeting on Population Distribution and Migration held in Santa Cruz, Bolivia from 18 to 23 January 1993.

13 Satterthwaite, David (1993). *Op. cit.*

Urbanization: A Report From Around The World

1   The regional divisions used in this paper are those used by the United Nations in making estimates and projections. See United Nations (1993). *World Urbanization Prospects: The 1992 Revision* (New York).

2   Berry, B.J.L. (1976). "The Counterurbanization Process: Urban America Since 1970." In *Urbanization and Counterurbanization*, B. J. L. Berry, ed. (Beverly Hills, CA: Sage).

3   Beale, C.L. (1975). *The Revival of Population Growth in Non-metropolitan America.* Washington, D.C.: Economic Research Service, U.S. Department of Agriculture.
4   Frey, W.H. (1992). "Perspectives on Recent Demographic Change in Metropolitan and Non-metropolitan America." In *Population Change and the Future of Rural America*, D. L. Brown and L. Swanson, eds. (Washington, D.C.: Economic Research Service, U.S. Department of Agriculture.)
5   Champion, Anthony G., ed. (1989). *Counterurbanization.* (London: Edward Arnold).
6   Champion, Anthony G. (1993). "Population Distribution Patterns in Developed Countries." Paper presented at the Expert Group Meeting on Population Distribution and Migration held in Santa Cruz, Bolivia from 18 to 23 January 1993; and United Nations (1993), op. cit.
7   Lattes, Alfredo (1993). "Distribución de la Población y Desarrollo en América Latina." Paper presented at the Expert Group Meeting on Population Distribution and Migration held in Santa Cruz, Bolivia from 18 to 23 January 1993.
8   United Nations (1993). *World Urbanization Prospects: The 1992 Revision* (New York).
9   Economic Commission for Latin America and the Caribbean (1993). Population dynamics in the large cities of Latin America and the Caribbean. Paper presented at the Expert Group Meeting on Population Distribution and Migration held in Santa Cruz, Bolivia from 18 to 23 January 1993.

Hania Zlotnik, "International Migration: Causes and Effects"

1   This term denotes persons entering the country of destination as migrants. The term "immigrant" is reserved to denote foreigners granted the right to permanent residence in the country of destination.
2   In this paper, developed countries include Australia, Canada, Japan, New Zealand, the United States, all countries in Europe and the former USSR, that is, the successor states of the former USSR.
3   The regional groupings used in this paper are consistent with those used by the United Nations. See United Nations (1992) *World Population 1992* (New York: United Nations, Sales No. E.92.XIII.12)
4   Although the indicator used here is total population growth, similar results would be obtained if the growth rates considered were those of the population of working age, namely, the number of persons aged 15 to 64. The growth rates of the population of working age are highly correlated to the growth rates experienced by the whole population some 15 or 20 years before. In particular, the growth rates of the working age population during 1985-1989 are closer to those of the total population in 1965-1969 than to those estimated for 1985-1989, thus displaying a lower range of variation among developing regions than the growth rates of the total population during 1985-1989. In addition, they are more affected by migration (thus confounding the effects of immigration and natural increase) and they essentially lead to the same results as those reported here: Regions with relatively low growth rates of the population in working ages have generated more migrants to developed countries than those with higher growth rates.
5   Homze, Edward L. (1967). *Foreign Labor in Nazi Germany.* Princeton, New Jersey: Princeton University Press.
6   Hawkins, Freda (1989). *Critical Years in Immigration: Canada and Australia Compared. Kingston and Montreal:* McGill-Queen's University Press.
7   U.S. Immigration and Naturalization Service (1991). *1990 Statistical Yearbook of the Immigration and Naturalization Service.* Washington, D.C.: Government Printing Office.
8   Ibid.
9   United Nations (1990). *World Population Monitoring, 1989* (New York: United Nations publication, Sales No. E.89.XIII.12).
10 Zlotnik, Hania (1991). South-to-North migration since 1960: The view from the North. *Population Bulletin of the United Nations* (New York), vol. 31/32, pp. 17-37.
11 Zlotnik, Hania (1992). "Who is Moving and Why? A Comparative Overview of Policies and Migration Trends in the North American System." Paper presented at the Conference on Migration, Human Rights and Economic Integration held in North York, Ontario from 19 to 22 November 1992.
12 State of Kuwait (1985). *Annual Statistical Abstract, 1985.* Kuwait: Central Statistical Office.
13 Zlotnik, Hania (forthcoming). "Migration to and from developing regions: A review of past trends." In *Alternative Paths of Future World Population Growth: What Can We Assume Today?*, Wolfgang Lutz, ed.
14 United Nations (forthcoming). *World Population Monitoring, 1993.* New York: United Nations.
15 United Nations (1994). *Action of the United Nations to Implement the Recommendations of the World Population Conference, 1974: Monitoring of Population Trends and Policies, with Special Emphasis on Refugees* (E/CN.9/1994/2).

16 *United Nations (1985). World Population Trends, Population and Development Interrelations and Population Policies: 1983 Monitoring Report, vol. 1* (New York: United Nations publication, Sales No. E.84.XIII.10).

17 United Nations (1985), op. cit.; Republic of South Africa (1985). *1985 Chamber of Mines of South Africa: Statistical Tables.* Johannesburg: Chamber of Mines of South Africa.

18 Borjas, George J. (1990). *Friends or Strangers: The Impact of Immigrants on the U.S. Economy.* New York: Basic Books, Inc., p. 79.

19 Ibid.

20 Ibid.

21 Massey, Douglas, Rafael Alarçón, Jorqe Durand and Humberto González (1990). *Return to Aztlán.* Berkeley, California: University of California Press.

22 Sharon Stanton Russell, "Migrant Remittances and Development," *International Migration,* Vol.XXX, No.3-4, 1992.

23 Ibid.

24 Commission for the Study of International Migration and Cooperative Economic Development (1990). *Unauthorized Migration: An Economic Development Response.* Washington, D.C.: U.S. Government Printing Office.

25 Ibid.

26 International Organization for Migration (1993), op. cit.

27 Brubaker, W. R. (1992). Citizenship struggles in Soviet successor states. *International Migration Review* (Staten Island, New York), vol. 26, No. 2, pp. 269-291.

28 United Nations High Commissioner for Refugees (1993). *The State of the World's Refugees.* New York: Penguin Books.

29 Suhrke, Astri (1993). Safeguarding the right to asylum. Paper presented at the Expert Group Meeting on Population Distribution and Migration held in Santa Cruz, Bolivia from 18 to 22 January.

30 As a result of a court case, Salvadorians as a group were granted temporary protected status by the U.S. legislature.

31 U.S. Committee for Refugees (1992). *World Refugee Survey,* 1992. Washington, D.C.: U.S. Committee for Refugees.

32 According to the United Nations 1951 Convention Relating to the Status of Refugees, a refugee is a person who "owing to a well-founded fear of being persecuted for reasons of race, religion, nationality, membership of a particular social group or political opinion, is outside the country of his nationality and is unable, or owing to such fear, is unwilling to avail himself of the protection of that country." Asylum-seekers are persons who are outside their country of nationality and make an official application for asylum in the country in which they find themselves. Many of those who are granted asylum are also granted refugee status under the terms of the 1951 convention.

33 United Nations (forthcoming). *World Population Monitoring,* 1993. Op. cit.

34 United Nations High Commissioner for Refugees (1993). Op. cit.

35 Ibid.

36 United Nations High Commissioner for Refugees (1993). Op. cit., p. 22.

SECTION VIII: POPULATION AND NATIONAL SECURITY
Thomas Homer-Dixon, Jeffrey Boutwell, and George Rathjens, "Population Growth, Environmental Change, and Civil Strife"

REFERENCES

This chapter is based on research conducted by the Project on Environmental Change and Acute Conflict. Other references include:

William H. Durham, *Scarcity and Survival in Central America: Ecological Origins of the Soccer War* (Stanford: Stanford University Press, 1979).

Ted Gurr, "On the Political Consequences of Scarcity and Economic Decline," *The International Studies Quarterly,* Vol. 29, No.1, pp.51-75; March 1985.

Jeffrey H. Leonard, ed., *Environment and the Poor: Development Strategies for a Common Agenda* (Transaction Publishers, 1989).

Thomas Homer-Dixon, "On the Threshold: Environmental Changes as Causes of Acute Conflict," *International Security,* Vol. 16, No.2, pp.76-116; Fall 1991.

"World Population Datasheet, 1993," Washington, D.C.: Population Reference Bureau, April 1993.

Emma Rothschild, "Population and Common Security"

1   *Die Werke von Leibniz*, ed. Klopp (Hanover: 1873), Vol. IX, p.143.

2   The Independent Commission on Disarmament and Security Issues, *Common Security* (New York: Simon and Schuster, 1982).

3   J.M. Keynes, *The Economic Consequences of the Peace* (1920) (New York: Harper and Row, 1971) pp. 14-15.

4   T.R. Malthus, *An Essay on the Principle of Population* (1830) (London: Penguin, 1979), p. 75.

5   Report of the United Nations Conference on Environment and Development (United Nations: 1993), Vol. I, pp. 45-48.

6   Amartya Sen, "The Economics of Life and Death," *Scientific American,* May 1993.

7   Malthus, op. cit., pp. 260-261.

8   Marianne Heiberg and Geir Ovensen, *Palestinian Society* (FAFO: Oslo, 1993). Chapters 2, 9.

Index

....